COMPUTER METHODS IN
STRUCTURAL MASONRY - 4

COMPUTER METHODS IN STRUCTURAL MASONRY - 4

**Proceedings of the Fourth International Symposium on
Computer Methods in Structural Masonry**

3–5 September 1997
Florence, Italy

Edited by

**G. N. Pande
J. Middleton
B. Kralj**
*Department of Civil Engineering
University of Wales Swansea
Swansea SA2 8PP, UK*

Taylor & Francis
Taylor & Francis Group

LONDON AND NEW YORK

This edition published 1998
by Taylor & Francis, an imprint of Routledge
2 Park Square, Milton Park, Abingdon, Oxon, OX14 4RN

Simultaneously published in the USA and Canada
by Routledge
270 Madison Ave, New York NY 10016

Transferred to Digital Printing 2009

Publisher's Note
This book has been prepared from camera-ready copy provided by the editors.
The publisher has gone to great lengths to ensure the quality of this reprint
but points out that some imperfections in the original may be apparent.

British Library Cataloguing in Publication Data
A catalogue record for this book is available
from the British Library

Library of Congress Cataloging in Publication data
A catalog record for this book has been requested

ISBN 0 419 23540 X

Proceedings of the Fourth International Symposium on
Computer Methods in Structural Masonry held at the Hotel
Demidoff, Pratolino, Florence, Italy, 3–5 September 1997

Symposium Organisers

G. N. Pande
J. Middleton
B. Kralj
Department of Civil Engineering
University of Wales Swansea
Swansea SA2 8PP, UK

Technical Advisory Panel

L. Binda	ITALY
R. de Vekey	UK
T. Grimm	USA
A. Hendry	UK
T. G. Hughes	UK
S. Lawrence	AUSTRALIA
G. C. Manos	GREECE
J. G. Rots	THE NETHERLANDS
N. G. Shrive	CANADA
R. Zarnic	SLOVENIA

Endorsed and Sponsored by

University of Wales Swansea
University of Wales Cardiff
The International Union of Testing and Research Laboratories for
Materials & Structures (RILEM)

CONTENTS

PREFACE

In recent years, computer methods have been increasingly used for the analysis of contemporary masonry structures as well as historic buildings. In view of this, a symposium series on Structural Masonry (STRUMAS) was initiated and its first meeting was held in Swansea in 1991. Encouraged by its success, subsequent symposia were organised at Swansea (1993) and Lisbon (1995). The fourth Symposium on Computer Methods in Masonry (STRUMAS IV), sponsored by RILEM, was held in the Hotel Demidoff, Pratolino near Florence, Italy, 3–5 September 1997. It was attended by over seventy participants from across the world. Technical papers on a wide range of topics relating to the application of computer methods to structural masonry were presented and these form the proceedings of the meeting.

Constitutive models for masonry is a growing field of research, and this is reflected by the number of papers on this theme that were presented. Here, homogenisation techniques of composite materials are gaining popularity and a number of papers adopted these in various forms. Strain softening and its modelling was also the subject of in-depth discussion.

A number of papers on the validation of the behaviour of masonry through small- and large-scale experiments are also included in the proceedings. Many relate to observation of patterns of cracking under various configurations, and boundary and loading conditions. Also included are papers dealing with reinforced concrete single-storey and multi-storey frames with masonry in-filled panels. Static, dynamic and creep loading conditions for unreinforced and reinforced masonry are considered.

Papers on the analysis of stone masonry arches and domes are included along with presentations on historical structures in Greece and Italy. Computer-aided design and design codes and procedures are the themes of ten papers included in these proceedings. Due to the inherent variability of the properties of masonry, test results, be they physical model experiments or in-situ tests, vary widely and there is considerable scatter in the data. Statistics and reliability analysis play an important role in the interpretation of these results. It is appropriate that papers on this theme are presented which discuss the importance of reliability analysis in various situations.

The papers included are generally of a high standard and contain a wealth of information which will be useful for researchers as well as practising engineers.

We are grateful to the members of the Technical Advisory Committee for their help in organising STRUMAS IV and would like to thank authors and delegates for their participation.

G. N. Pande, J. Middleton and B. Kralj
Swansea, March 1998

AN ANISOTROPIC FAILURE CRITERION FOR MASONRY

G. Frunzio Researcher, **A. Gesualdo** PhD Student and **M. Monaco** PhD
*Department of Science of Construction, University of Naples
'Federico II', Piazzale Tecchio 80, 80125 Naples, Italy*

1. ABSTRACT

The case of biaxially stressed masonry occurs in a large number of walls subjected to complex systems of in-plane loads. Masonry is a material which exibits distinct directional properties, so a failure criterion, because of the material anisotropy, cannot be postulated in terms of principal stresses, like the failure of an isotropic material. In the case of biaxial stress it is necessary to take into account the reference to the effect of a third variable, the orientation of the bed joints relative to the direction of principal stresses.

Hence, to define a failure criterion for masonry in a plane stress state, a three-dimensional surface is required in terms of the principal stresses and their respective directions relative to the bed joints.

The main objective of this paper is to present and discuss a surface of this type, in which a Mohr-Coulomb behaviour is considered, involving a friction tensor, to take into account the material anisotropy.

2. INTRODUCTION

The strength characteristics of masonry are highly sensitive to the orientation of the principal stresses with respect to the joint direction: a sort of dependence of "stress geometry" on "material geometry"; thus, to define the masonry failure in a plane stress

Keywords: Failure criterion, plane stress, friction.

Computer Methods in Structural Masonry – 4, edited by G.N. Pande, J. Middleton and B. Kralj.
Published in 1998 by E & FN Spon, 11 New Fetter Lane, London EC4P 4EE, UK. ISBN: 0 419 23540 X

state, it is necessary to develop a failure domain in the space of the two nonzero principal stresses and the angle ϑ between these ones and the direction of the bed joints.

Most contributions towards establishing a rational biaxial failure theory of masonry are phenomenological: experimental results are interpreted on the basis of a criterion which is assumed to be suitable for the observed mode of failure. Differences between formulae are mainly due to different failure hypoteses.

Due to the difficulty of developing a representative biaxial test as well as the large number of tests necessary to the definition, few attempts have been made to obtain experimentally a complete in-plane failure criterion. The most reliable experiments are those performed on panels under globally homogeneous stress state since they are directly translable in terms of masonry strength [7, 12, 13, 15]. In general, tests have been carried out mainly in the biaxial compression range, while tests in the biaxial tension range have not been performed, because of the complexities involved.

Theoretical approaches have been proposed in recent years to model the ultimate behaviour of masonry, based on homogenization procedures [14] or considering particular brittle models [6]. In general, these approaches involve elegant mathematcal formulations, but often need sophisticated laboratory tests to determine material parameters for the identification of the failure domain.

This paper proposes a failure criterion for masonry in which the ultimate behaviour of masonry is described by the classical Mohr-Coulomb relation, taking into account the material anisotropy by means of a friction tensor.

The main advantage of this criterion is due to the easy determination of the material parameters, since the classical shear box test is sufficient for the identification, together with the unique analytical formulation of the failure surface.

Since the material geometry of brick masonry, it can be postulated the existence of a friction tensor, in which the principal directions are those of the bed and head joints, according to the following picture:

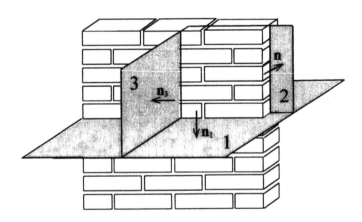

Fig. 1. *Principal friction frame*

This paper presents the analytical formulation of the failure criterion and discuss the structure of the limit domain thus obtained.

3. THE FAILURE CRITERION

3.1. The general limit condition

The criterion considers masonry as a brittle material in which the limit condition is given by the classical Mohr-Coulomb relation:

$$\left|\tau_{lim}\right| \le (c - \sigma)\, tg\, \varphi(\mathbf{n}) \tag{1}$$

in which the friction coefficient $tg\, \varphi(\mathbf{n})$ on a plane with unit normal $\mathbf{n}$, varies according to the following tensor law:

$$tg\varphi(\mathbf{n}) = \left\| \mathbf{Mn} \right\| = \left\| \begin{bmatrix} \mu_1 & 0 & 0 \\ 0 & \mu_2 & 0 \\ 0 & 0 & \mu_3 \end{bmatrix} \begin{bmatrix} n_1 \\ n_2 \\ n_3 \end{bmatrix} \right\| \tag{2}$$

where the symbol $\|\bullet\|$ is the Euclidean norm and:

$$\mathbf{M} = \begin{bmatrix} \mu_1 & 0 & 0 \\ 0 & \mu_2 & 0 \\ 0 & 0 & \mu_3 \end{bmatrix} \tag{3}$$

is the friction tensor in which $\mu_i = tg\varphi_i$ is the friction coefficient and φ_i is the friction angle. The cohesion $c\, tg\varphi(\mathbf{n})$ changes according to the variation of friction coefficient, while c is a material parameter.

It must be noted that (1), (2) and (3) must be written in the principal friction frame. In order to obtain the relation (1) in the principal friction frame, it is necessary to express the (1) in that frame.

If $\mathbf{S}$ is the diagonal stress tensor in the principal stress frame and $\mathbf{Q}$ is the orthogonal tensor expressing the reference frame change, the stress tensor in the principal friction frame is given by:

$$\mathbf{T} = \mathbf{Q}^T \mathbf{S} \mathbf{Q} .$$

If $\mathbf{l}$ and $\mathbf{m}$ is a couple of unit vectors forming with $\mathbf{n} = \mathbf{l} \wedge \mathbf{m}$ a cartesian reference frame, the limit condition (1) becomes:

$$\sqrt{\left(\mathbf{Tn} \cdot \mathbf{l}\right)^2 + \left(\mathbf{Tn} \cdot \mathbf{m}\right)^2} \le (c - \mathbf{Tn} \cdot \mathbf{n}) \|\mathbf{Mn}\| . \tag{4}$$

Note that the relation (4) depends on six variables: the three principal stresses in $\mathbf{S}$ and the three rotations in $\mathbf{Q}$, so that a geometrical representation of the domain in the principal stress space, like in the case of isotropic friction tensor, is impossible, because of the relation between the "stress geometry" and the "material geometry".

3.2. The case of plane stress

A parametrical representation of the failure surface is possible in the case of plane stress. Let σ_1 e σ_2 be the principal non zero stresses, laying in one of the principal friction planes, the failure domain can be obtained in the space $(\sigma_1, \sigma_2, \vartheta)$, where ϑ is the angle between the principal friction frame and the principal stress one.
In this case the tensor $\mathbf{Q}$ becomes:

$$\mathbf{Q} = \begin{bmatrix} \cos\vartheta & -\mathrm{sen}\vartheta & 0 \\ \mathrm{sen}\vartheta & \cos\vartheta & 0 \\ 0 & 0 & 1 \end{bmatrix} \tag{5}$$

while the stress tensor is given by:

$$\mathbf{S} = \begin{bmatrix} \sigma_1 & 0 & 0 \\ 0 & \sigma_2 & 0 \\ 0 & 0 & 0 \end{bmatrix} \tag{6}$$

The limit condition (4) provides, as n changes, a set of conic sections. The intersection gives, for every ϑ, the limit domain. The failure condition is the union of that hulls.

4. THE LIMIT DOMAIN IN THE CASE OF PLANE STRESS

In the following pictures, the limit domain is represented and discussed. The figure 2 shows five failure surface sections for five different values of the angle ϑ.
The failure domain, as a convex envelope of a set of conic sections, is convex [8]. As it can be noted in the figure 1, the shape of the domain is similar to the experimental one evaluated by Page [12, 13], not only in the biaxial compression range, but in the biaxial tension and in the tension-compression range.
A direct numerical comparison cannot be performed, since in literature the friction data are not available.

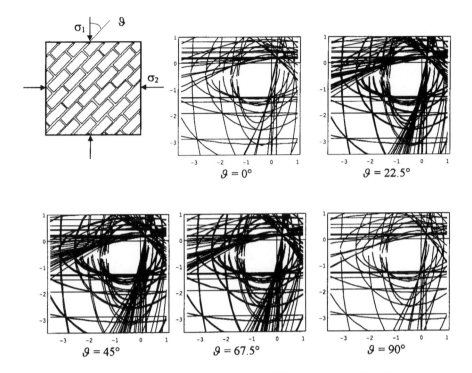

Fig. 2. *Failure surface section for different values of* ϑ *with*
$\mu_1 = 0.8$, $\mu_2 = 1.0$, $\mu_3 = 1.2$, $c = 0.2$ N/mm^2

The variation of the limit domain, reported in the figures 3-5, has been examined according the variation of the different material parameters adopted (for $\vartheta = 0$).

In the figure 3 and 4, the influence of the friction coefficient μ_1 and μ_3, respectively, can be observed. In the figure 3 the progressive symmetrization of the domain with respect to the bisection of I and IV quadrant, as the values of μ_1 approaches the value of μ_2 is shown.

The perfect symmetry represents an ultimate transversely isotropic behaviour, with symmetry axis orthogonal to the stress plane.

In the figure 4 it can be noted that the domain size, without changing of shape, is strongly increasing with the increase of μ_3, which is the friction coefficient on the stress plane, while in the figure 3 the size itself does not change sensibly.

This is a consequence of the strong influence of the friction coefficients in the plane orthogonal to the stress directions (the variation influence of μ_2 may be presumed similar, only by means of a simple change of frame).

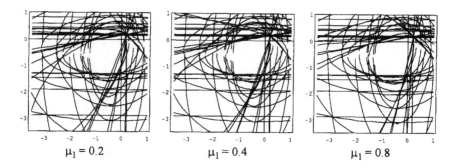

Fig. 3. *Variation of the limit domain according the variation of* μ_1 *with*
$\mu_2 = 1.0$, $\mu_3 = 1.2$, $c = 0.2$ N/mm^2 , $\vartheta = 0°$

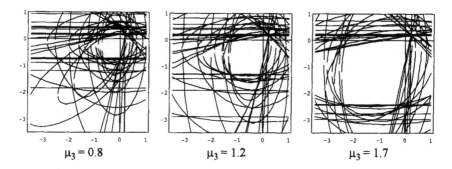

Fig. 4. *Variation of the limit domain according the variation of* μ_3 *with*
$\mu_1 = 0.8$, $\mu_2 = 1.0$, $c = 0.2$ N/mm^2 , $\vartheta = 0°$

The figure 5 shows the influence of the parameter *c* on the variation of the limit domain:
a strong increase in size, without change of shape.

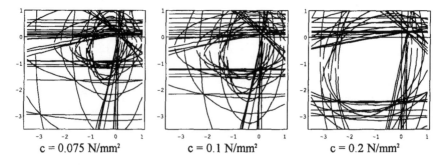

Fig. 5. *Influence of c on the variation of the limit domain with*
$\mu_1 = 0.8$, $\mu_2 = 1.0$, $\mu_3 = 1.7$, $\vartheta = 0°$

5. REMARKS

The hypothesis of a friction tensor in the Mohr-Coulomb limit condition provides, differently from the classical Mohr-Coulomb domain, a limited convex domain, which has easy application, due both to the simple formulation and to the simple laboratory tests involved, since the only test involved for the material identification is the classical shear box test to determine the friction coefficient, while the material parameter c can be evaluated performing a minimum of three shear tests for every direction.

The general relation (4) is useful in the computational methods for the analysis of triaxial stress states, though its geometrical representation is impossible.

The proposed failure criterion, which takes into account the different mechanical characteristics according to the material geometry, although the directional variation of friction coefficient postulated can have different analytical formulation, represents with great accuracy the ultimate behaviour of masonry. Nevertheless it is in general applicable to materials presenting ultimate behaviour variable according the direction (such as composite materials, anisotropic rocks), though the formulation should be specifically expressed.

REFERENCES

[1]   Aboudi, J., 'Mechanics of composite materials - A unified micromechanical approach', Elsevier, Amsterdam-Oxford-New York-Tokyo, 1991.

[2]   Anthoine, A. 'In-plane behaviour of masonry: a literature review', Report EUR 13840 EN, Commission of the European Communities, 1992.

[3]   Baldacci, R. 'Scienza delle costruzioni', U.T.E.T., Torino, 1983.

[4]   Dialer, C. 'Masonry and rock mechanics - An interdisciplinary look at two related materials', Proc. of The Sixth North American masonry conference, Philadelphia, Pennsylvania, U.S.A., 1993.

[5]   Frunzio, G., Gesualdo, A., Monaco, M. 'Stati limite di resistenza per materiali fragili con attrito anisotropo', Atti XII Congresso Nazionale A.I.M.E.T.A., Siena, Italy, 1997.

[6]   Gambarotta, L. 'Stati limite di resistenza per materiali fragili a danneggiamento isotropo', Atti XII Congresso Nazionale dell'Associazione Nazionale Meccanica Teorica ed Applicata, Napoli, Italia, 1995.

[7]   Ganz, H.R., 'Failure criteria for masonry', Proc. of the 5th Canadian Masonry Symposium, Vancouver B.C., Canada, 1989.

[8]   Jiang, J., Pietruszczak, S., 'Convexity of yield loci for pressure sensitive materials', Computers & Geotecnics, vol. 5, 1988.

[9]   Kupfer, H.B., Hilsdorf, H.K., Rush, H. 'Behaviour of concrete under biaxial stresses', Proc. ACI, vol. 66, 1969.

[10] Nemat Nasser, S. 'Micromechanics: overall properties of heterogeneous materials', Elsevier Science, 1993.

[11] Nova, R. 'An extended cam clay model for soft anisotropic rocks', Computers & Geotecnics, vol. 2, 1986.

[12] Page, A.W., 'An experimental investigation of the biaxial strength of brick masonry', Proc. of the 6th International Brick Conference, Rome, Italy, 1982.

[13] Page, A.W., 'The biaxial compressive strength of brick masonry', Proc. of Institution of Civil Engineers, Part 2, 1981.

[14] Pietruszczak, S., Jiang, J., Mirza, F.A., 'An elastoplastic constitutive model for concrete', Int. Journal of Solids and Structures, vol. 24, 1988.

[15] Samarasinghe, W., Hendry, A.W., 'Strength of brickwork under biaxial stress', Proc. of the International Symposium on Loadbearing Brickwork, London, U.K., 1980.

[16] Samarasinghe, W., Page, A.W., Hendry, A. W 'A finite element model for the in-plane behaviour of brickwork', Proc. of Institution of Civil Engineers, Part 2, 1982.

[17] Wolfram, S. 'Mathematica - A system for doing mathematics by computer', Addison-Wesley Publishing Company, Inc.

NUMERICAL MODELLING AND EXPERIMENTAL VERIFICATION OF THE MECHANICAL BEHAVIOUR OF MASONRY WALLETS

K. Chudyba Research Assistant, **K. Piszczek** Research Assistant,
K. Stępień Research Assistant, **J. Szarliński** Professor and
A. Urbański Reader
Cracow University of Technology (CUT), Faculty of Environmental Engineering, ul. Warszawska 24, 31-155 Cracow, Poland

1. ABSTRACT

The method of determining constitutive properties of homogenised masonry media based on consistent FEM formulation of the periodic media homogenisation problem is presented together with its experimental verification carried out on the series of in-plane loaded masonry wallets .

2. INTRODUCTION

Masonry is a typical representative of a multi-phase periodic micro-structured media whose distinct and common feature is that its microstructure regarding material non-homogenity and/or geometry is a periodic function of a place. Even though with the aid of the FEM the problem of determining all the states of those structures important for designers, i.e. their state of strains, stresses/internal forces and also limit states, is theoretically possible, if only the mechanical properties of the structural materials and their geometry are properly defined, some essential practical barriers arise, if a standard FEM - with discretisation over the areas of particular constituents - is used, since a large number of finite elements and D.o.F. makes such an approach, in principle, economically not justified. This disadvantage can be avoided in case of a structure built of a periodic composite medium by the use the so-called homogenisation concept, in which a fictitious homogeneous medium, equivalent - in a sense of macroscopically observed behaviour - to the given micro-structural medium, is introduced. In order to determine the constitutive properties of that composite as homogeneous medium, which can be then applied for MACRO-level FEM analysis of the structure, a MICRO-level non-standard FEM analysis of representative cell of a media is proposed. The application of the homogenisation technique for masonry, was the subject of Copernicus Joint Research Project „ATEM" (Advanced Testing of Masonry).

Keywords: Masonry, Homogenisation, Composites, Finite Element, Testing

Computer Methods in Structural Masonry – 4, edited by G.N. Pande, J. Middleton and B. Kralj.
Published in 1998 by E & FN Spon, 11 New Fetter Lane, London EC4P 4EE, UK. ISBN: 0 419 23540 X

Within that framework, the research team from CUT performed: 1) development of a theoretical basis of the presented homogenisation method, 2) implementation of the homogenisation tools to the existing 3D-FEM code enabling the homogenisation of arbitrary micro-structured periodic media, 3) modelling of in-plane loaded masonry walls 0.12m thick (half-brick walls) made with brick of different classes and mortar of approximately the same class, 4) laboratory tests for series of masonry panels with two different sorts of mortar to verify results obtained from the numerical calculations. Collecting the results of the laboratory tests for the data bank was an additional task foreseen in the project.

3. NUMERICAL MODELLING

3.1 Formulation and finite element solution of the generalised homogenisation problem

Let us consider a medium, with its micro-structure given within the representative cell being a cube in R^{NMACRO} space, submitted to a certain macroscopically controlled stress or strain state . We do not admit any other action in the medium, such as body or surface forces. The problem to be solved is: A) to find the relation between averaged macroscopic strains and corresponding averaged macroscopic stresses assuming a control by strains (strain-controlled homogenisation), by stress (stress-controlled h.) or by any compatible selection from both (mixed control h.); B) to find the distribution of strains and stresses over the microstructure. Note, that the standard FE analysis of the cell can not be used to solve A) or B) as both forces and displacement patterns are in general unknown (except some cases exhibiting symmetry) on the boundary of the cell, and the control of the loading / deformation process is imposed only in the sense of averaged field values.

The fundamental concept of presented homogenisation method lies in the split of the total deformation on the part related to the macroscopically controlled deformation of the medium and perturbation resulting from the presence of the microstructure. The deformation compatibility requirement results in condition $u(x^+)=u(x^-)$ to be fulfilled by perturbation displacement field u at corresponding points on opposite boundaries x^+, x^- of the cell. The word *generalised* concerns the concept of including other than standard Cauchy-type continua strains to the description of the deformation at the level of homogenised media. The possibilities are: different beam and shell -type models as well as higher order continua. A generalised formula of the deformation split may be put as follows , using the vector / matrix notation:

$$\varepsilon(x) = L_p(x)E + \mathcal{B}u(x), \quad u_{TOT}(x) = U(X) + C_p(x)E + u(x), \quad L_p(x) = \mathcal{B}C_p(x), \quad L_p(x^+) = L_p(x^-) \quad (1)$$

where: $\mathcal{B}$ - a strain -displacement differential operator, $U(x)$ displacement of the centre of the cell co-ordinate system $\{x\}$. In the present formulation we limit our-selves to the simplified version of a GHM, where E are generalised averaged strains controlling only strain periodic deformation. Above include the cases of standard continua , Kirchoff - Bernoulli beam and shell models and also the torsion of the beam, (see Ref. 5). In extended formulation, (see Ref. 6,7), additional (dependent) parameters, giving strain anti-periodic contribution to the deformation, are introduced.. These are the cases of transversal shear in beams and panels, (see Ref. 7) for details, or Cosserat continua. Matrices $L(x)$, $C(x)$ relate strains and displacements at the microstructure level to the parameters of assumed model of homogenised media. The case where anti - periodic components are absent corresponds to the situation when periodicity of the medium together with the uniformity of macroscopic fields result in the periodicity conditions for micro-level strains and stresses, and periodic stresses alone are in equilibrium. In the case of standard continua $L(x) = I$ and (1) represents the deformation split as described in earlier works (see Refs 1,2,3). Generalised averaged (g.a.) stresses are defined as:

$$\Sigma = \frac{1}{V_C} \int_V L_p^T(x) \, \sigma(x) \, dV \tag{2}$$

The formulation of GHM problem is based on virtual work principle (VWP), written for a representative cell of a medium. On the right side of VWP variational equation, the work of cell boundary traction on virtual displacement field should appear. It however vanishes in all GHM cases to which anti-periodic deformation is not introduced, due to the fact that displacements are then periodic, while traction are anti-periodic on the opposite boundaries of the cell.

$$\int_{\partial V} \tilde{u}^T t \, d\partial V = 0 \;, \tag{3}$$

For a given loading process, i.e. for a given path of control parameters, the response of the medium in terms of remaining g.a. strains and stress is investigated, as well as the displacement, strain and stress pattern at the level of microstructure. A complete

---

BOX 1. B.V.P of GHM (mixed stress & strain control)

Given : V- representative cell of the periodic media, controlling strains and stresses : $\hat{E} = \hat{E}(\lambda)$, $\hat{E} = \{E_I\}, I \in J_E$

$\hat{\Sigma} = \hat{\Sigma}(\lambda)$, $\hat{\Sigma} = \{\Sigma_I\}, I \in J_\Sigma$

such that: $J_E \cap J_\Sigma = \varnothing$, $J_E \cup J_\Sigma = J = \{1,2,\dots,NPAR\}$

Find: $\forall \lambda \in [0, \lambda_{MAX}]$, $\forall x \in V$, λ $\quad$ – pseudo -time

i. $u(x) \in P$ - perturbation displacement, $u(x^+) = u(x^-)$,

ii. $E = \{E_I\}, I \in J_\Sigma$ -generalised averaged strains

such that, $\forall \tilde{u}(x) \in P, \forall \tilde{E} : \int_V \left(\mathcal{B} \, \tilde{u}(x) + L_{p\Sigma} \tilde{E} \right)^T \sigma(\varepsilon) d\,V = 0$

where: $\varepsilon(x) = L_{pE}(x)\hat{E} + L_{p\Sigma}(x)E + \mathcal{B} \, u(x)$

$L_{pE} = \{L_{ij}\}, i = 1,\dots,6 \quad j \in J_E$, $\quad L_{p\Sigma} = \{L_{ij}\}, i = 1,\dots,6 \quad j \in J_\Sigma$

$C_{pE} = \{C_{ij}\}, i = 1,\dots,3 \quad j \in J_E$, $\quad C_{p\Sigma} = \{C_{ij}\}, i = 1,\dots,3 \quad j \in J_\Sigma$

iii. $\Sigma = \{\Sigma_I\}, I \in J_E$ - unknown generalised averaged stresses:

$$\Sigma = \frac{1}{V_C} \int_V L_{pE}^T(x) \sigma(x) dV$$

---

weak statement of the boundary value problem is given in the BOX 1. To the governing equation set, Eqns (5,6), expressing internal equilibrium, stress control (if $\#J_\Sigma > 0$) respectively, standard FE discretisation of perturbation displacement field u has been introduced. This leads to a non-standard FE problem, with $\{E\}$ as additional global D.o.Fs, given in BOX 2. In Eqn (6) stress control parameter $\hat{\Sigma}$ has been introduced as the traction integrated over the cell surface, such that:

$$\hat{\Sigma} = \frac{1}{V_C} \sum_{I=1}^{NMACRO} \int_{\partial V^I} C_{p\Sigma}^T(\underline{\sigma} n^I) d\partial V \tag{4}$$

The displacement compatibility condition is enforced by appropriate equation numbering. The FE model should have an additionally imposed set of isostatic constraints eliminating rigid body movements. Note that in the case of stress control deformation the banded structure of the global stiffness matrix is lost, but only on a small number of global D.o.Fs, to which all elements contribute.

BOX.2   Discrete form of GHM problem . Mixed control.

Given: FE-representation of a cell,

displacement approximation: $u(x) = C_{p\Sigma}(x)E + C_{pE}(x)\hat{E}(\lambda) + N(x)u^e$

strain approximation: $\varepsilon(x) = L_{p\Sigma}(x)E + L_{pE}(x)\hat{E}(\lambda) + B(x)u^e$

constitutive function: $\dot{\sigma} = \phi(\dot{\varepsilon}, q)$,   control by $\hat{\Sigma}(\lambda), \hat{E}(\lambda)$ like in BOX 1

Find $\forall \lambda \in [0, \lambda_{MAX}]$, λ – pseudo -time :

i.   **displacements** u and g.a. **strains** E on stress controlled directions, fulfilling:

internal equilibrium: $f(u, E, \hat{E}) \equiv \overset{Nele}{\underset{e=1}{A}} \left(\int_{V^e} B^T \sigma d V \right) = 0$ $\qquad$ (5)

if $\#J_\Sigma > 0$, stress control: $p(u, E, \hat{E}) \equiv \overset{Nele}{\underset{e=1}{\sum}} \int_{V^e} L_{p\Sigma}^T \sigma d V - \hat{\Sigma}(\lambda) V_C = 0$ $\qquad$ (6)

ii.   g.a. **stresses** on strain -controlled directions:

$\Sigma = \overset{Nele}{\underset{e=1}{\sum}} \int_{V^e} L_{pE}^T \sigma dV$,

---

The adopted non-linear problem solution scheme, based on Newton-Raphson method, requires tangent stiffness matrix, given by Eqn (7), where **D** is the current point tangent constitutive matrix.

$$K = \begin{bmatrix} K_{uu} & K_{uE} \\ K_{Eu} & K_{EE} \end{bmatrix} = \begin{bmatrix} \dfrac{\partial f}{\partial u} & \dfrac{\partial f}{\partial E} \\ \dfrac{\partial p}{\partial u} & \dfrac{\partial p}{\partial E} \end{bmatrix} = \overset{Nele}{\underset{e=1}{A}} \int_{V^e} \begin{bmatrix} B^T DB & B^T DL_{p\Sigma} \\ L_{p\Sigma}^T DB & L_{p\Sigma}^T DL_{p\Sigma} \end{bmatrix} dV . \qquad (7)$$

As the additional gain of the method, the tangent constitutive matrix of the composite Δ, relating averaged stresses and strains increments as $d\Sigma = \Delta \cdot dE$, may be obtained, by static condensation of the incremental equilibrium equations system, Eqn (8), written at each load step, after reaching an equilibrium state by appropriate iterative procedure ,

$$\begin{bmatrix} K_{uu} & K_{uE} \\ K_{Eu} & K_{EE} \end{bmatrix} \cdot \begin{bmatrix} du \\ dE \end{bmatrix} = \begin{bmatrix} 0 \\ d\Sigma \cdot V_C \end{bmatrix} , \qquad (8)$$

$$\Delta = \frac{1}{V_C} \left(K_{EE} - K_{Eu} K_{uu}^{-1} K_{uE} \right). \qquad (9)$$

3.2 Numerical modelling of masonry

Simulations of the macroscopic behaviour of the single leaf masonry wall with typical polish brick dimensions, submitted to in-plane loads according to the experiment schedule given in p. 4 were performed by means of FE analysis of the representative cell. Its 3D FE model is shown on Fig. 3.1. Mixed strain /stress control was applied with macroscopic generalised strains and stresses of Bernouli-type panel , i.e.

$$\mathbf{E} = \left\{ E_{XX}, \quad E_{YY}, \quad \Gamma_{XY}, \quad K_{XX}, \quad K_{YY}, \quad K_{XY} \right\}^T \text{-stretches ,curvatures}$$

$$\Sigma = \left\{ N_{XX}, \quad N_{YY}, \quad N_{XY}, \quad M_{XX}, \quad M_{YY}, \quad M_{XY} \right\}^T \text{-membrane forces ,bending moments.}$$

Active control variables used for each load case are given in Tab. 3.1, while for remaining generalised directions stresses are forced to be =0 by the control. Loading process was continued until the divergence of iterative procedure occurred, what was assumed to be a numerical image of the ultimate load state.

Tab 3.1.

load case:	$P_v/P_h = 1/0$	$P_v/P_h = 0/1$	$P_v/P_h = 1/1$	$P_v/P_h = 1/2$	$P_v/P_h = 2/1$
control:	Strain E_{YY}	Strain E_{XX}	Stress N_{XX}, N_{YY}	Stress N_{XX}, N_{YY}	Stress N_{XX}, N_{YY}

For both brick and mortar softening elasto-plastic model of Menetrey-Willam were used basing on Ref. 4. Necessary constitutive data of both constituents -i.e. brick and mortar which should be thoroughly identified include Young moduli , Poisson ratios, tensile and compressive strengths, dillatancy angles to model non-associated plastic flow. Moreover, ultimate crack width modelling tensile strain controlled softening de-cohesion process were used, what might be seen as regularised form of the description of the brittle behaviour.

3.3 Elasticity constants of orthotropic media equivalent to in-plane masonry

Study of elastic properties of masonry 1/2 brick thick wall , again with typical polish dimensions, treated as plane stress orthotropic medium has been done. The ratios between each component of composite matrix and brick stiffness are shown as functions of brick/mortar stiffness ratio η. Despite D_{XXYY} they prove to be almost independent of Poisson ratio of brick and mortar.

Fig. 3.1   Fig. 3.2

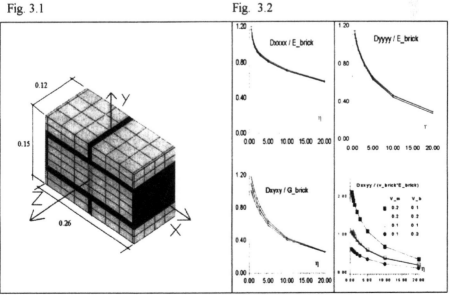

4. TESTING

The testing programme worked out for masonry wallets consisted of two parts:
1) testing constituents (brick and mortar) , 2) testing the masonry wallets.
The aim of *constituents' tests* was to determine the most important mechanical properties of mortar and brick. The following characteristics of constituents were determined: compressive strength (f_{mc} , f_{bc}), tensile strength (f_{mt} , f_{bt}), modulus of elasticity (E_m , E_b) and Poisson's ratio (v_m , v_b). In general, those tests were carried out according to appropriate Polish Codes, but in some cases not covered by codes regulations (tensile strength of brick , modulus of elasticity and Poisson's ratio for both constituents) special procedures were proposed,verified and carried out.
The aim of *testing masonry wallets* was to verify the numerical modelling.
Masonry wallets were square panels having thickness 12 cm and other dimensions 65 cm, and were made of constituents having had the properties as determined by testing the samples (p.1) of testing programme). Variable parameter was path of loading, i.e. the ratio of loads in both perpendicular directions: parallel to bed joints and perpendicular to bed joints. The following paths of loading were applied to masonry wallets:
(1) uni-axial vertical compression (P_v/P_h=1/0), (2) uni-axial horizontal compression (P_v/P_h=0/1), bi-axial compression with the ratio of vertical to horizontal load with P_v/P_h equal to: (3) 1:1, (4)1:2, (5) 2:1.
Two series of wallets were prepared and tested. The first one consisted of 20 wallets (four wallets for each path of loading) and was prepared from the sand-cement mortar of class no. M5 and solid clay brick of class no. 15. The second one consisted of 15 wallets (three wallets for each path of loading) and was prepared from the same mortar, but with solid clay brick of class no. 10.
The laboratory stand used for testing the wallets is presented in Fig. 4.1. , where:
1 - masonry wallet in testing position , 2 - press , 3 - steel brushes ,
4 - oil pressure sensor , 5 - steel frame.

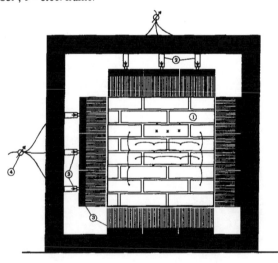

Fig.4.1

The contact surfaces between brushes and testing element were made from the very thin layer of epoxy resin and the layer of PVC foil. Brushes joints were hinged.

In the course of testing each of the masonry wallets, for a given level of external load (stress controlled testing procedure), the measurements of strains were made. Horizontal and vertical strains were measured by the means of „Demec" strain gauge, and transverse strains were determined using Hottinger's inductive sensors (datum and measuring points' arrangement is shown in Fig.4.1.). Besides strain measurements, the maximum load resulting in failure was recorded for each wallet.

5. COMPARISON AND DISCUSSION OF THE RESULTS

Graphs in Fig 5.1-5.10 represent strains in both vertical and horizontal directions measured as an average gauge elongation, registered at the subsequent load levels during experiment for: each sample, averaged over all samples subjected to the same path of load and, finally, results of corresponding numerical simulations , see p. 3.2.

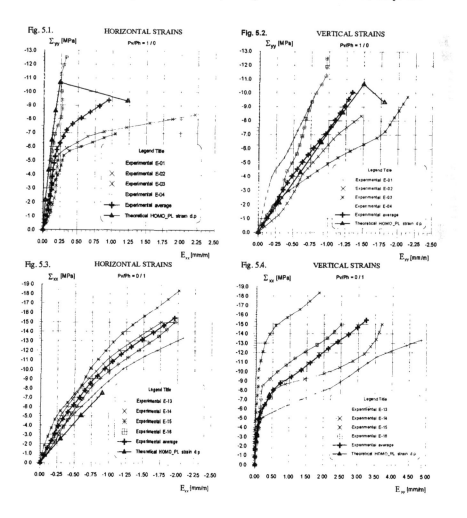

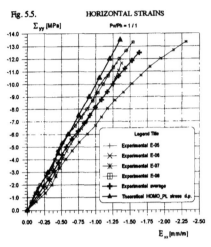

Fig. 5.5. HORIZONTAL STRAINS

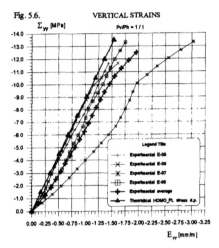

Fig. 5.6. VERTICAL STRAINS

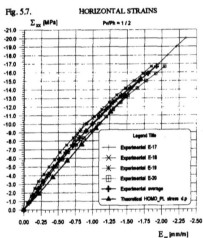

Fig. 5.7. HORIZONTAL STRAINS

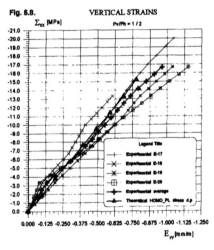

Fig. 5.8. VERTICAL STRAINS

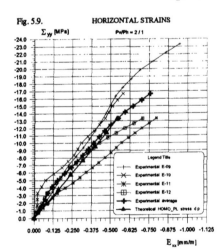

Fig. 5.9. HORIZONTAL STRAINS

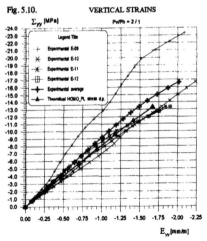

Fig. 5.10. VERTICAL STRAINS

Although substantial dispersion of the measured strains are observed within each population of wallets , in most of cases averaged values of measured strains correspond satisfactorily to numerical results . The reason for this dispersion is lack of homogenity in quality of used bricks, i.e. initial cracks and other faults .

It should be pointed out that the numerical results presented above were obtained after several corrections of the constituent materials' data, especially those of mortar. The first FEM calculations with the mortar mechanical properties, i.e. E_m, f_{mc}, f_{mt}, taken from the testing of the samples, see p.4, led to so big discrepancies (100-300%) between the results of the FEM calculations and testing the panels for various paths of loading that they were far beyond acceptable limits. They confirmed the well-known phenomenon that the properties of mortar „in situ", viz. inside the joints of the masonry panels, were by far different from those, which were exhibited by the mortar samples tested separately. Therefore, in the next FEM simulations, mortar properties, particularly E_m, were significantly reduced to values, which were derived analytically from the testing results obtained for the vertically loaded panels ($P_v/P_h=1/0$). Those results were much closer to the experimental ones for the remaining paths of loading, i.e. 1/1, 1/2 and 2/1, with the distinct exception in this respect for the path 0:1 (for horizontal compression), where too early failure of the FEM calculations occurred and it was due to unjustified numerical instability of the FEM process.

It has been stated that applying the same reduction factor for the mortar properties of all the panels on the basis of that defined for the panels vertically loaded is in most of the cases not correct, since the properties are very susceptible to many environmental factors and should be defined more rigorously for a given masonry. To find realistic properties of the mortar „in situ" the authors suggest to apply compression tests on three layered half-brick prisms with mortar joints made simultaneously with the given panel.

Another factor affecting deformation and strength of the masonry panels was the interface bond strength between brick and mortar, which was assumed arbitrarily, because of lack of more reliable data, as a half of the value of tensile strength of the mortar.

6. CONCLUSIONS

Even though the overall investigations have not yet been completed, it seems that on the basis of those, which have been hitherto carried out, some preliminary conclusions can be already formulated as follows.

1. The FEM homogenisation technique worked out under the „ATEM" project is an advanced tool for determining averaged constitutive properties of composite materials with periodic structures on the whole and, particularly, such as masonry. The formulation is open, in the sense that other constitutive models than that of Willam-Menetrey, e.g. with damage-brittle features, for masonry constituents, brick and mortar, may also be used . Introducing those models should allow to obtain a better numerical response of masonry subjected to every loading within the range of the 2D states and, as a consequence, to avoid too early failure of the computational process and such discrepancies between FEM modelling and lab testing as for example those, which were observed for the horizontal compression ($P_v/P_h=0/1$) and mentioned above in the point 6 (see also Fig.5.3 and 5.4).

As the next step, results of the numerical simulation will be used for the creation of MACRO -level non-linear constitutive model of in-plane masonry wall (identification of the linear part of such a model is completed already, see p. 3.3).

2. Defining realistic „in situ" properties of mortar plays a decisive role in determining proper mechanical characteristics of masonry, since these properties, especially modulus of elasticity, both compressive and tensile strength, as well as bond strength, are much lower than defined on the mortar separate samples; it would be very useful to find better methods to define these properties more accurately by testing, e.g. those suggested in the foregoing point 6 (with the half-brick prisms). Also, some analytical methods to derive proper formulas for them could be considered and worked out.

3. It has been proved by the FEM computations that dimensions of the panels/wallets used for the laboratory tests (with brushes) are sufficient to verify average stress and strain relationships of the masonry obtained from FEM homogenisation calculations.

7. AKNOWLEDGMENT

The authors wish to express their thanks to the European Commission in Brussels for financing the research within the framework of the Copernicus JERP „ATEM", under the contract No CIPA CT94-0174.

8. REFERENCES

1. F.Devries, H.Dumonet, G.Duvaut, F.Lene, 'Homogenization and damage for composite structures' , Int.J.Num.Meth..in Eng. ,1989, 27, pp. 285-298.

2. P.Pegon, A.Anthoine , 'Numerical strategies for solving continuum damage problems involving softening : application to the homogenization of masonry', Advances in Non-Linear Finite Element Methods 143(eds. Topping *et al*), CIVIL-COMP Ltd., Edinburgh Scotland, 1994.

3. C.C.Swan, A.S.Cakmak,' Homogenization and effective elastoplasticity models for periodic composites', Communications in Numerical Method In Engineering, 1994, 10, pp.257-256.

4. P.Menetrey, 'Numerical analysis of punching failure in reinforced concrete structures', Ph.D. Thesis , EPFL Lausanne ,1994.

5. A.Urbański, J.Szarliński, Z.Kordecki.,'Finite element modelling of the behaviour of the masonry walls and columns by homogenisation approach', Proc. of 3rd Int. Symposium on Structural Masonry, (Eds.G.N. Pande *et al*), Swansea, ,1995

6. A.Urbański, 'Finite element formulation of generalised homogenisation method of periodic composite media and its application to the analysis of masonry', Proceedings of XIII PCCMM, Poznań, Poland,1997, p. 1323-1330.

7. A.Urbański, 'Numerical analysis of the cross - sectional behaviour of an arbitrary composite beam or panel as a case of generalised homogenisation theory', Proceedings of XIII PCCMM, Poznań, Poland,1997, p. 1331-1338.

AN ELASTIC-PLASTIC MODEL WITH DAMAGE FOR CYCLIC ANALYSIS OF MASONRY PANELS

A. Callerio and E. Papa[1]

[1]*Politecnico of Milan, Department of Structural Engineering, Piazza Leonardo da Vinci 32, 20133 Milano, Italy*

1. ABSTRACT

A local model with damage is introduced to describe the behavior of in-plane loaded masonry, viewed in terms of an orthotropic equivalent continuum. The model is based on plasticity and damage theory. The yield domain, and its hardening and softening law, is defined according to piecewise linear functions. The evaluation of stress in the material is then reached by solving a Linear Complementarity Problem (LCP). The material damage is taken into account considering the change of the stiffness matrix as a function of the inelastic deformation. The preliminary analyses show a good agreement between experimental and numerical results obtained from two dimensional cyclic load tests.
Keywords: Masonry, Plasticity, Softening, Damage, Cyclic Analysis

2. INTRODUCTION

The aim of the model is to study the behavior of masonry panels subjected to cyclic plane loads adopting the finite element method. The analysis performed are oriented toward seismic vulnerability evaluation of masonry structures.

The constitutive law introduced is based on the plasticity theory of piecewise-linear yield functions [1] [2]. Here softening behavior of the material is considered, coupled with a stiffness reduction, related to inelastic strains. The effective stress state in the material is obtained by solving a Linear Complementarity Problem (LCP), adopting iterative procedures [3].

The theory is developed starting from some previous works [4] [5] in which a piecewise-linear model is introduced and results are carried out under elastic perfectly plastic material hypothesys. In [4] a theoretical approach to softening problems is followed,

Computer Methods in Structural Masonry – 4, edited by G.N. Pande, J. Middleton and B. Kralj.
Published in 1998 by E & FN Spon, 11 New Fetter Lane, London EC4P 4EE, UK. ISBN: 0 419 23540 X

investigating for the stability of the solution algorithm in the presence of such unstable materials.

A yield function Γ, needed by the theory of plasticity, is discretized by means of N planes Φ^k limiting in such a way the elastic stress domain Ω. In plane stress analysis the Γ function can be defined as a surface in the $(\sigma_{11}, \sigma_{22}, \sigma_{12})$ space, where index 1 means the horizontal bed joint direction and index 2 the vertical one.

The admissible stress must verify the inequality:

$$\Gamma(\boldsymbol{\sigma}, \mathbf{p}) = \bigcup_{k=1,N} \Phi^k \leq 0 \tag{1}$$

where $\mathbf{p}$ is a vector of history-dependent parameters and, according to the Drucker stability postulate, Γ has to be a convex function.

The yield domain has been defined according to experimental results obtained from tests on multi axial loaded masonry panels [6] [7]. The domain Ω limited by Γ is symmetric respect to $\sigma_{12} = 0$ plane and a Mohr-Coulomb criterion (showed in fig. 1a) is introduced. Plane 9 is here introduced for the first time, in order to limit the maximum τ stress in the mortar joint. In fig. 1b the solid line shows the cutting

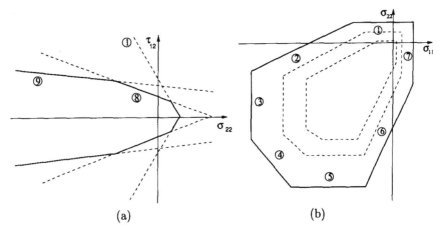

(a)   (b)

Figure 1: Yield planes

of the limit domain for $\sigma_{12} = 0$, and the dotted ones represent the contour lines at different σ_{12} values. Can be seen that the compression strength is modeled quite greater than the tensile one: the planes 1 and 7 represent the cut-off in tensile stresses, when the others limit the compressive stresses.

For each plane an hardening rule is defined by means of a piecewise linear function $\mathcal{H}$, (fig. 2) in terms of the plane distance by the origin of the axes (r). The shape of the hardening function may vary depending on which plane is considered: a compression plane (i.e., plane 4) has a smoother shape than a tensile one (plane 1 or 7), or plane 8 and 9, considering in such a way the different class of damage mechanism.

Regarding the shape of the hardening function, different level of ductility of the material can be then modeled, due to compression or tensile stresses. Usually two shapes of the $\mathcal{H}$ function are considered: one grouping planes 1, 7, 8, 9 and one describing the others.

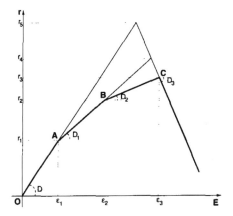

Figure 2: Plane hardening law.

3. MODEL THEORY

In plane stress analysis the non zero components of stress tensor, and the respective components of strain, are related by the ortothropic stiffness matrix $\mathbf{D}$. Its rank is equal to 3 and $\mathbf{D}$ is defined by the Young modulus in the vertical and horizontal direction (E_1 and E_2), the Poisson ratio ν_{12} and the shear modulus G_{12}. The evaluation of the stress follows a Predictor-Corrector scheme:

Prediction:

$$\sigma^* = \sigma^0 + \mathbf{D}\,\Delta\varepsilon \tag{2}$$

Correction:

$$\sigma = \sigma^* - \mathbf{D}\,\Delta\varepsilon_i \tag{3}$$

in which σ^0 is the initial stress, σ^* represents the predicted stress, σ is the corrected stress, $\Delta\varepsilon$ is the total strain increment and $\Delta\varepsilon_i$ the variation of its inelastic part (unknown).
Inelastic strains occur only when eq. (1) is violated by the predicted stress. The eq. (1) can be rewritten for the k-th plane as:

$$\mathbf{n}^{k^T} \times \sigma \leq \mathbf{r}^k \tag{4}$$

where $\mathbf{n}^k$ is the normal to the plane and $\mathbf{r}^k$ the distance between the plane and the origin at the beginning of the analysis step. The predicted stress σ^* can stay outside of one or more planes. If more than one plane has been violated by the predicted stress σ^* (saying M the number of those planes), the inelastic strain variation needs to be evaluated by a sum over the contribution coming from each plane.
Introducing an associated plasticity criterion with $\Delta\varepsilon_i$ normal to the yield surface,

one has:

$$\Delta\varepsilon_i = \sum_{k=1,M} \Delta\lambda^k \, \mathbf{n}^k \tag{5}$$

Considering $M = 1$ (only the k-th plane has been violated), given the hardening rule showed in fig. 2, an hardening matrix $\mathbf{H}^k$ and the $\mathbf{r}^k$ vector need to be introduced as follows:

$$\mathbf{H}^k = \begin{bmatrix} h_1^k & 0 & 0 & 0 & 0 \\ 0 & h_2^k & 0 & 0 & 0 \\ 0 & 0 & h_3^k & h_4^k & h_5^k \\ 0 & 0 & h_4^k & h_4^k & h_5^k \\ 0 & 0 & h_5^k & h_5^k & h_5^k \end{bmatrix} \quad \mathbf{r}^k = \begin{bmatrix} r_1^k \\ r_2^k \\ r_3^k \\ r_4^k \\ r_5^k \end{bmatrix} \tag{6}$$

The h_i^k are hardening factors evaluated on the basis of the hardening function shape, and they depend on the stiffness reduction during the hardening and softening phases. The r_i^k coefficients are the distance between plane and origin at different hardening states (see fig. 2). The plastic multiplier $\Delta\lambda^k$ can be decomposed into different contributions coming from the different levels of the hardening law:

$$\Delta\lambda^k = \sum_{i=1,5} \Delta\lambda_i^k \tag{7}$$

Introducing a vector $\mathbf{\Delta\lambda}$, which components are the $\Delta\lambda_i^k$, one can write eq. (1) in terms of a Linear Complementarity Problem:

$$\phi^k = \left(\mathbf{n}^{k^T}(\sigma^0 + \mathbf{D}\,\Delta\varepsilon)\right)\mathbf{1} - \mathbf{n}^{k^T}\mathbf{D}\,\mathbf{n}^k\,\Delta\lambda^k\,\mathbf{1} \tag{8}$$
$$-\mathbf{H}^k\,\Delta\lambda^k - \mathbf{r}^k \le 0$$

under the conditions:

$$\phi\,\mathbf{\Delta\lambda} = 0 \;\; ; \;\; \mathbf{\Delta\lambda} \ge 0 \tag{9}$$

The scalar coefficient $d^k = \mathbf{n}^{k^T}\mathbf{D}\,\mathbf{n}^k$ can be seen as a kind of "projected initial stiffness".

The rank of the LCP problem in this case is five. During the analysis the rank decreases to one when the material is damaged, and consequently the plane considered is totally collapsed to the origin of the axes, or to a minimum distance (if defined). More than one plane violated require an assembling of interacting LCP problems, leading one to write the entire problem as the solution of:

$$\Phi = \left(\mathbf{N}^T(\sigma^0 + \mathbf{D}\,\Delta\varepsilon)\right)\mathbf{1} - \mathbf{N}^T\mathbf{D}\,\mathbf{N}\,\Delta\Lambda\,\mathbf{1} \tag{10}$$
$$-\mathbf{H}\,\Delta\Lambda - \mathbf{R} \le 0$$

The solution in terms of $\mathbf{\Delta\lambda}$, assembling of the λ^k vectors, is iterative [3]. With a backward substitution in eq. (5), one finally obtains $\Delta\varepsilon_i$ that leads to the evaluation of the corrected stress through eq. (3).

3.1 Damage Evolution Law

An isotropic damage law is here adopted in order to model the material decreasing stiffness. In [8] a brittle material damage is modeled considering a part of the total inelastic deformation as irreversible. Here one starts from a plastic material and a ratio α (with $0 \leq \alpha \leq 1$) is introduced to obtain the inelastic strain tensor ε_d (considered as reversible) and the plastic strain tensor ε_p (non-reversible), through coaxial laws:

$$\Delta \varepsilon_d = \alpha \, \Delta \varepsilon_i \quad ; \quad \Delta \varepsilon_p = (1 - \alpha) \, \Delta \varepsilon_i \tag{11}$$

At the end of the step, after the evaluation of the corrected stress σ, one has the following expression for the total strain:

$$\varepsilon = \mathbf{C_0} \, \sigma + \varepsilon_p^0 + \Delta \varepsilon_p + \Delta \varepsilon_d \tag{12}$$

in which $\mathbf{C_0} = \mathbf{D_0^{-1}}$ is the compliance matrix at the beginning of the solution step. The increment of material compliance during the step is then related to the reversible part of the inelastic strain variation as follows:

$$\Delta \varepsilon_d = \Delta \mathbf{C} \, \sigma \tag{13}$$

Then from eq. (12) one obtains:

$$\varepsilon = [\mathbf{C_0} + \Delta \mathbf{C}] \, \sigma + \varepsilon_p^0 + \Delta \varepsilon_p = \mathbf{C} \, \sigma + \varepsilon_p^0 + \Delta \varepsilon_p \tag{14}$$

where $\mathbf{C}$ is the final compliance matrix. Introducing two unit vectors $\mathbf{n}$ and $\mathbf{m}$, coaxial respectively to stress and damage strain variations, as:

$$\Delta \varepsilon_d = \|\Delta \varepsilon_d\| \, \mathbf{n} \quad ; \quad \sigma = \|\sigma\| \, \mathbf{m} \tag{15}$$

and a scalar quantity ρ:

$$\rho = \frac{\|\Delta \varepsilon_d\|}{\|\sigma\|} > 0 \tag{16}$$

the expression of the variation $\Delta \mathbf{C}$ of the compliance matrix is:

$$\Delta \mathbf{C} = \frac{\rho}{\mathbf{m^T n}} \, \mathbf{n \, n^T} \tag{17}$$

4. NUMERICAL ANALYSES

In order to check the reliability of the model proposed, some simple analyses have been performed.

Imposing a cyclic strain history along direction 2 on a single finite element, the plot of fig. 3 is obtained. It is possible to note that the model describe correctly the different compression and tensile strength of the material. In fig. 4 the response due to increasing level of σ_{12} strain in two direction is showed. In fig. 4b the plane 9 (shear stress limit) has been made inactive: can be seen how the maximum shear stress increases, towards high incompatible values. A first conclusion is that the

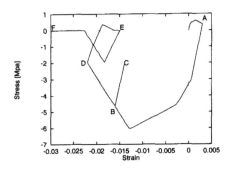

Figure 3: σ_{22} vs. ε_{22} plot.

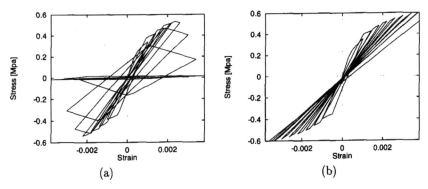

Figure 4: σ_{12} vs. ε_{12} plot: (a) nine planes (b) eight planes.

introduction of plane 9 is necessary in order to obtain a realistic behavior.

After the implementation of the model in a finite element code, shear tests on a base fixed masonry panel have been performed. A constant vertical load is applied, and monotonic or cyclic displacement at the top have been imposed. The analyses have been carried out under displacement control and the results obtained in terms of force-displacement and contour plots. In fig. 5a is showed the force displacement plot from the monotonic test: the final strength (260 KN) is similar to the one obtained from experimental tests on masonry panels [9]. In fig. 5b the plot shows the results gathered from cyclic tests: decreasing stiffness of the model is clear showed, with also the reduction of the strength due to the same displacement amplitude. The contour plots in fig. 6a, obtained from the cyclic analysis, show the position of the plane one respect to the origin, moving from the initial undamaged state, following the given hardening rule, toward the final position. Fig. 6b concerns about plane six, that limit compressive stresses: a smaller area is interested by damage, due to the greater material strength in compression. Finally fig. 6(c) shows the damage occurred to plane eight (Mohr-Coulomb criterion).

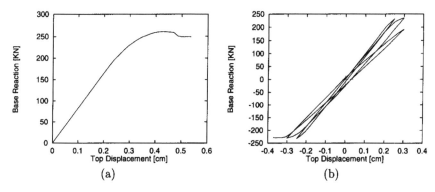

Figure 5: Shear test. Horizontal force vs. displacement plot: (a) monotonic test, (b) cyclic test.

5. CONCLUSIONS

A local model for masonry based on experimental results has been presented. The preliminary results obtained show good agreement respect to experimental results. More computational and theoretical efforts are needed to extend the model introducing different hardening rules. Another further development will be the implementation of more realistic damage laws depending on the different damage mechanism occurred in the material. An extension of the model will be the considering of more complex stress states, taking into account the effects of out-of-plane displacements on brick-work damage under seismic actions.

6. ACKNOWLEDGEMENT

A grant from the Italian Ministry of University and of Scientific and Technological Research is gratefully acknowledged.

7. REFERENCES

1. Maier, G. 'A matrix structural theory of piecewise linear elastoplasticity with interacting yield planes', *Meccanica*, 5, 1970, pp. 54–66.
2. De Donato, O. 'Fundamental of elastoplastic analysis', in *Engineering Plasticity by Mathematical Programming*, M.Z. Cohn and G. Maier eds., NATO Advanced Study Institute, Pergamon Press, 1977, pp. 325–349.
3. Mangasarian, O.L. 'Solutions of symmetric linear complementarity problems by iterative methods', *Journal of Optimization Theory and Applications*, 22, 1977, pp. 465–485.
4. Nappi, A., Facchin, G. and Marcuzzi, C. 'Structural dynamics: convergence properties in the presence of damage and applications to masonry structures', *Structural Engineering and Mechanics*, (to appear).

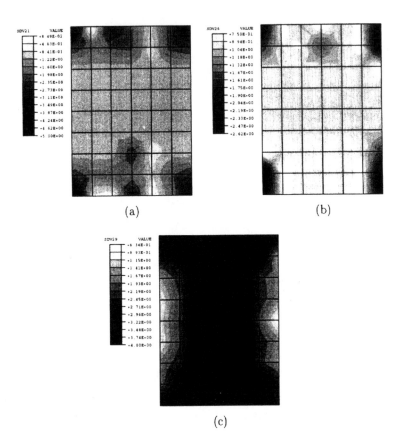

(a)   (b)

(c)

Figure 6: Cyclic shear test. (a) plane 1 failure (tension), (b) plane 6 failure (compression), (c) plane 9 failure (shear)

5. Papa, E. and Nappi, A. 'Modellazione numerica di strutture murarie soggette a carichi ciclici', in *La meccanica delle murature tra teoria e progetto*, L. Gambarotta ed., Università di Messina, Italy, 1996, pp. 441–450 (in Italian).

6. Page, A.W. 'The biaxial compressive strength of brick masonry', in *Proc. Inst. Civ. Eng.*, 3, 1981, pp. 893–906.

7. Maier, G., Papa, E. and Nappi, A. 'On damage and failure of brick masonry', *Experimental and Numerical Methods in Earthquake Engineering*, 1991, pp. 223–245.

8. Ortiz, M. 'A constitutive theory for the inelastic behavior of concrete', *Mechanics of Material*, 4, 1985, pp. 67–93.

9. Magenes G. *Comportamento Sismico di Murature in Mattoni: Resistenza e Meccanismi di Rottura di Maschi Murari*, Ph.D. thesis, Università di Pavia, Italy, 1992.

NUMERICAL VALIDATION OF A CLOSE-FORM APPROACH TO LOCALIZATION IN LINEAR ELASTIC HOMOGENIZATION OF MASONRY

S. Briccoli Bati Associate Professor, **G. Ranocchiai** Technical Official and **L. Rovero** Doctorate Graduate
Dipartimento di Costruzioni, Università di Firenze, Piazza Brunelleschi 6, 50121 Firenze, Italy

1. ABSTRACT

The results of the numerical homogenization process of linear elastic simple-textured masonry, are compared against the results of the homogenization based on the micromechanical model, proposed by the authors, making use of the close-form solution due to Eshelby. The numerical process is realized with the finite element analysis of the representative cell, which describes the local elastic field. The results of numerical localization are employed to interpret the average elastic field obtained by means of the close-form localization of the proposed micromechanical model.
Macroscopic elastic constants determined by the numerical and close-form processes are compared against the elastic constants of masonry panels, determined by laboratory compression tests.

2. INTRODUCTION

In recent years homogenization techniques have been applied to the study of masonry-like materials, in order to introduce information about brickwork texture into structural analysis [1]. In fact homogenization techniques can determine the macroscopic behavior of a heterogeneous material, making use of information about its microstructure and about the properties of the constituent materials.

Keywords: Homogenization, Micromechanical model, Localization

Computer Methods in Structural Masonry – 4, edited by G.N. Pande, J. Middleton and B. Kralj.
Published in 1998 by E & FN Spon, 11 New Fetter Lane, London EC4P 4EE, UK. ISBN: 0 419 23540 X

A linear elastic homogenization process can be realized by means of a micromechanical model which localization problem can be solved either by close-form solution, or by finite element computation.

In this paper the results of the linear elastic homogenization process of masonry are reported, obtained while performing the localization process by means of a finite element analysis over the elementary cell. The local elastic field is reproduced by a proper discretization of brick and mortar in the representative volume element.

The homogenized elastic constants obtained by the numerical localization process are compared with the overall properties obtained by the micromechanical model proposed by the authors, that makes use of close-form solution. This model relies upon the exact solution due to Eshelby and describes brickwork as a mortar matrix with insertion of elliptic cylinder-shaped bricks. Making use of a close-form solution, only the average values of the local stress and strain fields, computed on every single phase, can be determined.

The results of numerical localization are then employed to interpret the average localization process of the close-form homogenization model. The results of both the homogenization and localization processes are compared with experimental results obtained by laboratory tests on masonry panels.

The homogenization techniques introduced in the present research were applied to masonry panels made of bricks and lime or cement mortar. The case example employed for the comparison are simple texture masonry panels made with clay bricks (UNI standard) and one centimeter thick mortar joints. The use of both lime and cement mortar enabled reproduction of composite materials where mortar represents respectively the less stiff phase (lime mortar) and the stiffer phase (cement mortar).

3. SUGGESTED MODEL

The micromechanical model suggested by the authors represents the brickwork solid as a matrix of mortar with the inclusion of elliptic cylinder-shaped bricks [2] [3] (fig. 1). Eshelby's solution is employed to determine the localization tensors of the phases.

Eshelby's problem [4] addresses the case of an infinitely extended material containing an ellipsoidal inhomogeneity. The solution can be employed to determine the average localization tensors and the homogenized stiffness matrix of an infinitely extended material containing scattered inclusions, provided that their concentration is low. If this is the case the stiffness matrix is:

$$\mathbf{K}^{hom} = c_1 \, \mathbf{A}_1 \, \mathbf{K}_1 + c_2 \, \mathbf{A}_2 \, \mathbf{K}_2 \qquad\qquad (1)$$

where $\mathbf{A}_1$ and $\mathbf{A}_2$ are the average localization tensors of phases 1 and 2 respectively, $\mathbf{K}_1$ and $\mathbf{K}_2$ are the stiffness matrices of the phases, and c_1 and c_2 their concentration factors.

If the concentration of inclusions is high, as is the case for brickwork, the reciprocal influence of the scattered phase cannot be neglected and it is necessary to employ Mori and Tanaka's theory [5] to extend Eshelby's solution.

4. NUMERICAL MODEL

Masonry material may be considered as a periodic composite medium, made up of an array of elementary cells (fig. 2). Suppose that a panel made of periodic material is subjected to a globally homogeneous stress state, that is every cell is subjected to the same stress conditions; the periodic texture of the panel can be assumed to produce the following properties in the elastic state [6]:

a)   periodicity of stress state

b)   periodicity of strain state, that is displacements are able to produce periodic strain.

The first condition, owing to the equilibrium of stress, causes the stress vectors Σ n to be equal and opposite on two opposite sides of the cell boundary; on the other hand, the second condition, owing to the compatibility of strain, causes the displacement field to assume the same value on two opposite sides of the cell boundary, up to a rigid displacement. That is, as a consequence of the periodicity condition b), the displacement field, producing periodic strain, can be written:

$$u(x) = \overline{E} \, x + u^p(x) \qquad (2)$$

where u^p is a periodic displacement field, and $\overline{E}$ is a constant symmetric tensor, representing (by virtue of Hill's theorem) the average strain field.

The elastic problem of a single cell, with strain controlled loading, reads:

$$\text{div}(K \, E(u^p)) + \text{div}(K\overline{E}) = 0 \qquad \text{on S}$$

$$\qquad (3)$$

$$K \, E(u^p) \, n \qquad \text{anti-periodic on } \partial S$$

$$u^p \qquad \text{periodic on } \partial S$$

where the u^p function has to be determined and can be evaluated via finite element analysis. The periodicity conditions permit to define the boundary conditions for the cell subjected to numerical analysis.

Due to the   uniformity of elastic tensors within each phase, the fictitious body forces $f = \text{div}$ $(K \, \overline{E})$ become surface forces applied at the interface between constituents:

$$f = (K^1 - K^2) \, E \, n \, \delta_{\partial s} \qquad (4)$$

Owing to the linearity of the problem, the solution u^p for any $\overline{E}$ can be calculated by linear recombination of the solutions corresponding to the   elementary   strain tensors.

By means of equation (2), the total strain can be evaluated:

$$E(x) = \overline{E} + E(u^p(x)) \qquad (5)$$

that represents the localization process of the strain field.

Similarly, the local stress field is the sum of two shares, one derived from the periodic displacement, and the second from the homogeneous strain field.

By knowing the average stress and strain fields, the macroscopic elastic constants can be easily evaluated:

$$\langle \Sigma \rangle = K^{\text{hom}} \langle E \rangle \qquad (6)$$

where

$$\langle \Sigma \rangle = \int_{V} \Sigma \, dV$$

$$\langle E \rangle = \int_{V} E \, dV$$

and where K^{hom} is the homogenized stiffness tensor.

In practice the homogenization process can be performed by means of elasticity problems defined on the representative cell, subjected to uniform unitary strain.

Mesh and constraints of the cell subject to numerical analysis, under the plane stress assumption, are shown in fig. 3; constraints depend on the periodicity conditions, while the surface forces applied at the interface between the phases were determined through eq. 4 ((a) scheme was solved twice, with different surface forces, able to produce the elastic state corresponding to the elastic constants of the ortotropic medium).

5. EXPERIMENTAL ANALYSIS

13 panels reproducing simple-leaf brickwork on a small scale, were realized with small bricks obtained from hand-pressed bricks.

Brickwork panels of dimensions 19.5 x 4.8 x 15.5 cm were constructed according to Rilem Lum B1. Lime mortar panels were cured for a period of 90 days before carrying out mechanical tests.

Brick samples of dimensions 2 x 2 x 4 cm were cut from real bricks, keeping the height of samples aligned with the load direction of bricks in brickwork. Mortar samples of dimensions 2 x 2 x 2 cm were cast; small dimension make the resulting mechanical properties closer to the properties of mortar joints in brickwork. Lime mortar was cured for a period of 90 days.

Mechanical uniaxial compression tests of both material samples and brickwork panels were carried out using a machine which permitted controlled displacement tests. A uniform load distribution was applied with loading rate equal to about 4 E-3 MPa/sec. Friction between the load plates and loaded surfaces of the specimens was reduced by a thin layer of talcum powder. Electric transducers were employed to measure load and displacement, and generate real-time load-displacement diagrams.

Global displacement measurements were taken in order to evaluate elastic constants: displacement transducers placed over the upper surface of samples furnished data for Young's modulus; displacement transducers whose gauge length was equal to the sample width were employed to record transverse deformation, through which transverse elastic constants were calculated. Global measurements were employed to determine elastic properties of materials, so as to cope with small dimension of samples, where local transducers could not be placed. To obtain comparable results, global measurements were employed for the determination of elastic properties of panels too. Moreover, displacement transducers with gauge length of the same order of magnitude of the sample were used to permit measurement of the overall deformations of the heterogeneous sample. Determination of elastic constants was referred to the linear section of the load-displacement diagram ascending branch. With regard to the lime mortar samples, the relatively short linear section usually occurs between 1/2 and 2/3 of the ascending brunch.

	E (Mpa)	ν
bricks	1545,35	0,117
cement mortar	5381,18	0,120
lime mortar	309,12	0,100

Tab. I: Young's modulus and Poisson coefficient for the constituent materials of masonry panels.

Experimental values of Young's modulus and Poisson's ratio of the constituent materials are reported in tab. I. Experimental values of Young's modulus and Poisson's ratio of masonry panels are shown in tab. II.

6. RESULTS AND CONCLUSIONS

Some observations can be made while comparing the results of the homogenization techniques against the results obtained with the laboratory tests.
The values of the elastic constants determined with the two homogenization techniques and by means of experimental analysis are shown in tab. II.

micromechanical model	lime mortar masonry			cement mortar masonry		
	E (MPa)	mE (MPa)	G (MPa)	E (MPa)	mE (MPa)	G (MPa)
numerical model	1020,06	9511,4	417,28	1792,6	17410	779,91
suggested model	887,47	8049,1	474,80	1867,3	16121	949,70
experimental values	807,62	5115,4	-	1884,0	25372	-

Tab. II: Young's modulus, transverse elastic constant and shear modulus for masonry panels, as determined by means of the numerical homogenization procedure, the proposed close-form procedure, and the experimental analysis.

With regard to cement mortar masonry, overall elastic constants, determined by means of both the homogenization procedures, fit to experimental results. On the contrary, close-form methods results are closer to experimental data for lime mortar panels; in this case the numerical prediction overestimates Young's modulus by more than 20% of its value. However, both homogenization techniques exceed experimental results, probably due to the real mechanical behavior of lime mortar, roughly approximated by linear elasticity and by the perfect interface bonding hypothesis.
Some observations can be made about the localization process of lime mortar panels; in fact, the ratio of the phases stiffness in this case is closer to the typical masonry one (bricks are stiffer than mortar).
The values of the average stress of the phases produced by a uniaxial unitary vertical compression (fig. 4), and calculated with the average localization process for lime mortar masonry, are shown in tab. III.
The local stress state (fig. 5) is well describer by means of the numerical localization process : nothing can be added to common knowledge about the distribution of stress in masonry; however, we can notice the absence of high stress gradient in mortar phase, so that the average localization of mortar stress is quite representative of the real one. On the contrary, the numerical localization of brick stress shows large variation of the stress function, so that it can be asserted that, when the ratio of the phase stiffness is significant, the average localization process is not able to give any useful information about the local state.

	brick	mortar
σ_x	0,0014745	-0,005898
σ_y	-3,050037	-2,799858

Tab. III: Average stress localization as obtained by means of the close-form solution of the micromechanical model proposed by the authors; the local stress state is produced by a vertical compression of 3 MPa, applied to a lime mortar masonry wall.

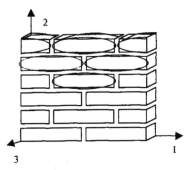

Fig. 1 Schematic representation of a masonry wall subjected to the analysis, with elliptic cylinders representing bricks (proposed micromechanical model).

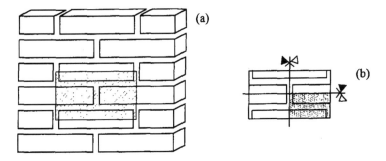

Fig. 2 (a) a masonry wall with a representative cell, characterized by two symmetry axes; (b) the region subjected to the numerical analysis (screen).

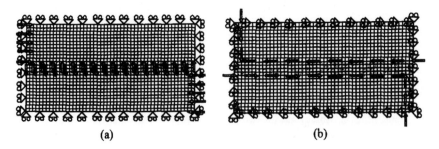

(a)

(b)

Fig. 3 The panels subjected to the numerical analysis. (a) Constrained panel corresponding to unitary uniform extensional strain; (b) Constrained panel corresponding to unitary uniform shear strain.

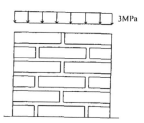

Fig. 4 The case example for the comparison of the localization processes.

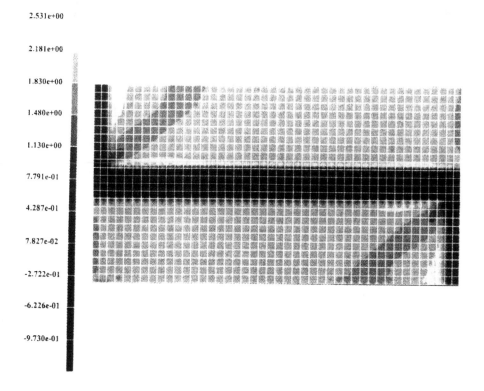

Fig. 5 Results of the numerical localization process: horizontal stress σx produced by a vertical compression of 3 Mpa.

REFERENCES

[1] Pande, G. N., Lang, J. X., Middleton, J., "Equivalent elastic moduli for brick masonry", Computers and Geotechnics, Vol. 8, 243-265, (1989).

[2] Ranocchiai, G., Rovero, L., "Un modello micromeccanico per la modellazione del materiale muratura", atti del Convegno Nazionale La meccanica delle murature tra teoria e progetto, Messina, 285-294, (1996).

[3] Briccoli Bati, S., Ranocchiai, G., Rovero, L., "A micromechanical model for linear homogenization of brick masonry", subjected to RILEM, Materials and Structures.

[4] Eshelby, J. D., "The determination of elastic field of an ellipsoidal inclusion and related problems", Proc. Roy. Soc. London, vol. 241 (A), 376-396, (1957).

[5] Mori, T., Tanaka, K., "Average stress in matrix and average elastic energy of materials with misfitting inclusions", Acta Metallurgica, Vol. 21, 571-574, (1973).

[6] Suquet, P.M., "Elements of homogenization for inelastic solid mechanics", in Sanchez-Palencia, Zaoui (Ed.), Homogenization techniques for composite media, 193-279, (1987).

ACKNOWLEDGEMENTS:
this work was partially supported by Ministry of University and of Scientific and Technological Research (MURST 40%).

FAILURE PROCESS OF VERTICAL JOINTS BETWEEN COMPOSITE PREFABRICATED MASONRY WALL ELEMENTS

D. Pume PhD, Senior Research Worker
Klokner Institute, Czech Technical University in Prague, Solínova 7, 166 08 Praha 6, Czech Republic

1. ABSTRACT

The aim of the paper is to present the simple analysis of the failure process of the storey-high vertical joints between the composite prefabricated masonry wall elements. The structural lay-out of elements consists of the planar subtle reinforced concrete frame and the mortarless confined masonry. The lateral faces of the elements are not provided with concrete shear keys. The shear resistance of the vertical joint results from the structural behaviour of two special tie-beams at the level of each floor slab. The simplified calculation procedure of the shear stiffness of the vertical joints is given.

2. INTRODUCTION

The prefabricated wall elements JCKD made from concrete and clay brick masonry were designed in the former Design and construction institute of clay brick industry (PKÚ CVKP) in Brno [1]. The elements were used in the external walls of apartment houses. The dimensions of the elements: the height is 2.8 m to 3.0 m, the width is 1.2 m to 2.4 m, the thickness is 0.4 m. The structural lay-out of elements consists of the planar subtle reinforced concrete frame and the mortarless confined masonry made from vertically perforated clay masonry units (fig. 1). The upper and lower edges of the elements are made from concrete. The lateral faces of the elements are made either from concrete (type A, figs. 1 and 2, upper part) or from masonry (type B, fig. 2, lower part).

Keywords: Vertical shear joints, Failure process, Prefabricated wall elements, Mortarless masonry, Clay masonry units

Computer Methods in Structural Masonry – 4, edited by G.N. Pande, J. Middleton and B. Kralj.
Published in 1998 by E & FN Spon, 11 New Fetter Lane, London EC4P 4EE, UK. ISBN: 0 419 23540 X

The composite action of reinforced concrete slender columns and mortarless masonry under vertical compressive loads, resulting from the lateral restraint produced by the subtle reinforced concrete frame was studied very carefully [2]. The design formulae which resulted from these studies were included into the Czech Standard ČSN 73 1103 Design of the prefabricated clay brick wall elements.

3. THE STRUCTURAL LAYOUT OF THE VERTICAL JOINTS

The vertical joint under consideration consists of three parts. The first part is the unreinforced vertical mortar joint between two prefabricated elements along the clear storey-height. The second part is the prefabricated floor beam (*el* in fig. 3). The third part is the tie-beam cast in situ (*fv* in fig 3). Both beams are provided with the longitudinal horizontal reinforcing bars. The reinforcement ratio of the total vertical joint is very low (see next chapter). The shape and the material exposition in the lateral surfaces of wall elements depend on their structural layout (designated as the type A or type B). No concrete shear keys were used in the vertical joints.

4. TEST SPECIMENS

The methodology of experimental investigations was built up in the Klokner Institute, in the institute PKÚ CVKP and in the Technical institute of testing in construction in the town České Budějovice (TZÚS). The loading tests of six specimens were performed in TZÚS [3].

The test specimens (fig. 3) with the total height h_{tot} = 1580 mm and with the total width w_{tot} = 1220 mm involved the full scale vertical joint (either type A or type B) with the height h_j = 1130 mm. The vertical cross section of the joint was A_{bj} = 1130 . 400 = = 452 000 mm². Three specimens were designated as A1, A2 and A3, three specimens as B1, B2 and B3. Before placing the infill concrete, the contact surfaces of the elements *a*1, *a*2 and a3 were painted with the oil-colour in order to initiate the separation of wall elements from the mortar in the vertical joints.

The total cross section area of the floor beam and of the tie-beam was A_{bel} = 89 700 mm² and A_{bfv} = 6 300 mm², respectively. The reinforcing bars in the beams were 5 ∅ E10 (A_{sel} = 393 mm²) and 1 ∅ E16 (A_{sfv} = 201 mm²), respectively. The reinforcement ratio was μ_{el} = 0.0044 and μ_{fv} = 0.0319, respectively.

The vertically perforated clay bricks with the dimensions 360 x 220 x 145 mm, the volume density ρ_{vol} = 940 kg.m^{-3} and the compressive strength f_{uc} = 13.4 N.mm^{-2} were used. The mean cube compressive strength of the concrete in the floor beams was $f_{200. el}$ = 23.0 N.mm^{-2}. The mean cube compressive strength of the concrete in the tie-beams was $f_{200,fv}$ = 8.4 N.mm^{-2}.

5. TESTING PROCEDURE AND MEASURED QUANTITIES

Two loading forces Q were applied to the upper and lower steel beams *su* and *sl* which were fixed to the concrete elements by means of steel bolts (fig. 3). The test specimens were subjected to the monotonously increasing load $Q \in <Q = 0; Q_{obs}>$. Within this interval, the course of loading was interrupted by regularly distributed pauses at the load levels Q_i in order to register the measured quantities and to monitor the formation and development of cracks and other failures. The horizontal and vertical displacement components (w_i, δ_i) were measured at levels U, M and L (fig. 3) and the mean values of w_{iU}, w_{iM}, w_{iL} and δ_{iU}, δ_{iM}, δ_{iL} were considered as w_i and δ_i, respectively.

6. FAILURE PROCESS OF VERTICAL JOINTS BETWEEN THE ELEMENTS JCKD

The failure process of the vertical joints within the interval $<0; Q_{obs}>$ involves the set of six successive particular failure intervals: $<0; Q_{vj}>$, $< Q_{vj}; Q_{hel}^{(1)} >$, $<Q_{hel}^{(1)}; Q_{hel}^{(2)} >$, $<Q_{hel}^{(2)}; Q_{bj}>$, $<Q_{bj}; Q_{cr}>$, $< Q_{cr}; Q_{obs}>$,

where:

Q_{vj} is the shear force at which the separation of the infill mortar from the element $a1$ or $a2$ occurs,

$Q_{hel}^{(1)}$ is the shear force at which the first vertical crack appears in the floor beam in the distance of 150 to 250 mm from the axis of symmetry of the joint,

$Q_{hel}^{(2)}$ is the shear force at which the second vertical crack, symmetrical with the first one, appears,

Q_{bj} is the shear force at which the crack appears in any horizontal mortar joint,

Q_{cr} is the shear force at which the vertical crack appears in any wall element.

If $Q \geq Q_{hel}^{(1)}$, only the floor beam el and the cast in-situ tie-beam fv resist the shear force Q. The values of the boundaries between the particular intervals of all joints are given in table 1. The values were estimated from the course of the curves describing the relationships $Q = Q (w_U)$, $Q = Q (w_M)$, $Q = Q (w_L)$, $Q = Q (\delta_U)$, $Q = Q (\delta_M)$, $Q = Q (\delta_L)$ and from the crack development records (fig. 4). The fictitious shear stress τ_{fi} in table 1 relates to the fictitious vertical cross-section area of the joint $A_b = A_{bel} + A_{bfv} = 96\,000$ mm^2.

7. THE STRUCTURAL MODEL OF THE BEAMS *el* AND *fv* OVER-LAPPING THE VERTICAL JOINT

The couple of both floor-beams is modelled as the beam with the constant stiffness EI, resting on the elastic half-plane, produced by the composite wall elements JCKD. The beam is loaded by the couple of forces (fig. 5). The points of the application of forces Q coincide with the positions of the cracks estimated by testing.

The complete solution is prevented by the unknown value of support stiffness coefficient k. Therefore, the solution consists in the comparison of the structural behaviour of the both types A and B. The loading case with the single force Q (fig. 5b) is considered instead of the loading case with two forces Q (fig. 5a), since the ratio between the deflections $\kappa_{\delta BA} = \delta (0)_B : \delta (0)_A$ is to be estimated. It is obvious that the conclusion based on the value $\kappa_{\delta BA}$ derived for the loading case shown in fig. 5b will be valid also for the loading case shown in fig. 5a. The notation $\delta\,(0)_B$ and $\delta\,(0)_A$ denotes the deflection of the couple of beams el and fv in the section (0) in the test specimens B and A, respectively. The bending moment in the section (x) with the co-ordinate x is given by

$$M(x) = M(0)\,F_M^{(x)} = \gamma_M\;\lambda\,Q\,F_M^{(x)}, \tag{1}$$

the deflection of the beam in the section (x) is given by

$$\delta\,(x) = (0)\,F_\gamma^{(x)} = \gamma_\delta\;\lambda^3\,Q\,F_\delta^{(x)}, \tag{2}$$

where $M(0)$ is the bending moment in the section (0), $\delta(0)$ is the deflection in the section (0), $F_M^{(x)}$ and $F_\delta^{(x)}$ is the function expressing the course of the bending moment and of the deflection along the beam length, respectively, γ_M and γ_δ is the constant resulting from the solution of the beam, respectively, λ is the characteristic of the beam on the elastic embedment, given by

$$\lambda = \left(\frac{4EI}{k}\right)^{\frac{1}{4}}. \tag{3}$$

The first cracks in the beams in the test specimens A and B appeared under the identical $M(0)$, caused by $\overline{Q}_{hel,A}^{(1)} \neq \overline{Q}_{hel,B}^{(1)}$, since $(EI)_A = (EI)_B$ and $k_A \neq k_B$. This feature may be expressed by

$$\gamma_M\,(4EI)_A^{\frac{1}{4}}\,(k_A)^{-\frac{1}{4}}\,\overline{Q}_{hel,A}^{(1)} = \gamma_M(4EI)^{\frac{1}{4}}(k_B)^{-\frac{1}{4}}\,\overline{Q}_{hel,B}^{(1)}, \tag{4}$$

from which it follows

$$\left(\frac{k_A}{k_B}\right)^{\frac{1}{4}} = \frac{\overline{Q}_{hel,A}^{(1)}}{\overline{Q}_{hel,B}^{(1)}}. \tag{5}$$

The ratio $\kappa_{\delta BA}^{(1)}$ and $\kappa_{\delta BA}^{(2)}$ between the deflections in the section (0) in the test specimens B and A, when the first or the second crack appears, is given by

$$\kappa_{\delta BA}^{(1)} = \frac{\delta(0)_B}{\delta(0)_A} = \left(\frac{k_A}{k_B}\right)^{\frac{1}{2}} = \left(\frac{\overline{Q}_{hel,A}^{(1)}}{\overline{Q}_{hel,B}^{(1)}}\right)^2 = 1.33, \tag{6}$$

and by

$$\kappa_{\delta BA}^{(2)} = \left(\frac{\overline{Q}_{hel,A}^{(2)}}{\overline{Q}_{hel,B}^{(2)}}\right)^2 = 1.21. \tag{7}$$

respectively.

The observed value $\kappa_{\delta BA,obs}^{(2)} = 1.20$ coincides with $\kappa_{\delta BA}^{(2)} = 1.21$ very well.

8. THE SHEAR STIFFNESS OF THE VERTICAL JOINT

The shear stiffness C of the vertical joint within the elastic, almost linear interval, i.e. within the first three intervals of the failure process $<Q = 0; Q_{hel}^{(2)}>$, is given by

$$C = \frac{G_{bel}\, A_{bel} + G_{bfv}\, A_{bfv}}{l_G},$$

(3)

where

G_{bel} is the shear modulus of elasticity of the concrete in the floor beam *el*,

G_{bfv} is the shear modulus of elasticity of the concrete in the tie-beam cast in situ *fv*,

l_G is the fictitious horizontal dimension of the shear area; according to the test results, the value $l_{GA} = 800$ mm and $l_{GB} = 2080$ mm related to the type A and the type B, respectively, should be considered.

It is supposed that the shear modulus G is the 0.42-multiple of the basic modulus of elasticity E. The mean values of the shear stiffness C within the elastic interval $<Q = 0; Q_{hel}^{(2)}>$ estimated by testing are $C_A = 1350$ kN.mm^{-1} and $C_B = 510$ kN.mm^{-1}.

CONCLUSIONS

1. The failure processes of vertical joints between the composite prefabricated masonry wall elements (types A and B) are treated in this paper. The shear resistance of the joint results from the structural behaviour of the two beams at the level of each floor slab.
2. The measurement of strains and displacements enabled to establish the set of six successive particular failure intervals within the total failure process.
3. The structural behaviour of the assemblage of both beams strongly depends on the structural layout of the wall elements.
4. The simplified calculation procedure of the shear stiffness of both types of vertical joints verified by tests has been proposed.

References

[1] Prokeš, B. - Šamonil, K. - Krubová, A.: Jednovrstvý celokeramický dílec z cihelných tvarovek CD-IDA. Zásady pro navrhování a výrobu (in Czech). (One-wythe composite concrete masonry wall element with clay bricks CD-IDA. Rules for design and production.) Ed. PKÚ CVKP. Brno 1984.
[2] Pume, D.: Stěnové dílce ze zdiva a železobetonu (in Czech). (Composite concrete masonry wall elements.) Stavebnícky časopis (Structural Journal), 36 (1988), 4, 297 - 319.
[3] Pume, D.: Proces porušování svislých styků keramických stěnových dílců (in Czech). (Failure process of vertical joints between the composite concrete masonry wall elements.) Stavebnícky časopis (Structural Journal), 36 (1988), 12, 889 - 904.

Table 1. The values Q_i of boundaries between the successive particular intervals of the failure process

The joints A

Notation of the boundary	The value Q_i of the joint in the test specimen			$\overline{Q}_i$	τ_{fi}
	A1	A2	A3		
-	kN	kN	kN	kN	N.mm^{-2}
Q_{obs}	164	195	180	180	1.87
Q_{cr}	153	153	153	153	-
Q_{bj}	119	136	136	130	-
$Q_{hel,A}^{(2)}$	102	119	136	119	1.24
$Q_{hel,A}^{(1)}$	85	85	102	91	0.95
Q_{vj}	17	34	34	28	-

The joints B

Notation of the boundary	The value Q_i of the joint in the test specimen			$\overline{Q}_i$	τ_{fi}
	B1	B2	B3		
-	kN	kN	kN	kN	N.mm^{-2}
Q_{obs}	160	153	136	150	1.56
Q_{cr}	136	136	119	130	-
Q_{bj}	119	119	102	113	-
$Q_{hel,B}^{(2)}$	119	102	102	108	1.12
$Q_{hel,B}^{(1)}$	85	68	85	79	0.83
Q_{vj}	34	17	34	28	-

Fig. 1 Scheme of composite concrete masonry prefabricated wall element JCKD.

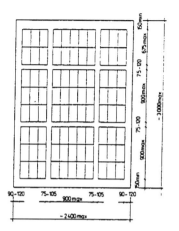

Fig. 2 Horizontal sections through two types A and B of vertical joint. The connected surfaces of wall panels $(el)^1$ and $(el)^r$ are either from concrete (type A) or from the vertically perforated clay bricks CD-IDA (type B).

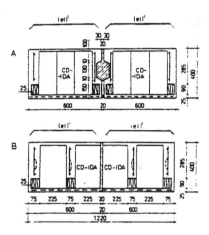

Fig. 3 Test specimen under the forces Q, with the vertical joint along the axis T - T; al, $a2$ - the parts of wall panels; $a3$ - parts of floor slabs; el - L-shaped beam; fv - tie-beam; su - upper steel beam; sl - lower steel beam; U, M, L - points where horizontal components w_U , w_M , w_L and vertical components δ_U , δ_M , δ_L were measured.

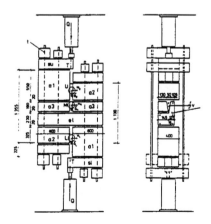

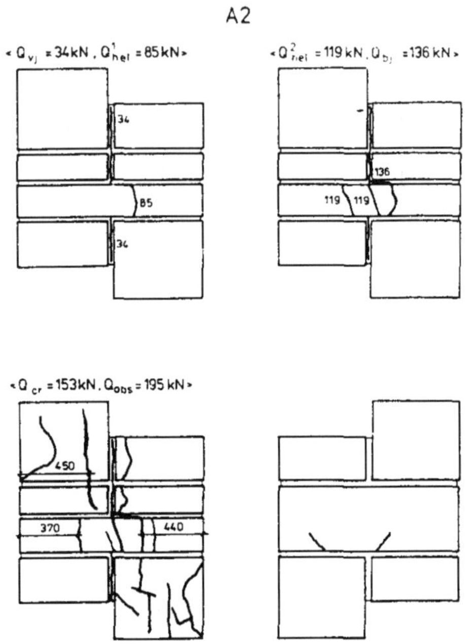

A2

Fig. 4 Cracks on the surface of the specimen A2 during the second interval (interval $<Q_{vj};\ Q_{hel}^{(1)}>$), the fourth interval (interval $<Q_{hel}^{(2)};\ Q_{bj}>$) and the sixth interval of the failure process of joint (interval $<Q_{cr};\ Q_{obs}>$). The number nearby the crack indicates the value of shear force Q [kN], when the crack has appeared.

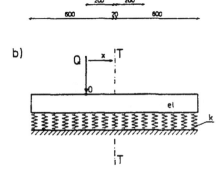

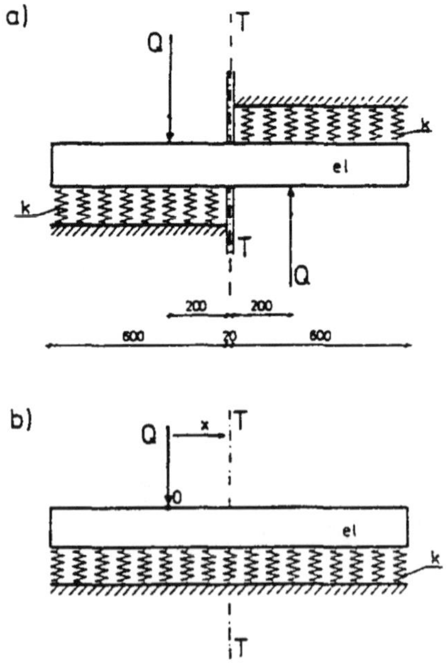

Fig. 5 The concrete beams *el* overlapping the plane *T* - *T* of vertical joint on the elastic embedments with the stiffness coefficient k: a) two-sided embedded beam loaded by two equal forces Q, b) one-sided embedded beam loaded by one force Q.

LONGITUDINAL VIBRATIONS OF MASONRY BEAMS

C. Casarosa Professor[1], **M. Lucchesi** Professor[2],
C. Padovani Researcher[3], **A. Pagni** Researcher[3] and
G. Pasquinelli Researcher[3]

[1]*Dipartimento di Energetica, University of Pisa, Via Diotisalvi 2, Pisa, Italy;* [2]*Dipartimento di Costruzioni, University of Florence, Piazza Brunelleschi 6, Florence, Italy and* [3]*CNUCE-C.N.R., Via Santa Maria 36, Pisa, Italy*

1. ABSTRACT

The explicit solution to a problem of free longitudinal vibrations of a beam made of a bimodular material is determined. The solution corresponding to a no-tension material is then obtained as a special case.

2. INTRODUCTION

In many equilibrium problems the constitutive behaviour of masonry is described in a sufficiently realistic way by the no-tension model [1]. In that model it is assumed that the infinitesimal strain is the sum of an *elastic* part and a *fracture* part, and that the stress, negative semi-definite, depends linearly on the former and is orthogonal to the latter. Thus, the actual cracks are represented in the model by a non-zero fracture strain, without resorting to the introduction of displacement field discontinuities. Although this constitutive law has allowed study of problems relevant to real-world application, it has yet to be employed in dynamic analysis.

In the present paper we examine the longitudinal vibrations of a fixed-ended beam made of a no-tension material. The beam is displaced initially at its midpoint and then left to oscillate freely. In keeping with the adopted constitutive assumptions, we imagine that the portion under tension extends in such a way as to maintain the continuity of the displacement field; this, in such a one-dimensional example, is tantamount to assuming a nil Young's modulus under tension.

We first consider a material whose in-tension Young's modulus, E_t, is a fraction of that

Keywords: Non-linear dynamics, no-tension materials, shock waves.

Computer Methods in Structural Masonry – 4, edited by G.N. Pande, J. Middleton and B. Kralj.
Published in 1998 by E & FN Spon, 11 New Fetter Lane, London EC4P 4EE, UK. ISBN: 0 419 23540 X

under compression, E_c, $E_t = \beta^2 E_c$, with $\beta \in [0,1]$; we are dealing with a bimodular material conforming to a no-tension material for $\beta = 0$, and to a linear elastic one for $\beta = 1$. The solution is then determined up to a certain instant as a function of β, and finally, its limit for β tending to zero is calculated. The explicit solution can be achieved by virtue of the fact that, despite the non-linearity of the material, the characteristics are straight lines.

For each value of β, the strain and velocity are constant in a certain number of regions in the plane (x,t), where x is the abscissa along the beam and t the time, and are discontinuous along the curves separating the different regions.

A feature of the solution determined in this way is that the mechanical energy of the beam (the sum of the potential and kinetic energies) is not constant with respect to time. In fact, a shock wave results at the interface between the taut and compressed parts of the beam, which determines a loss of mechanical energy and leads to progressive decay of the solution. The energy dissipated per unit time is a decreasing function of β and, of course, vanishing for $\beta = 1$.

Although the formation of shock waves in non-linear elastic solids, with the consequent loss of mechanical energy, is a well-know phenomenon [2], [3], it has, to our knowledge, never been encountered in no-tension solids in motion.

3. FORMULATION OF THE PROBLEM

Let us consider a fixed-ended beam with length l, made of a bimodular material. For $u(x,t)$, the longitudinal displacement of the point having abscissa x at time t, free vibrations of the beam are governed by the partial differential equation

$$u_{tt}(x,t) - \left(\kappa *(u_x) \right)^2 u_{xx}(x,t) = 0, \tag{1}$$

where $\kappa* = \sqrt{E*/\rho}$, with ρ the material density, and

$$E*(u_x) = E_c \quad \text{if } u_x < 0, \qquad E*(u_x) = \beta^2 E_c \quad \text{if } u_x > 0, \tag{2}$$

with the assigned boundary conditions

$$u(0,t) = u(l,t) = 0, \quad \text{for } t \geq 0. \tag{3}$$

In this paper we limit ourselves to considering cases in which the initial velocity is zero and thus can write

$$u(x,0) = a(x) \quad \text{and} \quad u_t(x,0) = 0, \quad \text{for } x \in [0, l], \tag{4}$$

with a a given continuous function of x.

The solution of equation (1) can be determined by using the characteristics method. To this end, it is necessary to detect the shocks, i.e. the curves in plane (x,t) where velocity u_t and strain u_x are discontinuous, and determine the corresponding jumps. This can be done with the help of the Rankine-Hugoniot conditions [4]

$$s[u_x] = -[u_t] \quad \text{and} \quad s[u_t] = -\frac{1}{\rho}[E * u_x], \tag{5}$$

where s is the speed of shock propagation and square brackets [ ] denote the jump (right minus left) of the enclosed quantities across the shock.

In general, the solution is not unique but the set of solutions can nonetheless be reduced by imposing the so-called Lax's E-condition [4]. This requires that when the characteristic beginning on either side of the discontinuity curve is continued in the direction of positive t, it will intersect the line of discontinuity. When the line of discontinuity constitutes the interface between the taut and compressed portions of the beam, the slopes $\dfrac{dx}{dt}$ of the characteristics are $\pm\beta\kappa$ and $\pm\kappa$, respectively, before and after traversing the shock; therefore, Lax's condition takes the form

$$-\beta\kappa > s > -\kappa, \quad \text{for } s < 0, \quad \text{and} \quad \kappa > s > \beta\kappa, \quad \text{for } s > 0. \tag{6}$$

It can be proved that condition (6) implies that the mechanical energy of the beam

$$\Phi(t) = \int_0^l \left(\frac{1}{2}\rho u_t^2 + \frac{1}{2}E * u_x^2 \right) dx \tag{7}$$

is a non-increasing function of time. More precisely, it can be verified that

$$\frac{d\Phi}{dt} = \frac{\rho s \kappa^2}{2}\left\{ \left(\frac{\kappa^2}{s^2}-1\right)(u_x^+)^2 + \beta^2\left(1-\frac{\kappa^2\beta^2}{s^2}\right)(u_x^-)^2 \right\}, \quad \text{for } s < 0 \tag{8}$$

$$\frac{d\Phi}{dt} = \frac{\rho s \kappa^2}{2}\left\{ \beta^2\left(\frac{\kappa^2\beta^2}{s^2}-1\right)(u_x^+)^2 + \left(1-\frac{\kappa^2}{s^2}\right)(u_x^-)^2 \right\}, \quad \text{for } s > 0, \tag{9}$$

from which, in view of (6), the desired result follows.

4. AN EXAMPLE

In this section we propose to explicitly solve equation (1) with boundary conditions (3) and initial conditions (4), in the case in which $\hat{u}$ is the piecewise linear function

$$\hat{u}(x) = 2ax, \quad \text{for } 0 \le x \le \tfrac{1}{2}l \quad \text{and} \quad \hat{u}(x) = 2a(l-x), \quad \text{for } \tfrac{1}{2}l \le x \le l, \tag{10}$$

with $a > 0$; thereby, at the initial time the left half of the beam is subjected to tension, and the right one, compression. In the plane (λ, τ), with $\lambda = \dfrac{x}{l}$ and $\tau = \kappa t$, let us consider the strip $0 \le \lambda \le 1$, $\tau \ge 0$, where the characteristics curves are drawn (Figure 1). Let V_1, V_2, A and C be the points having co-ordinates $(0,0)$, $(1,0)$, $(\tfrac{1}{2},0)$ and $(1, \tfrac{1}{2})$, respectively. The segment AC coincides with the characteristic issuing from point A and having slope κ. Each point of the triangular region AV_2C is crossed by two characteristics of slope κ and $-\kappa$, starting at segment AV_2. Therefore, recalling that the value of $u_t - \kappa u_x$ is constant along the first characteristic, while that of $u_t + \kappa u_x$ is constant along the second, it can be immediately deduced from (5)₂ and (10) that in this region we have

$$u_x(x,t) = -2a \quad \text{and} \quad u_t(x,t) = 0.$$ (11)

By using a similar procedure, it is possible to deduce that in the triangular region V_1AB it holds that

$$u_x(x,t) = 2a \quad \text{and} \quad u_t(x,t) = 0.$$ (12)

Now, we need to determine the shock and the values of u_x^+ and u_t^+. To this end, let us observe that relations (12) provide the values of u_x^- and u_t^-, whereas conditions (10) and (5) allow us to write $u_t^+ + \kappa u_x^+ = -2\kappa a$. Therefore, again from (5), we deduce that the shock is a straight line with slope

$$s = -\tfrac{1}{2}\kappa(1+\beta^2)$$ (13)

and that

$$u_x^+ = \frac{-2a(1-\beta^2)}{(3+\beta^2)}, \quad u_t^+ = \frac{-4a\kappa(1+\beta^2)}{(3+\beta^2)}.$$ (14)

By means of the foregoing we have determined the solution for $\tau \le 1.75$. The values of u_x and u_t are constant in the regions bounded by the discontinuity lines (shown in bold type). The co-ordinates of the vertices of these regions are listed below.

$$B = \left(0, \frac{1}{1+\beta^2}\right); \quad D = \left(\frac{1+3\beta^2}{4(1+\beta^2)}, \frac{5+3\beta^2}{4(1+\beta^2)}\right); \quad E = \left(0, \frac{3}{2}\right);$$

$$F = \left(\frac{\beta^3 + 2\beta^2 - \beta + 2}{2(1+\beta)(1+\beta^2)}, \frac{\beta^3 + 2\beta^2 + \beta + 4}{2(1+\beta)(1+\beta^2)}\right); \quad G = \left(0, \frac{\beta^3 + 2\beta^2 + 3}{(1+\beta)(1+\beta^2)}\right);$$

$$H = \left(1, \frac{3\beta^5 + 5\beta^4 + 8\beta^3 + 20\beta^2 + \beta + 15}{2(1+\beta)(1+\beta^2)(3\beta^2 - 2\beta + 3)}\right); \quad L = \left(1, \frac{5}{2}\right);$$

$$M = \left(\frac{(1-\beta)(3+\beta^2)}{4(1+\beta)(1+\beta^2)}, \frac{5\beta^3 + 7\beta^2 + 3\beta + 9}{4(1+\beta)(1+\beta^2)}\right).$$ (15)

The values of u_x and u_t in the different regions are summarised in the following Table.

Region	u_x	u_t
ACDB	$-\dfrac{2a\left(1-\beta^2\right)}{3+\beta^2}$	$-\dfrac{4a\kappa\left(1+\beta^2\right)}{3+\beta^2}$
BDE	$-2a$	0
CFD	0	$-\dfrac{2a\kappa\left(1+3\beta^2\right)}{3+\beta^2}$
DFME	$-\dfrac{4a\left(1+\beta^2\right)}{3+\beta^2}$	$\dfrac{2a\kappa\left(1-\beta^2\right)}{3+\beta^2}$
CHF	$\dfrac{2a\left(1+3\beta^2\right)}{\beta\left(3+\beta^2\right)}$	0
CHF	$\dfrac{2a\left(1+3\beta^2\right)}{\beta\left(3+\beta^2\right)}$	0
EMG	$-\dfrac{2a\left(1+3\beta^2\right)}{\left(3+\beta^2\right)}$	0
FHNM	$-\dfrac{16a\beta(1-\beta)\left(1+\beta^2\right)}{\left(3+\beta^2\right)\left(3\beta^3-\beta^2+5\beta+1\right)}$	$-\dfrac{2a\kappa\left(3\beta^2-2\beta+3\right)(1+\beta)^3}{\left(3+\beta^2\right)\left(3\beta^3-\beta^2+5\beta+1\right)}$

Table 1

The only discontinuity lines whose crossing causes a loss of mechanical energy are AB and FH; their traversal in the direction of increasing τ determines the passage from a zone where the beam is under tension to one where it is under compression. The mechanical energy dissipated per unit time through the shocks AB and FH can be calculated by means of relations (8) and (9), respectively, by using the corresponding values of u_x^- and u_x^+. For AB it is a simple matter to arrive at

$$\frac{d\Phi}{dt} = -\rho\kappa^3 a^2 \frac{\left(1-\beta^2\right)^2\left(1+\beta^2\right)}{\left(3+\beta^2\right)} \; ; \tag{16}$$

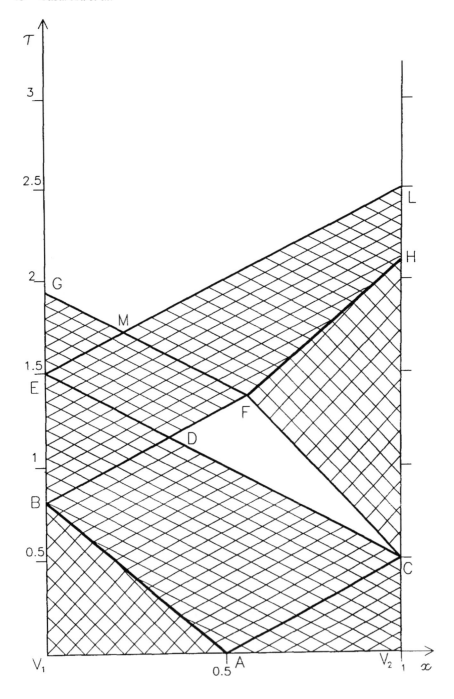

Figure 1. Behaviour of the characteristics in plane (λ, τ), for $\beta \neq 0$.

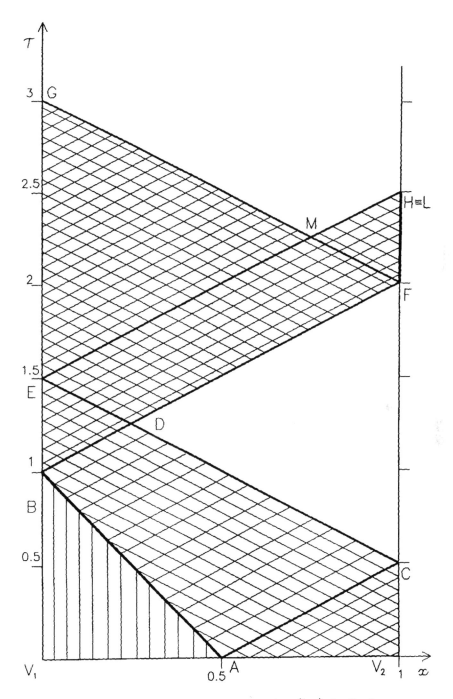

Figure 2. Behaviour of the characteristics in plane (λ, τ), for $\beta = 0$.

for FH, on the other hand, the calculation is quite complex and has been omitted here.

In region CFD it holds that $u_x = 0$, and the characteristics are therefore undefined; the presence of this region enables the transition from ACDB, where the beam is compressed, to CHF, where it is taut, and can therefore be thought of as a rarefaction wave.

Figure 2 presents the solution corresponding to $\beta = 0$. First of all, it can be seen that in region ABV_1, because the in-tension Young's modulus is vanishing, the two families of characteristics degenerate into a unique family with slope equal to zero, whereas in region AV_2C they remain unchanged. Moreover, as can be easily deduced from relations (15) and Table 1, for $\beta = 0$, points H and L coincide and point F belongs to the straight line $\lambda = 1$. Therefore, the rarefaction wave spreads, and region CHF degenerates to a segment where the strain increases indefinitely and the velocity is discontinuous. This fact also explains why the velocity is non-zero in the region EHM, even though this region borders the straight line $\lambda = 1$, in apparent violation of boundary condition (3).

The loss of mechanical energy brings about progressive decay of the solution. Considering, for example, the beam midpoint, it can be seen that its velocity is negative from the initial moment up to $\tau_0 = \dfrac{3}{2} + \dfrac{1}{1+\beta^2}$, and that the ratio between the displacement reached at this instant and that at the start equals

$$\frac{1}{a} u\left(\frac{1}{2}, \tau_0\right) = \frac{-2(1+3\beta^2)}{(3+\beta^2)(1+\beta^2)}. \tag{17}$$

Thus, in the particular case of $\beta = 0$, after the first half oscillation, the maximum displacement is two thirds the initial one.

As singular as these results may seem, they are a direct consequence of applying the constitutive laws of no-tension materials. However, whether this model can realistically be applied to solving dynamic problems of masonry must as yet remain an open question.

ACKNOWLEDGEMENT. The financial support of Progetto Finalizzato Beni Culturali of C.N.R. is gratefully acknowledged.

REFERENCES

1. Del Piero G., "Constitutive equations and compatibility of the external loads for linear elastic masonry-like materials", *Meccanica* 24, 1989, pp. 150-162.
2. Lax P. D., "Development of singularities of solutions of nonlinear hyperbolic partial differential equations", *J. Math. Phys.* 5, 1964, pp. 611-614.
3. Silhavý M., The Mechanics and Thermodynamics of Continuous Media, Springer-Verlag, 1997.
4. Lax P. D., "Hyperbolic systems of conservations laws and the mathematical theory of shock waves", Society for Industrial and Applied Mathematics, Philadelphia, 1972.

STRESS–STRAIN RELATIONSHIPS OF MASONRY MADE FROM CZECH CLAY UNITS PRODUCED BETWEEN 1955 AND 1992

D. Pume PhD, Senior Research Worker
Klokner Institute, Czech Technical University in Prague, Solínova 7, 166 08 Praha 6, Czech Republic

1 ABSTRACT

The aim of the paper is to present in a very comprehensive form the significant results of load tests of 275 masonry specimens (small walls) under compression, made from clay units produced between 1955 and 1992 The data are assigned to be exploited in both the Czech and European standardization studies and in the verification of structural models of masonry walls under compression.

2 INTRODUCTION

In the Czech Republic, many types of clay units have been produced between 1955 and 1997. The changes of their mechanical and other technical parameters have been caused by the increasing requirements upon masonry walls in this country. Undoubtedly, the load tests of masonry made from new types of clay units should be the permanent component of the technical development activities in that field.

The author has been able to supervise the majority of the consistent load tests under compression of all important types of masonry used between 1955 and 1997 in his country 334 masonry small walls have been tested.

Keywords. Clay masonry units, Masonry made from clay units, Stress-strain relationship, Structural model of masonry wall

Computer Methods in Structural Masonry – 4, edited by G.N. Pande, J. Middleton and B. Kralj.
Published in 1998 by E & FN Spon, 11 New Fetter Lane, London EC4P 4EE, UK. ISBN: 0 419 23540 X

3 DEVELOPMENT OF CLAY UNIT PARAMETERS IN THE CZECH REPUBLIC

Thermal insulation requirements imposed on external walls influence the perforation pattern, dimensions and also compressive strength of clay units. The required thermal resistance value for external walls was considered to be $R_w = 0.6$ m² K/W for many years (from 1920 through 1980) This was raised to $R_w = 1.0$ m² K/W in 1980 and to $R_w = 2.0$ m² K/W in 1993 Until 1980, the thickness of external walls made from solid clay bricks was $t_w = 440$ mm (usually designed without calculation of thermal resistance R_w). If the external wall was made from vertically perforated clay units, the thickness $t_w < 440$ mm was estimated through the appropriate calculation.

Before 1955, 290 140 65 mm solid clay bricks were prevalent. After 1955, more consistent use was made of vertically perforated clay units. At first, 240.115.113 mm vertically perforated bricks with a bulk density of $\rho_{vol} \in$ <1300 kg.m⁻³; 1550 kg.m⁻³> were produced. This brought about the reduction of the thickness of external walls from $t_w = 440$ mm to $t_w = 365$ mm All values of t_w indicated in this paper express the thickness of the wall without plaster, though almost all external walls are plastered, since the sufficiently frostresistant clay masonry units are produced in this country in a very limited extent.

Since about 1960, external walls have in most cases been designed as single wythe walls in header bond. During the period 1980-88, the thickness of external single wythe walls varied $t_w \in$ <365 mm; 390 mm>. From 1993 onwards, the thickness of external single wythe walls should be $t_w = 490$ mm, if the value of ρ_{vol} varies within the narrow interval, $\rho_{vol} \in$ <900 kg.m⁻³; 950 kg.m⁻³> or $t_w \in$ <390 mm, when $\rho_{vol} = 800$ kg.m⁻³. Internal walls are erected either from solid clay units, or from special vertically perforated units of length $l_h \leq 300$ mm, or from vertically perforated units (produced for external walls) with a length of $l_h > 300$ mm laid in stretcher bond, therefore $t_w = w_u$.

At present, in most cases, the compressive strength of clay units does not exceed the value $f_b = 25$ MPa. Only frostresistant clay units used in unrendered walls have a compressive strength of $f_b = 20$ to 35 MPa.

4. TEST RESULTS

All test specimens (tables 1 and 2) were small walls. Their dimensions and testing procedure corresponded with the provisions of RILEM BW 24 recommendations and subsequent standardization documents. The main results of testing each specimen were the compressive strength of masonry f_{obs} and the stress-strain relationship $\sigma = \sigma(\varepsilon)$ expressed by

$$\frac{\sigma}{f_{obs}} = \frac{K_s \, \eta - \eta^2}{1 + (K_s - 2) \, \eta},$$

(1)

where $\eta = \varepsilon/\varepsilon_{fu}$, ε_{fu} - ultimate strain of masonry; K_s - empirical coefficient estimated by the regression analysis.

5 EFFECT OF VERTICAL PERFORATION OF MASONRY UNITS

For many years, till 1970, the phenomenological approach to the determination of masonry compressive strength f_m, prevailed. Empirical equations expressing the values f_m, of many types of masonry, involving up to 5 empirical physical parameters were proposed and successfully applied to the structural design. The theoretical and experimental studies performed by E. Amrein, H. K. Hilsdorf, G. Schellbach and others caused a principal reversal in this point. Their studies resulted, among others, in the structural model of the single wythe masonry wall under compression. The model enables to express numerically the effect of vertical perforation pattern of clay masonry units on the masonry compressive strength. In this chapter, the model will be verified by author's test results [1]

The single wythe masonry wall is a wall without any longitudinal vertical masonry joint. If the thickness t_w of the wall is equal to the greater dimension, i.e to the length l_u of the unit ($t_w = l_u$), header bond was applied. If the thickness t_w is equal to the smaller dimension, i e to the width w_u ($t_w = w_u$), the stretcher bond was applied.

Let us consider the single wythe masonry wall (fig. 1) under the vertical uniformly distributed compressive stress σ_1. The values of compressive stress in any horizontal cross section of the wall, both in the masonry units (the stress σ_{uv}) and in the mortar bed joints (the stress σ_{mv}) are identical ($\sigma_1 = \sigma_{uv} = \sigma_{mv}$). In the Czech Republic, the masonry units with the compressive strength $f_u \in \,<10$ N mm^{-2} ; 20 N mm^{-2} > and the cement-lime mortar with the compressive strength $f_m \in \,<1\,0$ N.mm^{-2} , 2 0 N.mm^{-2} > are used in most cases. The value E_m of the secant modulus of elasticity of mortar is lower than that of masonry units E_u , if both of them are related to the identical value of stress σ_1. If in the masonry wall under compression no bond or friction existed in the interfaces between masonry units and mortar bed joints, the unrestrained lateral strain ε_{mxl} of the mortar bed joints would exceed that of bricks ε_{uxl} (fig 1a). As a consequence of the existing bond and friction, the lateral strains of mortar and bricks in the vicinity of contact interfaces are equal, $\varepsilon_{ux} = \varepsilon_{mx}$ (fig. 1b). It follows that $(\varepsilon_{ux} - \varepsilon_{mxl}) > 0$ and $(\varepsilon_{mx} - \varepsilon_{mxl}) < 0$.

The additional horizontal normal stress σ_{ux} in masonry units and σ_{mx} in mortar bed joints is given by $\sigma_{ux} = (\varepsilon_{ux} - \varepsilon_{mxl})\, E_u$, and $\sigma_{mx} = (\varepsilon_{mx} - \varepsilon_{mxl})\, E_m$, respectively. Both masonry units and mortar bed joints in the wall under consideration are under the planar state of stress, with the x-axis perpendicular to the wall faces and the vertical y-axis placed in the central plane of the wall. Thus, the masonry units are under axial compression and lateral tension, and the mortar bed joints under axial and lateral compression. Clay brick masonry under compression fails   in most cases as a consequence of vertical longitudinal splitting of bricks. Consequently, the masonry compressive strength is equal to the compressive stress which is related to the resistance of masonry units under the stress state mentioned above. The value of this compressive stress is equal to the negative co-ordinate of the point on the strength curve (or straight line) of the masonry unit

The masonry compressive strength increases with the increasing resistance of bricks and the decreasing lateral stress intensity in bricks. The resistance of solid bricks under "compression-tension" depends upon the tensile strength of the brick material in the critical middle vertical cross section passing through the centre of gravity of the brick parallel to the wall face. and upon the dimensions of bricks The lateral stress intensity in bricks decreases with the increasing ratio of brick height h_u to the thickness t_j of mortar bed joints and with decreasing lateral strain of mortar, i e with the increasing mortar strength

The resistance of vertically perforated clay units depends moreover upon the perforation pattern In this case the critical middle section is the section in the inner part of the masonry unit. which is weakened through perforation to the greatest extent Examination of test results shows however, that this critical section does not pass through the inner great cores in masonry units, destined for handling (fig. 2). In coincidence with the basic assumption of the proposed model, the ratio $k_{bo} = f_{ms,IID}/f_{ms,ST}$ of the masonry compressive strength in stretcher bond $f_{ms,ST}$ to the masonry compressive strength in header bond $f_{ms,IID}$ should be compared with the coefficient k_{IIV} determined as

$$k_{IIV} = \frac{A_{II',ef} \cdot A_v}{A_{I',ef} \cdot A_w},$$ (2)

where $A_{I',ef}$ and $A_{II,ef}$ is the effective area of section $V\text{-}V$ and $W\text{-}W$, respectively,

A_I and A_w is the total area of section $I'\text{-}V$ and $W\text{-}W$, respectively.

According to the test results, the value of ratio k_{bo} of masonry made from the perforated bricks CDK - 36 - B and CDK - 32 was $k_{bo} = 0.68$ and $k_{bo} = 0.73$, respectively. and the value of ratio k_{IIV} was $k_{IIV} = 0.72$ and $k_{IIV} = 0.68$, respectively. The deviations of the values k_{bo} from the relevant values k_{IIV} were - 5 6 % and +7 4 %

6 CONCLUSIONS

(1) The set of design parameters of masonry made from the Czech clay units, both solid and vertically perforated, produced between 1955 and 1992 (chapter 4, tables 1 and 2) is assigned to be exploited in both the Czech and European standardization studies

(2) The simple structural model of the single wythe wall under compression made from the vertically perforated masonry units (chapter 5, figs. 1 and 2) was verified successfully by the comparative load tests of masonry walls in the stretcher and header bond.

(3) The presented study is a part of results of the European research project COPERNICUS-CT 94-0174

7 REFERENCES

1   PUME D Effect of brick geometry on masonry strength. International Journal of Masonry Construction 1 (1980), 1, 11-17

2   PUME D Compressive strength of masonry made from Czech clay units produced between 1955 and 1992 Acta Polytechnica, 34 (1994), 3, 19-36

Table 1 Design parameters of masonry made from clay bricks

	Clay masonry units							Design parameters of masonry							
								mortar M 2.5				mortar M 5.0			
Type of units, volume density, total sum of test specimens	l_u	w_u	h_u	V_{rel}	f_u	l_w	bond	f_m	f_{obs}	K_s	ε_{fu}	f_m	f_{obs}	K_s	ε_{fu}
-	mm	mm	mm	%	N.mm^{-2}	mm	-	N.mm^{-2}	N.mm^{-2}	-	-	N.mm^{-2}	N.mm^{-2}	-	-
1	2	3	4	5	6	7	8	9	10	11	12	13	14	15	16
Solid bricks $\rho_{vol}=1600$ kg.m^{-3} 84 specimens	290 240	140 or 115	65	0	14.0	$1.5l_u$	long. vert. joints	3.2	2.84	2.30	0.00250	5.9	4.20	2.10	0.00220
Vert. perf. bricks $\rho_{vol}=1360$ kg.m^{-3} 120 specimens	240 240	115 or 115	113 140	16	15 7	$1.5l_u$	long. vert. joints	3.1	3.52	2.45	0.00230	6.6	4.32	2.35	0.00195
Vert. perf. bricks $\rho_{vol}=1450$ kg.m^{-3} 6 specimens	240 240	115 115	113 113	16 16	23.5 23 5	115 240	ST HD	2 9 2.9	6.17 5.14	2.00 2.00	0.00230 0 00240	-	-	-	-
Vert. perf. bricks $\rho_{vol}=1300$ kg.m^{-3} 7 specimens	290 290	90 140	188 188	20 24	19.1 14.9	90 140	ST ST	2.4 2.4	6.53 5.43	2.40 2 10	0.00240 0.00200	-	-	-	-
Vert. perf. units $\rho_{vol}=1300$ kg.m^{-3} 8 specimens	290 290	190 190	188 188	32 32	12.9 12.9	190 290	ST HD	2.4 2.4	4 45 4 92	2 20 2 30	0.00250 0.00220	-	-	-	-

Notations:

l_u, w_u, h_u - dimensions of masonry units; l'_{rel} - volume of perforation; f_u - compressive strength of masonry units; f_m - compressive strength of mortar, f_{obs} - compressive strength of masonry estimated by tests; K_s - coefficient in the relationship $\sigma=\sigma(\varepsilon)$, ε_{fu} - ultimate strain of masonry, long vert joints - masonry wall with longitudinal vertical joints; ST, HD - single wythe wall in stretcher and header bond, respectively

Table 2 - Part 1. Design parameters of masonry made from clay vertically perforated masonry units.

Type of units	Clay masonry units									Design parameters of masonry mortar M 2.5				
	l_u	w_u	h_u	ρ_{vol}	V_{rel}	f_u	t_w	bond	sum of spec.	f_m	f_{obs}	K_s	ε_{fu}
-	mm	mm	mm	kg.m^{-3}	%	mm	mm	-	-	N.mm^{-2}	N.mm^{-2}	-	-
1	2	3	4	5	6	7	8	9	10	11	12	13	14
CD - TI - M	394	291	139	908	48	9.2	394	HD	6	3.4	3.28	2.0	0.00095
CD -360	360	185	210	930	45	19.1	360	HD	3	2.7	3.95	2.0	0.00140
CD -TYN	285	189	213	970	50	11.8	285	HD	3	2.8	3.70	1.9	0.00115
CD K - 36 - B	364	237	116	1050	43	7.2	364	HD	6	3.0	2.68	-	-
CD - INA	362	240	141	1055	42	9.1	362	HD	-	-	-	-	-
CD - 32	321	235	148	1092	41	8.9	321	HD	-	-	-	-	-
CD - TI - 49K	491	249	146	892	42	11.2	491	HD	4	2.8	4.89	1.9	0.00100
CD - TI - 49S	491	243	139	917	42	13.1	491	HD	3	2.8	4.37	2.2	0.00120

Table 2 - part 2 Design parameters of masonry made from clay vertically perforated masonry units

Type of units	Design parameters of masonry														
	mortar M5					mortar M10					mortar M15				
	sum of spec.	f_m	f_{obs}	K_s	ε_{iu}	sum of spec.	f_m	f_{obs}	K_s	ε_{iu}	sum of spec.	f_m	f_{obs}	K_s	ε_{iu}
-	-	N.mm^{-2}	N.mm^{-2}			-	N.mm^{-2}	N.mm^{-2}				N.mm^{-2}	N.mm^{-2}		
1	15	16	17	18	19	20	21	22	23	24	25	26	27	28	29
CD - T1 - M	-	-	-	-	-	-	-	-	-	-	-	-	-	-	-
CD -360	-	-	-	-	-	-	-	-	-	-	-	-	-	-	-
CD -TYN	-	-	-	-	-	6	9.6	4.43	1.9	0.00135	-	-	-	-	-
CDK - 36 - B	-	-	-	-	-	5	9.8	2.88	-	-	5	15 0	3 37	-	-
CD - INA	3	4.1	3.98	2 2	0.00115	-	-	-	-	-	-	-	-	-	-
CD - 32	3	5.0	3.80	2 0	0.00110	-	-	-	-	-	3	18,0	4 13	2 1	0.00140
CD - T1 - 49K	-	-	-	-	-	-	-	-	-	-	-	-	-	-	-
CD - T1 - 49S	-	-	-	-	-	-	-	-	-	-	-	-	-	-	-

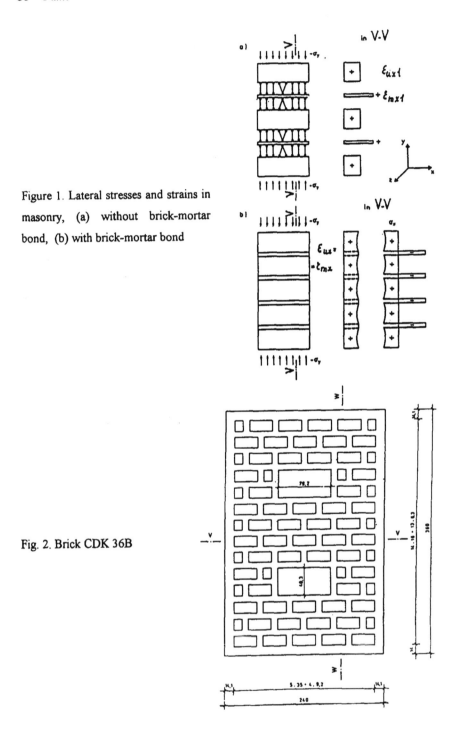

Figure 1. Lateral stresses and strains in masonry, (a) without brick-mortar bond, (b) with brick-mortar bond

Fig. 2. Brick CDK 36B

DAMAGE AND FRACTURE IN BRITTLE MATERIALS

L. Briseghella Professor, **P. Pavan** Engineer and **S. Secchi** Engineer
Dipartimento di Costruzioni e Trasporti, Università degli Studi di Padova, Via Marzolo 9, 35131 Padova, Italy

1. ABSTRACT

A scalar damage model and a crack opening model are applied to describe the behaviour of brittle materials. The damage laws take into account for different behaviours in tension and in compression. The definition of the damage constitutive model is suitable for extensions to more complex laws including tensorial damage. The crack opening model consists in a re-meshing which enable one to describe the progressing of the fracture in the material. The criteria for the opening of the cracks consist in an evaluation of the stress or strain state in the neighbourhood of the tip. The interaction between the two sides of the crack are not considered here but can be simply accommodated in the model. An application to the three points beam problem is presented.

2. INTRODUCTION

A finite element method, to investigate the softening behaviour and the propagation of cracks in brittle materials, is here presented. Large or finite strains and finite displacements problems can be described. Two constitutive models are considered: a scalar damage constitutive model and a crack opening model. The damage model is defined in accordance with the second law of thermodynamics. The Helmoltz free-energy function is obtained coupling in a multiplicative the stored energy function of the undamaged material and a scalar damage function. The latter, following the works of Mazars [1] and Simo [2], depends on equivalent measures of stress or strain. Different behaviours in tension and compression are possible by the use of "ramp" measures. It results possible to define different damage constitutive model. For example, anisotropic

Keywords: Damage, Fracture, Finite deformations

Computer Methods in Structural Masonry – 4, edited by G.N. Pande, J. Middleton and B. Kralj.
Published in 1998 by E & FN Spon, 11 New Fetter Lane, London EC4P 4EE, UK. ISBN: 0 419 23540 X

constitutive models can be considered, modifying in a suitable way the free-energy function. It is also possible to avoid the softening behaviour in compression taking into account for the presence of stiffness in the re-closure of micro-cracks phenomenon. These developments will be the object of future works. The discretization of the linear momentum balance weak form is obtained by the use of the Galerkin projections method. The non-linear system, which is obtained, is solved applying a Newton method. A consistent linearization enable one to have a quadratic rate of convergence. The crack opening model is defined in a very simple way. The stress or the strain states are evaluated in the neighbourhood of the crack tip. An elastic-perfect brittle law is used and the crack opening is related to the elastic threshold. The discretization of the model is obtained with the Delaunay triangulation method. Unstructured meshes are used with a refinement near the possible way of cracks. The progress of the fracture is obtained with a re-meshig of the model. At this stage of development the re-meshing consists simply in to redouble some nodes of the mesh. The direction of fracture propagation is chosen considering the principal directions of the deformation and progressing in the normal direction to the maximum principal stretch. The Delaunay topology structures, however, could be used to carry out a systematic refinement of the mesh in the tip of the crack. The interactions between the two sides of a crack are not considered her but can be easy accommodated in the model too. This will enable one to consider the re-closing phenomenon. The basic idea of the work is to obtain, in a next future, a constitutive model in which the damage is coupled with the crack opening model. In this way one will be able to consider, in the same time, the macro-cracks progress and the softening behaviour associated to the micro-cracks around the tip. A simple example regarding the three points beam problem, treated with the two methods, is presented.

3. KINEMATICS OF THE CONTINUUM

A lagrangian description of the motion is adopted. Let B be the reference configuration of a body with smooth boundary ∂B. A material point of B is labelled with X. The motion is a one-to-one map which relates the material point X of B with the spatial points $x = \varphi(X)$ of the current configuration $S = \varphi(B)$. The deformation is described by the deformation gradient, defined as the derivative of the spatial coordinates with respect to the material coordinates $F = \partial x / \partial X$. The local condition of regularity of the motion is $J = \det F > 0$. One defines the Green-Lagrange strain tensor as $E = \frac{1}{2}(C - I)$, where $C = F^T F$ is the right Cauchy-Green tensor. To obtain the stress and the strain measures for a plain problem the spectral decomposition theorem is adopted (see Simo & Taylor [3] for the three-dimensional case). The right Cauchy-Green tensor is written as

$$C = \sum_{A=1}^{3} \lambda_{(A)}^2 N^{(A)} \otimes N^{(A)}, \quad \left\| N^{(A)} \right\| = 1, \tag{1}$$

where $N^{(A)}$ are the three principal directions of the deformation and $\lambda_{(A)}$ the principal stretches. For plane stress and plane strain problems one can think to the right Cauchy-Green tensor as a two-two symmetric tensor. The square of the principal stretches are obtained by the following closed form

$$\lambda_{(1),(2)}^2 = \frac{I_1 \pm \sqrt{I_1^2 - 4I_3}}{2}, \tag{3}$$

where the principal invariants of the two-dimensional right Cauchy-Green tensor C are

$$I_1 = tr\,C = C_{11} + C_{22} \; ; \; I_3 = det\,C = C_{11}C_{22} - C_{12}C_{21}. \tag{4}$$

The principal material directions can be obtained in a closed form too and are represented by the tensors

$$M^{(A)} = \frac{1}{\lambda_{(A)}^2} N^{(A)} \otimes N^{(A)} = \frac{I + \left(\lambda_{(A)}^2 - I_1\right)C^{-1}}{2\lambda_{(A)}^2 - I_1} \quad ; \quad (A = 1,2) \tag{5}$$

where $N^{(A)}$ with $\left\|N^{(A)}\right\| = 1$, are two-dimensional vectors and I is the second order unit tensor. For equal principal stretches the latter equation presents a singularity, hence a little difference between the two stretches must be imposed. The derivative of the principal stretches with respect to the right Cauchy-Green tensor results

$$\frac{\partial \lambda_{(A)}}{\partial C} = \frac{1}{2}\lambda_{(A)}M^{(A)} \quad \text{for} \quad A = 1,2 \,; \lambda_{(1)} \neq \lambda_{(2)}, \tag{6}$$

$$\frac{\partial \lambda_{(A)}}{\partial C} = \frac{1}{2}\lambda_{(A)}C^{-1} \quad \text{for} \quad A = 1,2 \,; \lambda_{(A)} = \lambda_{(A)}. \tag{7}$$

4. BALANCE EQUATIONS

Following a lagrangian description of the continuum (Marsden & Hughes [3]), the balance of the linear and angular momenta, in the local form, are written as

$$DIV(P) + \rho_0 f = 0, \quad P^T F^{-T} = F^{-1}P. \tag{8}$$

being f the body forces for unit of mass, P the first (non symmetric) Piola-Kirchhoff stress tensor, ρ_0 the reference unit mass. The mass conservation law results as

$$\rho\,J = \rho_0. \tag{9}$$

The boundary conditions are defined as imposed traction $\bar{t}$ on $\partial_\sigma B$ and placements $\bar{\varphi}$ on $\partial_u B$, resulting $\partial_\sigma B \bigcup \partial_u B = \partial B$ and $\partial_\sigma B \bigcap \partial_u B = \varnothing$. The boundary values problem is completed specifying a constitutive relation for the stress tensor.

4.1 Discretization

The weak form of the linear momentum is obtained from (8)

$$\int_B P:GRAD(\eta)\,dB = \int_B f \cdot \eta\,dB + \int_{\partial_\sigma B} t \cdot \eta\,d\partial_\sigma B \tag{10}$$

where η are the admissible variations of the configuration. The method of Galerkin projections is adopted for the space discretization. The latter consists in the projection of the infinite dimension continuum system in a finite dimension system. The admissible configurations and their admissible variations are expressed in terms of the nodal values u^i, η^i:

$$\varphi^h(X) = \sum_{i=1}^{n_{node}} N^i(X)u^i \,, \quad \eta^h(X) = \sum_{i=1}^{n_{node}} N^i(X)\eta^i \tag{11}$$

being

$$N^i(X) \quad i = 1,2..n_{node} \quad ; \quad N^i(X_j) = \delta_{ij} \tag{12}$$

the shape functions. The discrete weak form obtained by the (10), is expressed in terms of the known displacements u_n^i and of the unknown displacements u_{n+1}^i. The residual is minimised by the application of a Newton method. The linearization of the (10) with respect to the current displacements gives the system to be solved in terms of u'_{n+1}.

5. DAMAGE CONSTITUTIVE MODELS

The elasto-damage models are obtained as combination of a stored energy function with suitable scalar damage functions. The employed models are characterised by a monotonic function g ($g \leq 1, \dot{g} \leq 0$), which can be used to estimate the damage degree of the material. The damage depends on the maximum value of a scalar measure attained during time history. The free energy function is expressed in the multiplicative way

$$\psi = gw(C), \tag{13}$$

where w is the stored energy. Considering the Clausius-Duhem inequality

$$-\dot{\psi} + \frac{1}{2}S:\dot{C} \geq 0, \tag{14}$$

one comes to the expressions for the symmetric Piola-Kirchhoff stress tensor and the internal dissipation

$$S = 2g\frac{\partial w(C)}{\partial C} \quad , \quad D^{int} = -w(C)\dot{g} \geq 0. \tag{15}$$

The so called De Saint Venant-Kirchhoff constitutive model is here used. However other hyperelastic constitutive models can be adopted. The De Saint Venant-Kirchhoff model is defined by the following stored energy function

$$w(E) = \frac{1}{2}E:C:E \tag{16}$$

being $C = 2\mu I_t + \lambda I \otimes I$. The stress response is simply given by $S = C:E$.

5.1 Constitutive tensor for pure hyperelastic models

In what follows, only the elastic response of the model is considered. The extension to the damage field will result very easy introducing only scalar factors. The constitutive tensors for the plane stress and the plane strain cases are well known in literature. Considering the spectral decomposition of the second Piola-Kirchhoff stress tensor

$$S = \sum_{A=1}^{3} S_{(A)}N^{(A)} \otimes N^{(A)} \tag{17}$$

and taking the contraction of the last form with the tensors $N^{(A)} \otimes N^{(A)}$ one obtains the principal stresses

$$S_{(A)} = S:N^{(A)} \otimes N^{(A)} \quad (A = 1,2). \tag{18}$$

The plane stress and the plane strain conditions are respectively obtained by the imposition of the condition $S_{(3)} \equiv 0$ and $\lambda_{(3)} \equiv 1$. In this way the constitutive tensor can be obtained starting from the stored energy function thought as depending on only two principal stretches since $\lambda_{(3)} \equiv cost \equiv 1$ for the plane strain condition and $\lambda_{(3)} = f(\lambda_{(1)}, \lambda_{(2)})$ for the plane stress one:

$$w = w(\lambda_{(1)}, \lambda_{(2)}). \tag{19}$$

The second Piola-Kirchhoff stress tensor is obtained by the application of the chain rule. For not equal principal stretches we have

$$S = \sum_{A=1}^{2} \lambda_{(A)} \frac{\partial w}{\partial \lambda_{(A)}} M^{(A)} \qquad \lambda_{(1)} \neq \lambda_{(2)}. \tag{20}$$

Remembering the expression for the tensors $M^{(A)}$, the principal stresses result

$$S_{(A)} = \lambda_{(A)}^{-1} \frac{\partial w}{\partial \lambda_{(A)}} \qquad (A = 1,2). \tag{21}$$

Again the singularity resulting in the previous relationships in the case of equal principal stretches is avoided forcing a difference between the latter. The material constitutive tensor, for the hyperelastic part, is

$$\mathbf{C} = 2\sum_{A=1}^{2} \lambda_{(A)} \frac{\partial w}{\partial \lambda_{(A)}} \frac{\partial M^{(A)}}{\partial C} + \sum_{A=1}^{2}\sum_{B=1}^{2} \lambda_{(B)} \frac{\partial}{\partial \lambda_{(B)}} \left(\lambda_{(A)} \frac{\partial w}{\partial \lambda_{(A)}} \right) M^{(A)} \otimes M^{(B)}, \tag{22}$$

with the obvious consideration in the case of equal stretches. The only difference between the constitutive tensor for the plane stress condition and the constitutive tensor of the plain strain one is due to the expressions of the scalar coefficients. The derivative of the principal directions in the material form with respect to the right Cauchy-Green tensor is

$$\frac{\partial M^{(A)}}{\partial C} = \frac{1}{2\lambda_{(A)}^{2} - I_{1}} \left[\left(I_{1} - \lambda_{(A)}^{2} \right) I_{C^{-1}} - \lambda_{(A)}^{2} M^{(A)} \otimes M^{(A)} - \lambda_{(B)}^{2} M^{(B)} \otimes M^{(A)} \right], \tag{23}$$

where $\lambda_{(A)} \neq \lambda_{(B)}$ and $I_{(\cdot)} = \frac{1}{2}\left[(\cdot)_{il}(\cdot)_{jk} + (\cdot)_{ik}(\cdot)_{jl} \right]$.

5.2 Basic damage model

For the scalar function g the well known form employed

$$g(\Xi) = \begin{cases} 1 & \text{if } \Xi \leq \Xi_0 \\ b + (1-b)\dfrac{1 - exp[-(\Xi - \Xi_0)/a]}{(\Xi - \Xi_0)/a} & \text{if } \Xi > \Xi_0 \end{cases} \tag{24}$$

where Ξ is the equivalent strain. The function g depends on maximum value of the latter in the strain history of the material, defined as

$$\Xi_t^{max} = \max_{\tau \in (-\infty, t]} \sqrt{2w(C(t))}. \tag{25}$$

The parameters a, b can take the values $b \in [0,1]$ and $a \in [0,\infty)$. The parameter b is dependent on the possible maximum degree of damage while the parameter a depends on the rapidity of its attainment. The evolution of damage phenomenon preserves in any case the inequality

$$f\left(\Xi_t^{max}, (C(t)) \right) = \sqrt{2w(C(t))} - \Xi^{max} \leq 0. \tag{26}$$

The evolution of damage can be take place only for loading from damage surface, that is

$$\begin{cases} \dot{g} < 0 & \text{as } f = 0, \quad \dot{f} > 0 \\ \dot{g} = 0 & \text{otherwise.} \end{cases} \tag{27}$$

The material constitutive tensor $\mathbf{C} = 2\,\partial S/\partial C$ has the following expression

$$\mathbf{C} = \begin{cases} 4g\left(\Xi_t^{max}\right)\dfrac{\partial^2 w(C)}{\partial C \partial C} + 4\dfrac{g'\left(\Xi_t^{max}\right)}{\Xi_t^{max}}\dfrac{\partial w(C)}{\partial C} \otimes \dfrac{\partial w(C)}{\partial C} & \text{as } f = 0, \quad \dot{f} > 0 \\ 4g\left(\Xi_t^{max}\right)\dfrac{\partial^2 w(C)}{\partial C \partial C} & \text{otherwise.} \end{cases} \tag{28}$$

Our proposal is to extend some concepts used in the damage models for small strains. See for example the work of Simo & Ju [5]. In this way, starting from the same definition of the Helmoltz free energy function, it is possible to take into account different behaviours in tension and compression. This can be obtained, for example, defining other measures of equivalent strains or stresses and maintaining the same function g. The constitutive tensors are again quite simple since the only difference is in the expression of the damage function g and its derivatives with respect to the right Cauchy-Green tensor. The proposed models however have some inconsistencies: for example they can not describe cyclic deformations which present a re-closing cracks phenomenon and a mesh dependence is not avoided. On the other hand this wants only to be a proposal for a way to be investigated and many improvements for the constitutive law are possible.

5.3 Damage as function of "ramp" principal stretches

One defines modified principal stretches in the following form

$$\lambda^*_{(i)} = \langle \lambda_{(i)} - 1 \rangle_+ + 1 \quad , \quad (i = 1,2,3) \tag{29}$$

being $\langle x \rangle_+ = (|x| + x)/2$ the positive ramp function of the variable x.
It is very easy to verify that the previous forms result as

$$\lambda^*_{(i)} = \begin{cases} \lambda_{(i)} & if \quad \lambda_{(i)} > 1 \\ 0 & if \quad \lambda_{(i)} \le 1 \end{cases} \quad ; \quad (i = 1,2,3) \tag{30}$$

that is only the principal stretches which are greater than one are taken in account.
The equivalent strain is now defined as

$$\Xi^*_+ = \sqrt{2\sum_{i=1}^{3}\lambda^*_{(i)}(t)} \tag{31}$$

and the damage function is calculated as $g = g(\Xi^{max}_+)$ where

$$\Xi^{max}_+ = \max_{\tau \in (-\infty,t]} \Xi_+(\tau) \tag{32}$$

is again the maximum value attained during the time history of the material. This definition could be in some way unrealistic because of its linearity. However we present it only for the simplicity of the related constitutive tensor. Equation (26) is now rewritten as

$$f(\Xi^{max}_+, \Xi^*_+) = \Xi^*_+ - \Xi^{max}_+ \le 0 \tag{33}$$

and the damage behaviour follows again the law described by the (24). The material constitutive tensor is modified by introducing the new form for the derivative of the function g and results clearly non symmetric. The expression for the material constitutive tensor results in the following form

$$\mathbf{C} = \begin{cases} 4g(\Xi^{max}_+)\dfrac{\partial^2 w(\mathbf{C})}{\partial\mathbf{C}\partial\mathbf{C}} + 4\dfrac{\partial w(\mathbf{C})}{\partial\mathbf{C}} \otimes \dfrac{\partial g(\Xi^{max}_+)}{\partial\mathbf{C}} & as \quad f = 0, \quad \dot{f} > 0 \\[2ex] 4g(\Xi^{max}_+)\dfrac{\partial^2 w(\mathbf{C})}{\partial\mathbf{C}\partial\mathbf{C}} & \text{otherwise.} \end{cases} \tag{34}$$

The expression of the derivative of g with respect to the right Cauchy-Green tensor is

$$\frac{\partial g(\Xi^{max}_+)}{\partial\mathbf{C}} = \frac{1}{\Xi^{max}_+} g(\Xi^{max}_+)\sum_{i=1}^{3}\frac{1}{2}\left[1 + \frac{\lambda_{(i)} - 1}{|\lambda_{(i)} - 1|}\right]\frac{\partial\lambda_{(i)}}{\partial\mathbf{C}} . \tag{35}$$

For the plane strain condition the third principal stretch has constant unit value and so the summation must be extended from 1 to 2. For a plane stress condition the summation is extended from 1 to 3 remembering however that $\lambda_{(3)} = f(\lambda_{(1)}, \lambda_{(2)})$.

5.4 Damage as function of "ramp" principal stresses

The equivalent stress is defined again with the use of the positive ramp function

$$\Xi_+^i = \sqrt{2 \sum_{i=1}^{3} \langle S_{(i)}(t) \rangle_+ }, \qquad (36)$$

where $S_{(i)}$ are the principal values of the second Piola-Kirchhoff stress tensor and the damage function is again calculated on its maximum value of the time history. For the presented stored energy functions the principal stresses are function of the principal stretches and so the constitutive tensor is obtained in a simple way by considering the chain rule. The summation in (36) is extended from 1 to 3 for the plain strain condition and from 1 to 2 for the plane stress one.

6. CRACK OPENING MODEL

A simple model is adopted to take into account for the possible opening of cracks. An elastic-perfect brittle material is considered. The principal stresses are evaluated in the material above all in the neighbourhood of, possible, already existing cracks. If the stress is grater than the elastic threshold, the crack progresses. The progress of the fracture is describe redoubling some nodes of the mesh. The direction of crack propagation is on one of the boundaries of the elements. The direction is the closest to the normal at the maximum principal stress. The stresses are hence calculated for the re-meshed structure and the method proceeds in an iterative way. The method of nodes redoubling is described in figure 1 for an external node.

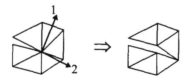

Fig. 1: Redoubling of the node and crack progress.

Thinking to apply also a damage constitutive model to describe the inelastic behaviours near the tip of the crack, the criterion to define the possible progress of the fracture is based on the evaluation of the strain state rather than on the evaluation of the stress one. In the presented model, the interaction between the two sides of the crack is not considered. The existence of stresses related to the cohesion or to the re-closure of the crack can be carried on. It must be underlined that the use of the Delaunay topology structures enable one to consider the possibility of a refinement of the model in a simply and fast way since only the zones around a crack must be re-meshed while the other parts of the model can be no modified.

7. EXAMPLES AND APPLICATIONS

The method and the constitutive models were implemented using an objects oriented language. The same problem is studied with the two methods, using both the damage constitutive model and the crack opening model

7.1 Three points beam problem.

The beam of figure 2 has length $l = 80$ cm and square section with edge $b = 10$ cm. It is subjected to a nodal force in the middle of the span. The De Saint Venant-Kirchhoff constitutive model with $\mu = 14870$ MPa and $\lambda = 3550$ MPa is coupled with the damage function (24) while the equivalent strain is defined with (31). The constants of the damage function are $a = 0.01$ and $b = 0.1$. The unstructured mesh of the beam is showed in figure 2. In the same figure, the damage contour and the opened crack obtained with the two approaches are also described.

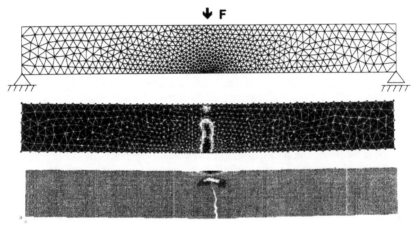

Fig. 2: Unstructured mesh, damage contour and propagation of crack.

8. REFERENCES

1. Mazars J., 'A model of unilateral elastic damageable material and its application to concrete', Proceedings of the International Conference on Fracture Mechanics of concrete, Lausanne, Switzerland, 1986.
2. Simo J.C., On a fully three-dimensional finite-strain visco-elastic damage model: Formulation and computational aspects, Comput. Meths. Appl. Mech. Engrg. 60, 1987, pp 153-157.
3. Simo J.C., Taylor R.L., Quasi-incompressible finite elasticity in principal stretches: Continuum basis and numerical algorithms, Comput. Meths. Appl. Mech. Engrg. 85, 1991, pp 273-310.
4. Marsden J.E., Hughes T.J.R., Mathematical Foundations of Elasticity, Prentice-Hall, Inc. Englewood Cliffs, 1983.
5. Simo J.C., Ju J.W., Strain- and stress-based continuum damage models-I. Formulation, J. Solids Structures 23, 1987, pp 821-840.

A MACROELEMENT DYNAMIC MODEL FOR MASONRY SHEAR WALLS

A. Brencich Graduate Research Assistant and
S. Lagomarsino Assistant Professor
Department of Structural and Geotechnical Engineering, University of Genoa, Via Montallegro 1, 16145 Genova, Italy

1. ABSTRACT

In this work a finite macroelement model for masonry panels is formulated, taking into account the main damage and dissipative mechanisms that had been observed in real cases and reproduced by means of detailed theoretical models. In particular, overturning phenomena are modelled by a monolateral elastic contact at the two ends of the macroelement, while shear cracking is described through an inelastic strain component that takes into account both damage and friction effects. The overall behaviour of masonry walls with openings is obtained by a suitable assembling of the macroelements used for both piers and architraves. The proposed model is then validated with reference to an example that had been carefully tested.

KEYWORDS: Macro-element model, masonry walls, friction, dissipation.

2. INTRODUCTION

The response of masonry structures to seismic actions is based upon the mutual contribution of the walls of the building. The behaviour of the structure is strongly dependent on the connection between orthogonal walls, on the presence of tie rods or, when some rehabilitation had been done, on the stiffening effect of reinforced concrete beams and slabs. In these cases the response of the structure to seismic actions up to final collapse is mainly due to the response of each wall in its own plane. This problem had been thoroughly studied following three approaches: *i*) observing the effects of earthquakes on real buildings; *ii*) carrying on experimental tests both on reduced models and on full scale buildings; *iii*) performing theoretical analyses by means of proper constitutive models. On these bases it is possible to define simplified and user-friendly procedures, known as macroelement models, characterised by a low number of unknowns and able of describing the wall response in its plane up to collapse.

Computer Methods in Structural Masonry – 4, edited by G.N. Pande, J. Middleton and B. Kralj.
Published in 1998 by E & FN Spon, 11 New Fetter Lane, London EC4P 4EE, UK. ISBN: 0 419 23540 X

Though a generic wall, containing an irregular distribution of openings, is on principle a bi-dimensional continuum, it can be well approximated by an assembly of vertical piers and horizontal architraves. This approximation is acceptable since the effect of horizontal loading on real scale buildings points out that the areas where piers and architraves are connected seldom experience any kind of damage. Various models have been proposed: in some of them piers and architraves are regarded as macroelements reproducing the main collapse mechanisms [1,2], while in others the wall is reduced to a frame which columns and beams are considered as elasto-plastic elements with stiffness and ultimate strength related to the constituent material and to the geometry [3]. Ancient buildings after structural rehabilitation or recent masonry structures present reinforced concrete floors that can be considered as rigid slabs; the resistance to horizontal actions can then be evaluated floor by floor considering a system of shear walls, each one characterised by its own elasto-plastic properties [4]; in this way the solution does not satisfy equilibrium conditions, since it does not consider the variation of the vertical load on each pier during the seismic actions.

The available simplified models are able of predicting the collapse load and the global ductility. Nevertheless, in the case of seismic actions it is very important to get information's on the cyclic response of the structure, which turns out to be more or less dissipative according to the prevalence in the piers of shear or overturning collapse mechanisms.

In this work a simplified macroelement model for masonry walls loaded in their own plane is presented, consisting of panels simulating piers and architraves connected by means of rigid blocks. The macroelement, which is a generalisation of the panel model proposed in [5], takes into account both the overturning mechanisms, related to cracking at the corners, and shear mechanism by means of a damage model with friction. In this way a generic wall with openings can be described by means of a limited number of unknowns involving few model parameters. The whole procedure is verified through a comparison with the experimental Pavia building prototype [6]: the results of the approximated model point out the flexibility of the procedure, able of reproducing the stiffness, ultimate load, ductility, dissipation and damage evolution that had been experimentally observed up to the final collapse.

3. THE MACROELEMENT MODEL

Masonry is characterised by non linear strain mechanisms resulting mainly from opening and sliding between constitutive elements (bricks, stones). In wall panels such phenomena are localised at the extremities, where open cracks originate as a consequence of low tensile strength of the joints and in the middle, mainly along the diagonals of the panel, where shear slidings are activated. Therefore, the formulation of a simplified macroelement needs, on one hand, the equilibrium conditions given by means of a limited number of unknowns, on the other hand, a kinematic model able of representing the overall mechanism of deformation, damage and dissipation.

Let us now consider the wall panel represented in figure 1.a and let assume that it can be considered as the assembly of three sub-structures: the axial compliance is supposed concentrated in the extremities ① and ③, while the shear deformability is represented by the central modulus ②.

A complete kinematic model needs to take into account three degrees of freedom for each node i and j at the extremities and for the two interface nodes, called 1 and 2; w stands for the axial displacements, u for the horizontal ones and rotations are named φ. The aforementioned hypothesis on the rigidity of the substructures allow some compatibility conditions at the interfaces between the modules ①, ② and ③ to be posed, so as to reduce the number of degrees of freedom:

$$w_1 = w_2 = \delta \ , \quad \varphi_1 = \varphi_2 = \phi \ , \quad u_1 = u_i \ , u_2 = u_j \ , \tag{1}$$

where δ and ϕ stand for the axial displacement and the rotation of the central zone ②.

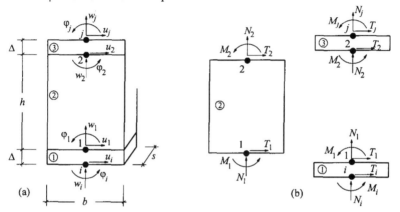

Fig. 1. Kinematic (a) and static (b) variables of the macroelement.

The kinematics is therefore described by eight displacements: six related to the nodes i and j (u_i, w_i, φ_i, u_j, w_j, φ_j) and two that can be regarded as internal variables (δ and ϕ). This kinematic model and compatibility conditions are equivalent to the hypothesis of extremities with vanishing thickness ($\Delta \rightarrow 0$).

3.1 Constitutive equations

The overturning mechanism, allowed by the no-tension assumption for the material, is modelled by means of an elastic monolateral contact at the interfaces ① and ③ (k is the axial stiffness per unit area). The constitutive equations between the kinematic variables w, φ and the corresponding static quantities N, M are linear and decoupled up to limit condition $|M/N| \leq b/6$, when the section is no longer entirely under compression. With reference to substructure ① they are given as:

$$N_i = kA(\delta - w_i) - N_i^* \ , \qquad M_i = \tfrac{1}{12} kAb^2(\varphi_i - \phi) - M_i^* \ , \tag{2.a, b}$$

where: $A = sb$ stands for the cross section of the panel, while the inelastic contributions N_i^*, M_i^* are given as:

$$N_i^* = \frac{kA}{8|\varphi_i - \phi|}\left[|\varphi_i - \phi|b + 2(\delta - w_i)\right]^2 H\left(|e_i| - \tfrac{1}{6}b\right) \ , \tag{3.a}$$

$$M_i^* = \frac{kA}{24b(\varphi_i - \phi)|\varphi_i - \phi|}\left[(\varphi_i - \phi)b - (\delta - w_i)\right]\left[|\varphi_i - \phi|b + 2(\delta - w_i)\right]^2 H\left(|e_i| - \tfrac{1}{6}b\right), \tag{3.b}$$

being H the Heaviside function.

The shear response of the panel is modelled considering a uniform stress distribution in the central modulus ② and imposing a relationship between the kinematic quantities u_i, u_j and ϕ and the shear forces $T_i = -T_j$. Cracking damage is usually localised on the diagonals, where slidings take place along the bed joints, and is represented by an inelastic strain component which is activated when a Coulomb friction limit condition is attained. Since the effective shear deformation of modulus ② is given as $u_T = u_i - u_j + \phi h$, and indicating as G the elastic shear modulus, the constitutive equations are given in the following terms:

$$T_i = \frac{GA}{h}\left(u_i - u_j + \phi h\right) - T_i^* , \qquad (4)$$

$$T_i^* = \frac{GA}{h} \frac{c\alpha}{1+c\alpha}\left(u_i - u_j + \phi h + \frac{h}{GA} F\right) . \qquad (5)$$

where the inelastic component takes into account the effect of the friction force F, opposite to sliding mechanisms. In the frame of the present model, friction plays the role of an internal variable ruled by the limit condition:

$$\Phi_s = |F| - \mu N_t \leq 0 , \qquad (6)$$

where μ is the friction coefficient. These constitutive equations are able of representing the variations of the panel strength due to changes in the axial forces $N_j = -N_i$. Damage, and its effect on the mechanical characteristics of the panel, is described by a damage variable α, which increases according to a fracture criterion [7]:

$$\Phi_d = Y(\alpha, Q) - R(\alpha) \leq 0 , \qquad (7)$$

where Y stands for the released free energy, R for a toughness function and $Q = \{N, T, M\}^T$ for the internal forces vector. Assuming R an increasing function of α up to the critical value $\alpha_c = 1$, and a decreasing function for higher values, the model is able of representing the stiffness degradation and the strength of the panel.

The shear model of the macroelement is a simplification of a more complex continuum model [8], which parameters are directly correlated to the mechanical properties of the masonry elements. However the parameters of the macroelements need to be considered as representative of an average behaviour. Besides its geometric characteristics, the macroelement is described by six parameters: G and k related to the elastic response, f_{vk0} as the shear strength of masonry; c that represents a non dimensional compliance coefficient governing the inelastic shear strain; μ standing for a global friction coefficient; β that is related to the toughness function R governing the softening phase.

The macroelement model so forth discussed is able of representing the transition from an overturning collapse mechanism, characteristic of slender walls, to a shear mechanism, typical of squat walls. If T_O and T_S are the overturning and shear limits, according to the preceding formulation they are respectively given by the following formulas:

$$T_O = Nb/h = N/\lambda , \qquad T_S = f_{vk0} A + \mu N, \qquad (8)$$

where λ is the wall slenderness. The diagonal collapse characteristic of masonry structures [9] is not considered in this model; in this case the resistance T_d to diagonal cracking is given by the formula:

$$T_d = \frac{f_{vk0} A}{\lambda} \sqrt{1 + \frac{N}{f_{vk0} A}} , \qquad (0 < \lambda < 1.5). \qquad (9)$$

Figure 2 shows a comparison between the limit conditions for the macroelement, represented by equations (8), and the collapse mechanism of equation (9), pointing out that the approximated formulation turns out to be admissible.

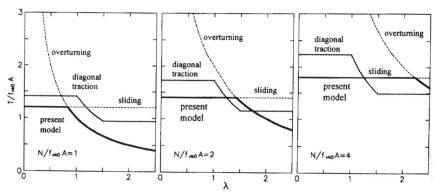

Fig. 2. Comparison among different collapse models for the masonry panels.

3.2 Formulation of the macroelement

The kinematic unknowns of the problem consist of the six external and the two internal displacements of the macroelement, and can be related to the static quantities (T_i, N_i, M_i, T_j, N_j, M_j) by the six equilibrium conditions at nodes i and j. The two missing conditions are given by the vertical and rotational equilibrium equations of the central modulus ② (figure 1.b).
The dead load is reduced to equivalent nodal forces, so that the constitutive equations can be written for a load free panel separating the elastic and the inelastic contribution:

$$
\begin{Bmatrix} T_i \\ N_i \\ M_i \\ T_j \\ N_j \\ M_j \\ 0 \\ 0 \end{Bmatrix} =
\begin{bmatrix}
GA/h & 0 & 0 & -GA/h & 0 & 0 & 0 & -GA \\
0 & kA & 0 & 0 & 0 & 0 & -kA & 0 \\
0 & 0 & \tfrac{1}{12}kAb^2 & 0 & 0 & 0 & 0 & -\tfrac{1}{12}kAb^2 \\
-GA/h & 0 & 0 & GA/h & 0 & 0 & 0 & GA \\
0 & 0 & 0 & 0 & kA & 0 & -kA & 0 \\
0 & 0 & 0 & 0 & 0 & \tfrac{1}{12}kAb^2 & 0 & -\tfrac{1}{12}kAb^2 \\
0 & -kA & 0 & 0 & -kA & 0 & 2kA & 0 \\
-GA & 0 & -\tfrac{1}{12}kAb^2 & GA & 0 & -\tfrac{1}{12}kAb^2 & 0 & GAh+\tfrac{1}{6}kAb^2
\end{bmatrix}
\begin{Bmatrix} u_i \\ w_i \\ \varphi_i \\ u_j \\ w_j \\ \varphi_j \\ \delta \\ \phi \end{Bmatrix} -
\begin{Bmatrix} T_i^* \\ N_i^* \\ M_i^* \\ T_j^* \\ N_j^* \\ M_j^* \\ N^* \\ M^* \end{Bmatrix}, \quad (10)
$$

where the non linear terms N^* and M^* depend on the quantities introduced in (3) and (5):

$$ N^* = N_j^* - N_i^*, \qquad M^* = -M_j^* - M_i^* + T_i^* h. \qquad (11) $$

4. THE MACROELEMENT MODEL OF A MASONRY WALL

The effect of earthquakes on real masonry structures, as well as experimental results on real scale prototypes, pointed out that the damage mechanisms are mainly located in the piers and in the architraves, those parts of the wall which connect the piers and the architraves being almost undamaged. From the point of view of a simplified model, it is acceptable to consider these areas as infinitely rigid; thus a general masonry wall is considered as the assembly of piers and architraves, all of them modelled by means of the macroelement discussed in the previous section, connected to the nodes of the model by rigid extremities. The unknowns of the problem consist of the nodal displacements and the internal variables δ and ϕ for each macroelement.

As an example, let us consider the wall represented in figure 3.a, which is one of the walls constituting the prototype building tested at the University of Pavia [6]. Once the vertical loads had been applied, representing the dead loads of the floors, horizontal displacements where imposed at the two levels in such a way so as to maintain equal reaction forces. The test has been conducted even after the peak load, pointing out the evolution of damage and the collapse mechanisms. Figure 3.b represents the macroelement model of the wall, showing that only 9 nodes and 10 elements are needed, summing up to 38 degrees of freedom. The eccentricity between the nodes and the axes of architraves ⑦ and ⑧ has been imposed so as to reproduce the exact experimental setup; such kind of eccentricity could be found also in other cases, when the wall presents irregular openings.

Once the contributions of all the macroelements have been assembled and the boundary conditions are given, the non linear system is solved by means of a step wise *predictor-corrector* algorithm; at each step the friction and damage variables F_e and α_e of all the elements are updated.

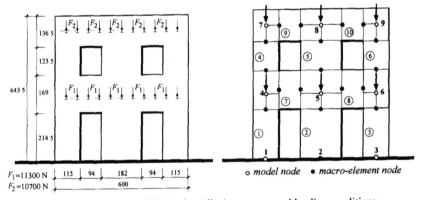

Fig. 3. The example of the Pavia wall: a) geometry and loading conditions;
b) macroelement mesh with rigid blocks (grey), piers and architraves.

5. A NUMERICAL APPLICATION

The wall of figure 3.a represents a challenging test since the experimental data allow a detailed knowledge either of the load-displacement cyclic response (stiffness degradation, softening branches and dissipation) and of the damaging evolution and collapse mechanisms. Figure 4.a shows the experimental diagram of the sum of the base shear as a function of the horizontal displacement at the top of the wall. A relevant

stiffness degradation and cyclic energy dissipation can be clearly observed; moreover the maximum load bearing capacity is progressively reduced by a softening response.
The model parameters, listed in table I, are derived from the continuum model [10] by means of appropriate averaging techniques, and therefore are deduced from the mechanical characteristics of the masonry; in particular k has been defined so as to approximate both the axial and bending stiffness of the wall panel ($k=3E/h$).
Figure 4.b shows the cyclic load-displacement diagram obtained by means of the macroelement model, pointing out that both the peak load and the softening branch of the diagram, as well as the hysteretic response, are well fitted by the simplified model.

Table I. Model parameters.

G (MPa)	E (MPa)	f_{vk0} (MPa)	c	μ	β
300	2000	0.15	1.2	0.20	0. 5

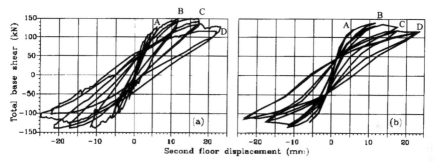

Fig. 4. Load-displacement curve: a) experimental results, b) macroelement model response.

Figure 5 presents a comparison between the state of damage which had been experimentally observed at the end of each loading cycle (points A, B, C and D in figure 4.a) and the damage distribution reproduced by the model through the variable α. It can be seen that the final collapse is the consequence of a two successive mechanisms: at the beginning (point A) only the architraves are broken while the base piers undergo a substantial overturning (the hatched areas represent the no tension parts of each macroelement as resulting from the overturning mechanisms). During the second load cycle (point B) damage in the architraves increases but also in the central base pier shear mechanisms take place. In the last two phases of the test (points C and D) the global mechanism changes: base piers no longer undergo overturning but are progressively damaged according to a shear mechanism, up to the final collapse when a weak store mechanisms (at the base of the wall) is responsible for the final collapse of the wall (point D). According to the experimental data, the transition from a mainly overturning to a shear collapse mechanism corresponds to an increase in energy dissipation, which is quite well reproduced by the simplified procedure outlined.

6. REFERENCES

1. D'Asdia P. and Viskovic A., 'Seismic analysis of masonry structures' (in italian), Ingegneria Sismica, **XI**, n. 1, 1994, pp. 32-42.
2. Braga F. and Liberatore D., 'Modeling of seismic behavior of masonry buildings', 9th I.B.Ma.C., Berlin, Germany, 1991.
3. Magenes G. and Calvi G. M., 'Perspectives for the calibration of simplified methods for the seismic analysis of masonry walls', Masonry Mechanics between theory and practice, L. Gambarotta ed., Messina, 1996, pp. 503-512.
4. Braga F. and Dolce M., 'A method for the analysis of seismic-safe multi-storey buildings', 6th I.B.Ma.C., Rome, 1982, pp. 1088-1099.
5. Gambarotta L. and Lagomarsino S., 'On the dynamic response of masonry walls', Masonry Mechanics between theory and practice, L. Gambarotta ed., Messina, 1996, pp. 451-462.
6. Magenes G., Kingsley G. R. and Calvi G. M., 'Static testing of a full-scale, two-story masonry building: test procedure and measured experimental response', in Numerical prediction of the experiment, CNR-GNDT, Report 3.0, 1995, pp. 1.1-41.
7. Gambarotta L. and Lagomarsino S., 'A microcrack damage model for brittle materials', International Journal Solids and Structures, **30**, 1993, pp. 177-198.
8. Gambarotta L. and Lagomarsino S., 'Damage models for the seismic response of brick masonry shear walls. Part I: the mortar joint model and its applications - Part II: the continuum model and its applications', Earthquake Engineering and Structural Dynamics, **26**, 1997, pp. 424-462.
9. Calvi G. M. and Magenes G., 'Seismic evaluation and rehabilitation of masonry buildings', Proc. of the U.S.-Italian workshp on seismic evaluation and retrofit, D. P. Abrams and G. M. Calvi ed.s, Technical Report NCEER-97-0003, march 1997, pp. 123-142.
10. Gambarotta L., Lagomarsino S. and Morbiducci R., 'Two-dimensional finite element simulation of a large scale brick masonry wall through a continuum damage model', Numerical prediction of the experiment, CNR-GNDT, Report 3.0, 1995, pp. 4.1-19.

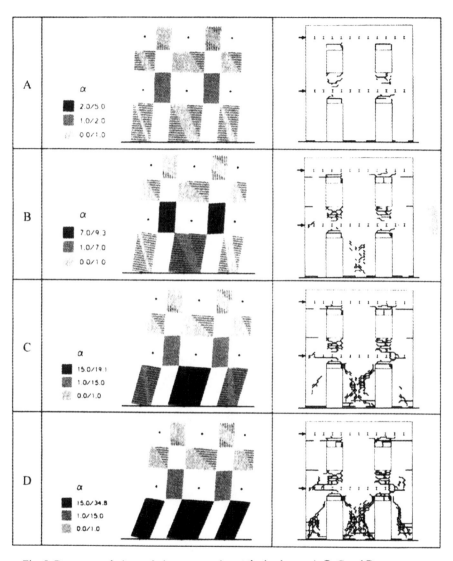

Fig. 5. Damage evolution and piers overturning at the load steps A, B, C and D:
comparison between the macroelement model results and the experimental ones.

ANALYSIS OF MASONRY STRUCTURES MODELLED BY A SET OF RIGID BLOCKS AND ELASTIC UNILATERAL CONTACT CONSTRAINTS

S. Briccoli Bati Associate Professor, **M. Paradiso** Associate Professor and **G. Tempesta** Associate Professor
Dipartimento di Costruzioni, Università degli Studi di Firenze, Firenze, Italy

1. ABSTRACT

This paper discusses a numerical procedure available for the analysis of masonry structures modelled as a discrete system of rigid blocks connected by elastic unilateral contact constraints. A convenient way to define the *contact device* which links the blocks, through which a mortar joint could be simulated, is to consider a curtain of elastic bars, orthogonal to the contact surface between two adjacent blocks, and an additional bar tangent to the same contact surface. In accordance with the assumptions of no tensile strength in the mortar joint, only compressive forces can be transmitted from one element to another. Moreover, reasonable hypotheses can be assumed for the bar tangent to the contact surface in order to calibrate both the shear behaviour and the influence of the friction between the blocks. Through the results of the numerical procedure it is possible both to define the cracking failure pattern, highlighting the actual reacting structure within the apparent one, and to evaluate the width of the cracks located in the mortar joints.

2. INTRODUCTION

The major problem in the analysis of any masonry structure is the different compressive and tensile strength born by the material. Such a circumstance makes it impossible to understand which is, in a pre-assigned configuration subject to assigned load conditions, the actual reacting structure. In fact such a structure does not necessarily correspond to the apparent one, but depends, from time to time, on the load conditions. In order to simplify, as much as possible, structural problems involving non-isoresistant materials, the assumption of no tensile strength for the masonry is appropriate. Such an hypothesis, that is equivalent to the statement that no tensile forces can be transmitted from one portion of the structure to another, although in some case may be unrealistic, can be

Computer Methods in Structural Masonry – 4, edited by G.N. Pande, J. Middleton and B. Kralj.
Published in 1998 by E & FN Spon, 11 New Fetter Lane, London EC4P 4EE, UK. ISBN: 0 419 23540 X

considered a *safe* assumption. Nevertheless, a no tensile strength assumption can be considered almost exactly true if, for instance, we deal with voussoir arches built with stone blocks assembled dry or with very weak mortar joints. Generally the analysis of structural problems involving unilateral constraints, expressed through systems of equations and inequalities, requires the use of Q.P. techniques. Otherwise, as an alternative, it is possible to obtain the solution by using a step by step procedure according to which the solution relative to the *standard material* (linear elastic and bilateral) is assumed as a starting point and is subsequently corrected according to the actual material skills. The final result is the individualization of the actual reacting structure contained within the structure shape. Such a method has already been practiced by Castigliano in 1879 [1]. The procedure presented in this paper, besides leading to the same results, is also less expensive in term of computational costs than those required by Q.P. techniques. Such a procedure is based on the introduction of appropriate *distortions*, whose potential has been firstly reported by S. Di Pasquale [2], allowing at the same time a more immediate mechanical interpretation of the necessary operations needed to reach the solution. The numerical procedure is based on the use of Moore-Penrose *generalized inverse* [3] and reduces the problem to the solution of systems of linear equations with the advantages of reaching available results also for large problems.

3. UNILATERAL ELASTIC CONTACT CONSTRAINT

Let's consider the general problem of a masonry structure consisting of rigid blocks linked through elastic mortar layers (Fig.1a). Assuming the aforementioned hypotheses are valid, it is possible to concentrate the no-tension behaviour totally in the mortar joint located in between two adjacent blocks. Such a joint can therefore be assumed as an unilateral elastic contact constraint. In such a model the rigid behaviour of the blocks is accompanied by the elastic-cracking behaviour of the joints. In particular, the mortar joint can be schematized through a sort of *interface device*, consisting of a curtain of unilateral bars, orthogonal to contact surface and capable of transmitting only compressive forces between the blocks, and an additional bar, parallel to the interface, through which the shear forces can be transmitted (Fig.1b). The behaviour of the orthogonal bars is assumed unilateral linear elastic, whereas for the parallel one further hypotheses can be added in order to specify both the shear strength and to calibrate, for instance, the influence of the friction between the blocks. In practice a reasonably low number of orthogonal bars is enough to describe with significant expressiveness the behaviour of the joint.

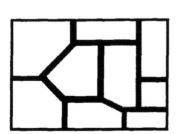
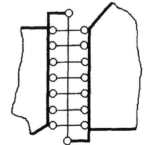

Fig. 1- a) Two-dimensional model of rigid blocks linked through elastic mortar layers.
b) *Interface device* consisting of unilateral linear elastic contact constraints.

4. GENERAL FORMULATION AND NUMERICAL PROCEDURE

Let's consider, therefore, the general problem of a masonry structure consisting of n rigid blocks linked through m unilateral elastic contact interfaces (Fig.2).

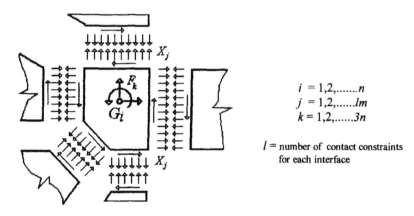

$$i = 1,2,.......n$$
$$j = 1,2,......lm$$
$$k = 1,2,......3n$$

l = number of contact constraints for each interface

Fig. 2- Rigid blocks linked through unilateral elastic contact constraints

Assuming that the structure is subject to the action of external loads represented by the vector $F \in \mathfrak{R}^{3n}$ and, eventually, to the influence of external inelastic displacements represented by the vector $\Delta_1 \in \mathfrak{R}^{lm}$, imposing equilibrium and elastic-kinematic conditions, the system of equations and inequalities that govern the problem, can be written in the following form :

$$\begin{cases} AX = F \\ A^T x + KX = \Delta_1 + \overline{\Delta}_2 \end{cases} \tag{1}$$

subject to
$$\begin{cases} X \leq 0 \\ \overline{\Delta}_2 \geq 0 \end{cases} \tag{2}$$

where $A \in \mathfrak{R}^{3n \times lm}$ is the geometrical configuration matrix; $X \in \mathfrak{R}^{lm}$ indicates the unknown vector of internal forces located in the contact joints. The value l, of course, depends on the number of contact constraints chosen to characterize the interface device and defines the degree of statically indeterminacy of the structure. The components of the unknown vector $x \in \mathfrak{R}^{3n}$ represent the displacements of the centroids of the blocks; $K \in \mathfrak{R}^{lm \times lm}$ is the diagonal stiffness matrix of the contact constraints.; $\Delta_1 \in \mathfrak{R}^{lm}$ represents the vector whose components are external inelastic displacements; $\overline{\Delta}_2 \in \mathfrak{R}^{lm}$ indicates the unknown vector whose components are *'distortions'* which need to be defined in order to obtain a solution capable of satisfying both the equilibrium equations, while respecting the sign conditions, and the kinematical compatibility of the problem. On this subject, it is convenient to distinguish, within the vector $\overline{\Delta}_2$, two types of distortions assuming for the former the notation Δ_2^* and for the latter Δ_2^{**}.

The general solution $X = X_0 + X_N$, that is able to satisfy the equilibrium problem and the first of the two inequalities (2), can be obtained assuming, as initial solution X_0, that relative to the bilateral linear elastic behaviour of the contact constraints:

$$X_0 = K^{-1}A^T(AK^{-1}A^T)^{-1}F + K^{-1}(I - A^T(AK^{-1}A^T)^{-1}AK^{-1})\Delta_1 \tag{3}$$

The initial vector solution X_0 can be suitably arranged in two sub-vectors. X_{0t}, whose components do not satisfy the sign conditions, and X_{0c} whose components satisfy the sign conditions:

$$X_0 = \begin{bmatrix} X_{0t} \\ X_{0c} \end{bmatrix} \tag{4}$$

Note that $X_{0t} \in \mathfrak{R}^t$, where t is the number of the contact constraints that, in the initial solution, prove to be stretched. Under no circumstances can t be greater than the degree of statically indeterminacy of the structure. Such an initial solution is then modified through the vector:

$$X_N = (I - K^{-1}A^T(AK^{-1}A^T)^{-1}A)\Delta_2^* \tag{5}$$

which, added to X_0, satisfies the first of the (1) while respecting the sign conditions.

The properties of the orthogonal projection matrix $C = (I - K^{-1}A^T(AK^{-1}A^T)^{-1}A)$ (see [4]) and the appropriate choice of the unknown vector Δ_2^*, are the keys to understanding the meaning of the procedure. In its turn also the matrix C can be suitably partitioned in four sub-matrices C_t, C_1, C_1^T, C_c:

$$C = \begin{bmatrix} C_t & C_1 \\ C_1^T & C_c \end{bmatrix} \tag{6}$$

where the sub-matrix $\overline{C}_t = [C_t \quad C_1] \in \mathfrak{R}^{t \times lm}$ has to be chosen as a full row rank matrix. On this subject the elimination of any linearly dependent row of the matrix $\overline{C}_t$, plays a key role in ascertaining the number of strictly necessary *distortions* to give back the compatibility in the sign conditions. Computing the Moore-Perrose generalized inverse of $\overline{C}_t$, it is easily possible to evaluate the vector Δ_2^*:

$$\Delta_2^* = \overline{C}_t^{-1}X_{0t} \tag{7}$$

If the solution of the unilateral problem exists, the vector solution which satisfies simultaneously the equilibrium equations and the first of the two inequalities (2), assumes the form :

$$X = \begin{bmatrix} 0 \\ X_c \end{bmatrix} \quad \text{with } X_c < 0 \tag{8}$$

Since the final vector X is different from the elastic solution vector X_0, it cannot satisfy, of course, the kinematical compatibility expressed through the second set of equations in the system (1).

A very easy way to build up again such a compatibility is to consider the second set of equations in the system (1) in the form $A^T \bar{x} + KX = 0$. Partitioning both the general matrix A^T in two sub-matrices A_t^T, A_c^T, and the constitutive matrix K in K_t, K_c, we obtain the solution:

$$\bar{x} = -(A_c A_c^T)^{-1} A_c K_c X_c \qquad (9)$$

which represents the vector of the displacements of the centroids of the blocks only due to the actual reacting structure. Finally we can determine the vector $\Delta_2^{\bullet\bullet}$, so that the compatibility of the second of the (1) is already reached :

$$\Delta_2^{\bullet\bullet} = A_t^T \bar{x} \qquad (10)$$

The components of the vector $\Delta_2^{\bullet\bullet} \neq 0$ give the position and width of the cracks located in the mortar joints:

$$\Delta_2^{\bullet\bullet} = \begin{bmatrix} \Delta_{2t}^{\bullet\bullet} \\ 0 \end{bmatrix} \qquad (11)$$

5. APPLICATION OF THE PROCEDURE TO THE ANALYSIS OF VOUSSOIR ARCHES

The voussoir arch can be analyzed as a simple but particularly significant type of masonry construction in order to test the effectiveness and the efficiency of the numerical procedure. Having established the general pattern of behaviour for such a case, the results can be applied to other structural elements. The term *masonry arch* is widely known as a conventionally monodimentional structure consisting of stone or brick elements assembled dry or with mortar joints. Since the difference between the mechanical parameters of stone and mortar deformability is so substantial, the latter case is suitable to be analyzed using the proposed model of rigid blocks linked through elastic mortar layers.

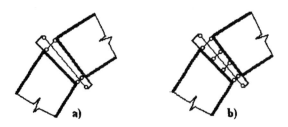

Fig 3- Two feasible *interface devices* between adjacent voussoirs

Even if three contact constraints would be strictly sufficient (Fig.3a), an *interface device*, consisting of four orthogonal bars (two of which located at the edges of block and the other two in the middle third position) and a parallel one, better describes the joint behaviour (Fig.3b).

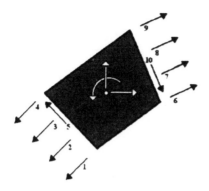

Fig 4- Reading order of the interactions acting on the block

The general behaviour of the structure depends mainly on the ratio between the stiffness value assumed for the contact constraints which are orthogonal to the interface surface, and the stiffness value of the contact constraint parallel to the interface surface. Moreover,

in addition to the no-tension behaviour of the orthogonal contact constraints, we can assume that a limited strength exists also for the shear forces. It is convenient to assume that the appropriate limit value of the tangential forces depends on the compressive value transmitted between the blocks. Nonetheless, it is very improbable that any sliding of one block upon another occurs in an arch under statical load conditions, and that the consequent behaviour is then that of opening hinges. The results obtained allow us to both locate the actual line of thrust and the cracked joints. In addition the corresponding width and depth of the cracks, measured in its radial direction, can be also evaluated

6. NUMERICAL EXAMPLES

In order to point out the efficiency and the versatility of the procedure in the cracking analysis of masonry arches, the structural behaviour of some simple cases has been investigated. The first example regards a simple arch, consisting of three blocks and four elastic-cracking joints, subject to self-weight loads and an asymmetrical load acting on the central block. The behaviour of the structure, after having located the actual line of thrust, becomes clearer through the amplification of its deformed configuration. In particular, through an appropriate graphic output, the crack width in one of the joints can be highlighted (Fig.5).

In the following examples the same procedure is applied in order to analyze the behaviour of arches with different geometry and different load conditions. In all the cases, it has been assumed that the friction between the blocks is high enough, and consequently, shear failure cannot occur.

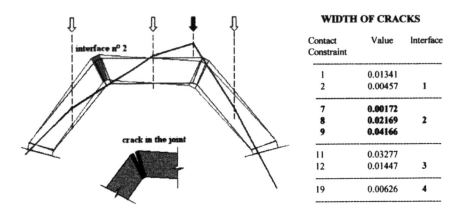

WIDTH OF CRACKS

Contact Constraint	Value	Interface
1	0.01341	
2	0.00457	1
7	**0.00172**	
8	**0.02169**	**2**
9	**0.04166**	
11	0.03277	
12	0.01447	3
19	0.00626	4

Fig 5- Structural behaviour and cracking analysis of a simple arch

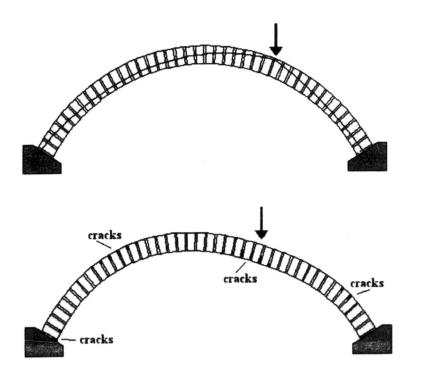

Fig 6- Incomplete circular arch subject to self-weight and single point load. Actual position of the thrust line, deformed shape and cracking pattern

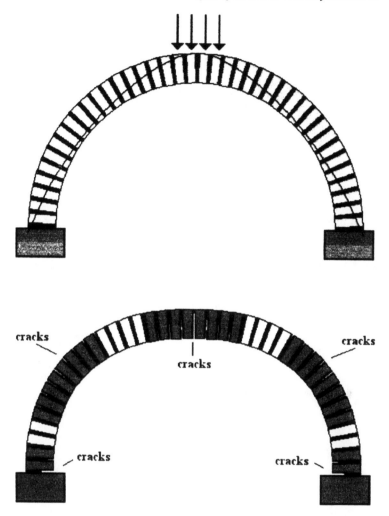

Fig 7- Circular arch subject to self-weight and central point loads. Actual position of the thrust line, deformed shape and cracking pattern

7. CONCLUSIONS

Based on the proposed numerical procedure, a computer program *MatLab* compatible was developed. This interactive software has proven to be very flexible for solving such a problem by using both the powerful of the matrix high-level programming language and the graphic one. Comparing this method with the iterative ones, usually used to solve this kind of problems, the solution, if it exists, is reached through a procedure with a low number of steps and without any particular numerical difficulty.

8. REFERENCES

1. Castigliano C.A.P., 'Théorie de l'équilibre des systèmes élastiques et ses applications', Torino 1879.
2. Di Pasquale S., 'Restauro dei Monumenti e questioni sismiche', Atti del Convegno di Ingegneria Antisismica e Protezione Civile, Ragusa 1981.
3. Ben Israel A. and Grenville T.N.E., 'Generalized inverses; theory and applications', Wiley Interscience Pubblication, London, 1974.
4. Briccoli Bati S., Paradiso M. and Tempesta G., ' A numerical procedure for the analysis of brickwork walls', IIIrd Int. Symposium on Structural Masonry, Lisbona 1995.
5. Kooharian A., 'Limit analysis of voussoir (segmental) and concrete arches', Journal of the American Concrete Institute, 1953.
6. Heyman J., 'The stone skeleton', Int. J. Solids Structures, vol. 2, pp. 249-279, 1966.
7. Heyman J., 'The masonry arch', Hellis Horwood, Chichester, 1982.
8. Di Pasquale S., "Questioni di meccanica dei solidi non reagenti a trazione", Atti VI Convegno nazioale AIMETA, Genova 7-9 Ottobre , 1982.
9. Franciosi V., 'The limit design of masonry arches under seismic loads' , Symposium Plasticity Today, CISM, Udine, 1983.
10. Di Pasquale S., 'Statica dei solidi murari: teoria ed esperienze', Pubb. n. 27, Dip. di Costruzioni, Universita' di Firenze, 1984.
11. Cristfield M.A., 'Computer methods for the analysis of masonry arches', Proc. 2nd Int. Conf. on Civil and Structural Engineering Computing, Londra 1985.
12. Franciosi, 'Su alcune questioni riguardanti la stabilita' delle strutture lapidee monodimensionali', Atti dell'Accademia Pontaniana, Nuova serie. vol.XXXIV, Giannini Napoli, 1986
13. Livesley R.K., 'A computational model for the analysis of three-dimensional masonry structures', Meccanica, vol. 27, pp. 161-172, 1992.
14. Briccoli Bati S., Paradiso M. and Tempesta G.,' Sul calcolo degli archi in muratura', Atti del Dipartimento di Costruzioni, Firenze 1992.
15. Briccoli Bati S., Paradiso M. and Tempesta G.,'Un modello numerico per l'analisi di strutture a vincoli unilateri', Atti del XII Congresso Nazionale AIMETA, Napoli 1995.

NUMERICAL ANALYSIS OF MASONRY STRUCTURES WITH A HOMOGENISED COSSERAT CONTINUUM MODEL

J. Sulem Chargé de Recherche[1], **M. Cerrolaza** Professor[2] and **A. El Bied** PhD Student[1]

[1]CERMES, Ecole Nationale de Ponts et Chaussées/LCPC, Cité Descartes, F-77455 Marne-la-Vallée, France and [2]Facultad de Ingenieria, Central University of Venezuela, PO Box 50361, Caracas 1050-A, Venezuela

1. ABSTRACT

In this paper the Cosserat continuum theory is used and validated to model the behaviour of blocky structures. For a typical benchmark problem, the results of finite elements computations for an equivalent Cosserat continuum are compared with the results obtained using a discrete approach. It is shown that the Cosserat homogenisation is applicable as soon as the size of the structure is bigger than 5 times the size of the block.

2. INTRODUCTION

The interest of developing continuous models for discrete structures is that discrete type analyses are very computer time intensive and, at least for periodic structures, one might argue that a homogenised continuum model would allow for a much more elegant and efficient solution. One could list the practical relevance of the development of continuum models : (a) it is extremely flexible when used with numerical methods since no interface elements are needed and since the topology of the finite element is independent of block size and geometry (one mesh can be used to study several different structures); (b) quite a number of analytical solutions can be provided that can be used as benchmarks for discrete codes; (c) unconditionally stable integration through implicit

Keywords: Blocky Structure, Cosserat Model, Finite Element Method

Computer Methods in Structural Masonry – 4, edited by G.N. Pande, J. Middleton and B. Kralj. Published in 1998 by E & FN Spon, 11 New Fetter Lane, London EC4P 4EE, UK. ISBN: 0 419 23540 X

algorithm can be used unlike discrete models where conditionally stable explicit integration schemes are used.

Averaging processes have been developed in order to describe the mechanical behaviour of inhomogeneous materials like blocky, jointed or layered rocks by considering an equivalent homogeneous continuum medium with averaged (effective) characteristics (Salamon, 1968, Bakhalov and Panasenko, 1989). These methods are based on the asymptotic developments technique and the validity of the approximation is restricted to the case where the characteristic size of the recurrent cell of the periodic medium (e.g. block size) is small as compared to the characteristic size of the problem (e.g. the wavelength of the deformation field). In other words when the characteristic length of the macroscopic deformation pattern is smaller than a certain multiple of the characteristic fabric length of the material, then the applicability of the continuum model has reached its limit. In order to overcome these limitations and to expand the domain of validity of the continuum approach one has to consider the salient features of the discontinuum within the frame of generalised continuum theories (e.g. Biot 1967, Herrmann and Achenbach 1968, Mühlhaus 1985, 1993, Vardoulakis and Sulem 1995).

An other important limitation of the homogenisation of layered or blocky structures with classical continuum theories is that they cannot account for elementary bending due to inter-layer or inter-block slip.

Consequently the framework of a Cosserat continuum theory is used here to model the behaviour of blocky rock. In Cosserat's theory a material point of the continuum has three additional rotational degrees of freedom as well as the three translations of a classical continuum. In a regular block structure, one can consider the influence of relative rotations between blocks by means of additional Cosserat rotations. The relative rotations cause moments and consequently, material parameters with dimension of length (here the block dimensions) appear in the constitutive relationship.

Numerical finite element modelling of layered or foliated materials has been recently presented by Adhikary and Dyskin (1996, 1997) and validated against explicit joint finite element model. For blocky structures with elastic joints, explicit determination of the material parameters of the equivalent Cosserat continuum has been presented in a previous paper of Sulem and Mühlhaus (1997). Analytical solutions for the dispersion function of the discrete structure and the corresponding continuous one have been presented and compared in order to establish the domain of validity of the Cosserat homogenisation. It was obtained that that the Cosserat model was appropriate for wavelengths greater than 5 times the size of the block.

In this paper a finite element analysis of blocky structures with the Cosserat model is presented. This numerical analysis is validated through the comparison of the response of a benchmark structure using the continuous approach and using a discrete approach . The discrete approach is developed by using the distinct element program UDEC. The key question is to evaluate the limit value of the ratio between the size of the model and the size of the individual block for which the applicability of the continuum model has reached its limit. It is obtained that for a structure with a characteristic dimension bigger than 5 to 6 times the size of the block the Cosserat approach is applicable.

3. THE COSSERAT MODEL

In a two-dimensional Cosserat continuum each material point has two translational degree of freedom (u, v) and one rotational degree of freedom ω^c. The index is used to distinguish the Cosserat rotation from the rotation

$$\omega = \frac{1}{2}\left(u_{2,1} - u_{1,2}\right) \quad ; \quad \left(.\right)_{,j} = \frac{\partial(.)}{\partial x_i} \quad i = 1,2 \tag{1}$$

For the formulation of the constitutive relationships we need deformation measures which are invariant with respect to rigid body motions which are the conventional strain tensor namely the strain tensor

$$\varepsilon_{ij} = \frac{1}{2}\left(u_{i,j} + u_{j,i}\right) \tag{2}$$

the relative rotation

$$\omega^{rel} = \omega - \omega^c \tag{3}$$

and the gradient of the Cosserat rotation which is called the curvature of the deformation

$$\kappa_i = \partial \omega^c / \partial x_i \tag{4}$$

It is usual to combine equations (2) and (3) to a single, tensorial deformation measure (see Mühlhaus 1993, Vardoulakis and Sulem 1995)

$$\begin{aligned} \gamma_{11} &= \partial u_1 / \partial x_1 \quad ; \quad \gamma_{12} = \partial u_1 / \partial x_2 + \omega^c \\ \gamma_{22} &= \partial u_2 / \partial x_2 \quad ; \quad \gamma_{21} = \partial u_2 / \partial x_1 - \omega^c \end{aligned} \tag{5}$$

The six deformation quantities (equations 4 and 5) are conjugate in energy to six stress quantities. First we have the four components of the <u>non symmetric</u> stress tensor σ_{ij} which is conjugate to the <u>non symmetric</u> deformation tensor γ_{ij} and second we have two moment stresses (moment per unit area) m_1 and m_2, which are conjugate to the two curvatures κ_1 and κ_2. Force and moment equilibrium at the element (dx_1, dx_2) lead to

$$\begin{aligned} \sigma_{11,1} + \sigma_{12,2} - \rho\ddot{u}_1 &= 0 \\ \sigma_{21,1} + \sigma_{22,2} - \rho\ddot{u}_2 &= 0 \\ m_{1,1} + m_{2,2} + \sigma_{21} - \sigma_{12} - I\ddot{\omega}^c &= 0 \end{aligned} \tag{6}$$

In the above equations dynamic effects are included through inertial forces and moment. The stress-strain relationships for a 2D anisotropic Cosserat continuum are (see Schaefer 1962)

$$\begin{aligned} \sigma_{11} &= C_{11}\gamma_{11} + C_{12}\gamma_{22} \\ \sigma_{22} &= C_{21}\gamma_{11} + C_{22}\gamma_{22} \\ \sigma_{12} &= \left[G + G_c(1-\alpha)\right]\gamma_{12} + \left[G - G_c\right]\gamma_{21} \\ \sigma_{21} &= \left[G - G_c\right]\gamma_{12} + \left[G + G_c(1+\alpha)\right]\gamma_{21} \\ m_1 &= M_1\kappa_1 \\ m_2 &= M_2\kappa_2 \end{aligned} \tag{7}$$

where α is a parameter of anisotropy.

4. BLOCKY STRUCTURES ANALYSIS

4.1 Discrete and continuous model

We consider here a simple model for a masonry wall (Figure 1). Each block is surrounded by six others. We are mainly concerned with the accuracy with which the continuum model reflects the domain of rigidity set by the size of the blocks. The elasticity of the blocks and the joints elasticities are lumped at the block edges for simplicity. We assume fully elastic joint behaviour. We assume that the interaction between the blocks is concentrated in six points of the edges as shown on figure 1. Normal and shear forces are written as

$$Q_{kl} = c_Q \Delta u_{kl}$$
$$N_{kl} = c_N \Delta v_{kl}$$

(8)

where c_Q and c_N are the elastic shear and normal stiffness respectively and Δu and Δv are the relative horizontal and vertical displacements at various contact points.

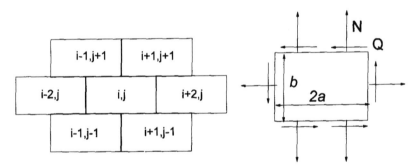

Figure 1 *Geometrical and statical configuration*

The homogenisation procedure is presented in the previous paper of Sulem and Mühlhaus (1997) for identification of the material parameters of the corresponding equivalent Cosserat continuum. We obtain

$$C_{11} = \left(c_Q + 2c_N\right)\frac{a}{b} \quad ; \quad C_{22} = c_N \frac{b}{a} \quad ; \quad C_{12} = C_{21} = 0$$

$$G = G_c = \frac{1}{4}\left[c_Q \frac{b}{a} + \left(c_N + 2c_Q\right)\frac{a}{b}\right]$$

$$\alpha = 2\frac{a^2\left(c_N + 2c_Q\right) - c_Q b^2}{a^2\left(c_N + 2c_Q\right) + c_Q b^2}$$

(9)

$$M_1 = \frac{a}{b}\left[c_N \frac{a^2}{4} + c_Q\left(\frac{b^2}{4} + 2a^2\right)\right] \quad ; \quad M_2 = \frac{b}{a}\left[c_N \frac{a^2}{4} + c_Q \frac{b^2}{4}\right]$$

4.2 Finite element formulation

For general boundary value problems numerical methods are used. The finite element method is a well established tool for these purposes. In the following we present the most important features of the extension of the method to a Cosserat continuum.
We define general displacement and traction pseudo vectors as

$$\mathbf{v} = \left\{ u_1, u_2, \omega_c^3 \right\} \quad ; \quad \mathbf{f} = \left\{ t_1, t_2, m_3 \right\} \tag{10}$$

by means of which the virtual work principal can be written as

$$\int_B \mathbf{s}^T \delta e dV = \int_{\partial B} \mathbf{f}^T \delta v dA \tag{11}$$

Above equation (11) looks formally the same as for the classical continuum.
Essentially the finite element method consists of specifying an assume distribution of the displacements and rotations within the domain B^e of a finite element. This can be written as

$$\mathbf{v} = \phi^M \mathbf{v}^M \quad ; \delta \mathbf{v} = \phi^M \delta \mathbf{v}^M \quad M = 1, 2 .... M^e \tag{12}$$

where ϕ^M are the so-called shape functions, M is the number of nodal points and M^e the total number of nodal points of each element.
The relation between the deformation vector **e** and the nodal variables is written as :

$$\mathbf{e} = \mathbf{B}^M \mathbf{v}^M \quad ; \quad \left[\mathbf{B}^M \right] = \begin{bmatrix} \phi_{,1}^M & 0 & 0 \\ 0 & \phi_{,2}^M & 0 \\ \phi_{,2}^M & 0 & \phi^M \\ 0 & \phi_{,1}^M & -\phi^M \\ 0 & 0 & \phi_{,1}^M \\ 0 & 0 & \phi_{,2}^M \end{bmatrix} \tag{13}$$

Inserting equation (13), equation (11) becomes

$$\left(\delta \mathbf{v}^M \right)^T \mathbf{K}^{MN} \Delta \mathbf{v}^N = \left(\delta \mathbf{v}^M \right)^T \left(\mathbf{F}_{ext}^M - \mathbf{F}_{int}^M \right) \tag{14}$$

where

$$\mathbf{K}^{MN} = \int_{B^e} \left(\mathbf{B}^M \right)^T \mathbf{C} \mathbf{B}^N dV \tag{15}$$

is the element tangent stiffness matrix,

$$\left[\mathbf{F}_{ext}^{M}\right] = \begin{bmatrix} \int\limits_{\partial_1 B^e} t\phi^M dA \\ \int\limits_{\partial_2 B^e} m\phi^M dA \end{bmatrix} \qquad (16)$$

is the generalised external load vector and

$$\left[\mathbf{F}_{int}^{M}\right] = \int\limits_{B^e}\left(\mathbf{B}^M\right)^T \mathbf{s}_t dV \qquad (17)$$

is the generalised initial stress vector.

5. BENCHMARK PROBLEM

5.1 Methodology

A specific finite element program for analysis of two-dimensional Cosserat continuum has been developed (Cerrolaza et al 1997). It is named the COSS program and allows static and dynamic analysis. It contains a graphic interactive pre-processor to generate and edit the geometry of the finite element mesh, a finite element processor to analyse the bidimensional model and a graphic interactive post-processor for the graphic displaying and interpretation of the finite element results.

The numerical analysis of blocky structure using the COSS finite element program with Cosserat structure is validated through the comparison of the response of a structure using the continuous approach and using a discrete approach. This has been done in the particular case of a two-dimensional structure with elastic joints under static loading. The discrete approach is developed by using the distinct element program UDEC. The key question is to evaluate the limit value of the ratio between the size of the model and the size of the individual block for which the applicability of the continuum model has reached its limit.

For elastic block interactions the dispersion relations of the discrete and the continuous models have been derived analytically (see Sulem et al 1996, 1997). The domain of validity of the continuous approach was discussed by comparing the dispersion function of the discrete and the continuous system. For dynamic loading, the Cosserat model gives an acceptable approximation for wave lengths λ larger than 5 times the size of the block.

5.2 Description of the problem

We consider a rectangular wall (length 12.6m, height 6.3m) made of identical rigid blocks (length 2a, height b) with elastic joints. For given length and height of the wall we consider four different models with different number of blocks (Figure 2):
Model MOD3 : b=2a=0.525m (3 horizontal ranks of blocks)
Model MOD6 : b=2a=0.7m (6 horizontal ranks of blocks)
Model MOD9 : b=2a=1.05m (9 horizontal ranks of blocks)
Model MOD12 : b=2a=2.1m (12 horizontal ranks of blocks)

The elastic joints are modelled as elastic springs in normal and tangential direction. We consider here the particular following values
$C_Q = 20$ MN/m and $C_N = 100$ MN/m.

The ratio b/2a is the same for the four models considered and consequently the elastic coefficients C_{ij}, G, G_c, and α of the equivalent Cosserat elastic continuum are the same and given by the following table (Table 1) (see equation 9)

C_{11} (MPa)	C_{22} (MPa)	C_{12} (MPa)	C_{21} (MPa)	G (MPa)	G_c (MPa)	α
110	200	0	0	27.5	27.5	0.55

Table 1: *Elastic constants for the equivalent Cosserat continuum*

The elastic constants M_1 and M_2 depend upon the size of the blocks (equation 9) and are given by

Model	MOD3	MOD6	MOD9	MOD12
M_1 (MPa.m^2)	46.86	11.71	5.21	2.93
M_2 (MPa.m^2)	99.22	24.80	11.02	6.2

Table 2: *Elastic bending stiffness for the various models*

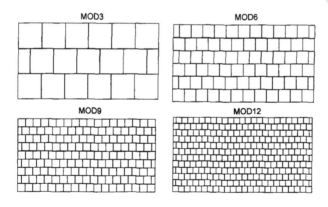

Figure 2 : *The benchmark problem*

5.3 Summary of the results

A uniform horizontal load of 220KN is applied on the left side of the structure. On the basement of the structures we assume elastic restraints. Typical deformed shape obtained with the finite element program is shown on figure 3.

We present on table 3 the average of the relative difference between the two sets of results obtained respectively with UDEC and with COSS for the displacements and for the stresses:

Model	MOD3	MOD6	MOD9	MOD12
difference in displacements	14.1%	8.9%	7.6%	5.5%
difference in stresses	10.4%	5.3%	3.4%	2.8%

Table 3: *Comparison of the results for COSS and UDEC*

As shown on the above table, for a wall which contains at least 6 ranks of blocks, the results of the two approaches differ from less that 10% for the displacements and less than 5% for the stresses. We can thus conclude that for a structure with a characteristic dimension bigger than 5 to 6 times the size of the block the Cosserat approach is applicable.

ACKNOWLEDGEMENTS

The authors want to acknowledge the Commission of the European Communities which supports this research through the ENVIRONMENT Project No EV5V-CT93-0300

REFERENCES

1. Salamon, M.D.G.(1968). Elastic moduli of stratified rock mass. *Int. J. Rock Mech. Min. Sci. and Geomech. Abstr.*, **5**, 519-527.
2. Bakhvalov, N. and Panasenko, G. (1989). *Homogenisation: Averaging processes in periodic media*. Kluwer Academic publ. Dordrecht, The Netherlands.
3. Biot, M. (1967). Rheological stability with couple stresses and its application to geological folding. *Proc. Roy. Soc. London*, **A2298**, 402-423.
4. Herrmann, G. and Achenbach, J.D. (1968). Applications of theories of generalised Cosserat continua to the dynamics of composite materials. *Proc. IUTAM Symp. Mechanics of generalised continua* (E. Kröner, ed.), 69-79, Springer-Verlag, Berlin.
5. Mühlhaus, H.B. (1985). Oberflächeninstabilität bei geschichtetem Halbraum mit Biegesteifigkeiten. *Ing. Arch.* **56**, 389-399.
6. Mühlhaus, H.-B. (1993). Continuum models for layered for layered and blocky rock. In: *Comprehensive Rock Engng.*, Vol. **2** (ed. Charles Fairhurst) Pergamon Press, 209-230.
7. Vardoulakis I, and Sulem J. (1995). *Bifurcation Analysis in Geomechanics*, Blackie Academic & Professional, Glasgow.
8. Adhikary, D.P., Dyskin, A.V. and Jewell, R.J. (1996). Numerical modelling of the flexural deformation of foliated rock slopes. *Int. J. Rock Mech. Min. Sci. and Geomech. Abstr.*. **33.6**. 595-606.

9. Adhikary, D.P. and Dyskin, A.V. (1997). A Cosserat continuum model for layered materials. *Computers and Geotechnics*, **20,1**, 15-45.

10.Sulem, J. and Mühlhaus, H.B. (1997). A continuum model for periodic two-dimensional block structures. *Mechanics of cohesive-frictional materials*, **2**, 31-46.

11.Schaefer, H., (1962). Versuch einer Elastizitätstheorie des zweidimensionalen ebenen Cosserat-Kontinuums. in: Miszellannenn der Angewandten Mekanik, Akademie Verlag, Berlin, 277-292.

12.Cerrolaza, M., Sulem, J., El Bied, A. (1997). A Cosserat non-linear finite element analysis software for blocky structures. *Int. J. of Advances in Eng. Soft*, Elsevier (submitted).

13.Sulem, J and Mühlhaus, H.B. (1996). A Cosserat continuum model for blocky rock under dynamic loading,, *ISRM Int. Symp. Prediction and performance in rock mechanics and rock engineering* (ed. G. Barla), Balkema *(EUROCK '96)*, Turin, Italy, September 2-5, 1996 Vol. 1, .359-366

TOWARDS NUMERICAL PREDICTION OF CRACKING IN MASONRY WALLS

G.P.A.G. van Zijl[1] and J.G. Rots[1,2]

[1]Department of Civil Engineering, Delft University of Technology, PO Box 5048, 2600 GA Delft, The Netherlands and [2]TNO Building and Construction Research, The Netherlands

1 ABSTRACT

As substitute for expensive, time consuming testing of large masonry structures, an accurate numerical predictive tool is of great value. Reference behaviour can be analysed as verification for more elegant, less computer intensive numerical modelling strategies. Here a discrete approach, capturing the most important failure modes by true, discontinuous deformation is reported. Available experimental results on the micro- and meso-scale are employed for verification of the strategy. The important large scale structural phenomenon of restrained shrinkage is analysed subsequently and the resulting behaviour employed as reference for a simplified modelling strategy to render large scale analyses viable.

2 INTRODUCTION

As part of the Dutch research programme *Structural Masonry* a brick-joint model was developed, capturing fracture, shear-slipping and crushing of masonry numerically [1]. The model is employed in a discrete approach, where the constituents and the interfaces between them are discretized. This approach is followed in recognition of the difficulty of numerical simulation of masonry structures, often exhibiting highly localized failure. By accurate, discrete modelling a substitute for large scale testing to determine behaviour of structural masonry, is envisaged. In this manner reference behaviour for verification of more elegant and less computational demanding computational strategies can be found.

Here enhancement and validation of the model are reported, after which prediction of large masonry wall behaviour is undertaken. The model is enhanced with a dilatancy formulation accurately capturing the smoothing of brick-mortar interfaces by compression and shear-slipping observed in micro-tests. Subsequently micro-experiments, as well as tests on wall parts on the meso-level are simulated numerically for verification of the model and validation of the strategy. With the confidence gained from the verified simulation capability, prediction of large masonry wall base-restrained shrinkage behaviour is undertaken. A simplified modelling approach for this shrinkage problem is illustrated next, employing the latter numerically predicted behaviour as reference.

Keywords: Masonry; Walls; Restrained; Shrinking; Numerical

Computer Methods in Structural Masonry – 4, edited by G.N. Pande, J. Middleton and B. Kralj. Published in 1998 by E & FN Spon, 11 New Fetter Lane, London EC4P 4EE, UK. ISBN: 0 419 23540 X

3 DISCRETE MODELLING APPROACH

The finite element (FE) method is employed and, because in-plane actions are of concern, masonry behaviour is studied in the two-dimensional framework of plane stress. As shown schematically in figure 1 either (a) bricks and mortar are discretized as elastic continua with nonlinear interfaces between the two constituents, or (b) only bricks are discretized and interfaces account for total joint behaviour. If, in addition, central, vertical potential cracks through bricks are modelled via interfaces - figure 1(c), the most important masonry failure modes can be captured.

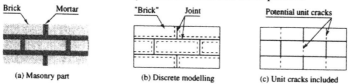

Figure 1. Discrete modelling strategy.

The interfaces account for tensile cracking, shear-slipping and compressive crushing via an interface material law set out in multi-surface computational plasticity, as shown in figure 2 [1].

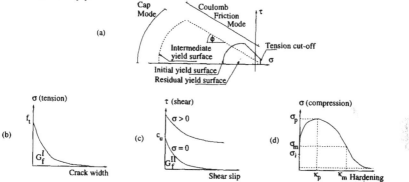

Figure 2. Interface material model (a) yield surface, (b) tensile- and (c) shear softening behaviour and (d) compressive hardening and eventual softening.

Complete behaviour of each material point from elastic through hardening to peak strength and eventual degradation to zero strength in tension, or some residual frictional strength in shear is captured. The parameters involved in the tension part are the tensile strength (f_t) and mode I fracture energy (G_f^I). For the Coulomb friction part the adhesion c_u, original (ϕ_o) and residual friction angle (ϕ_r), mode II fracture energy (G_f^{II}) and dilatancy angle (ψ) must be provided.

For proper dilatant description a non-associative formulation is employed. The interface plastic generalised strain rate is given by the flow rule:

$$\dot{\varepsilon}^p = \left\{ \begin{array}{c} \dot{\Delta u}^p \\ \dot{\Delta v}^p \end{array} \right\} = \dot{\lambda} \frac{\partial g}{\partial \sigma} \tag{1}$$

where Δu and Δv are the interface normal opening- and shear slipping displacement respectively. The gradient of a suitable potential function is defined as:

$$\frac{\partial g}{\partial \sigma} = \left\{ \begin{array}{c} \tan\psi^* \\ \text{sign}(\tau) \end{array} \right\} \tag{2}$$

ψ^* being the mobilised dilatancy angle. Now, following directly from the flow rule

$$\tan\psi^* = \frac{\dot{\Delta u}^p}{\dot{\Delta v}^p} \, \text{sign}(\tau) \qquad (3)$$

By integration the shear-slip induced crack width is found to be:

$$\Delta u^p = \int \tan\psi^* \, d|\Delta v^p| \qquad (4)$$

From measured normal uplift during shear-slipping [2] - figure 3 - it is clear that dilatancy is dependent on confining stress and shear-slipping displacement.

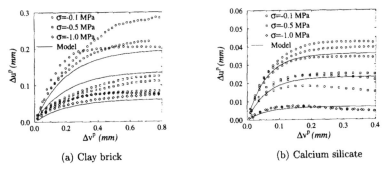

(a) Clay brick   (b) Calcium silicate

Figure 3. Measured joint normal opening vs. shear-slip of (a) clay brick and (b) calcium silicate specimen for different, constant confining pressures.

Curve fitting reveals that a separate variable function of the form

$$\Delta u^p = \frac{\tan\psi_o}{\beta} \left\langle 1 - \frac{\sigma}{\sigma_u} \right\rangle \left(1 - e^{-\beta \, \Delta v^p}\right) \qquad (5)$$

yielding after differentiation

$$\tan\psi^* = \tan\psi_o \left\langle 1 - \frac{\sigma}{\sigma_u} \right\rangle e^{-\beta \, \Delta v^p} \qquad (6)$$

captures the smoothing of a shearing interface with increased normal pressure and shearing displacement, as shown by the solid lines in figure 3. The dilatancy at zero normal confining stress and shear slip ($\tan\psi_o$), confining stress at which the dilatancy becomes zero (σ_u) and the dilatancy shear slip degradation coefficient (β) are parameters to be obtained by curve fitting of experimental measurements. The parameters for the specimens of figure 3 are given in table 1 and are employed in analyses throughout this paper. Also listed are the friction coefficients ($\tan\phi$).

Specimen	$\tan\phi$	$\tan\psi_o$	$\sigma_u (N/mm^2)$	β
Clay brick	1.01	1.06	-1.42	5
Calcium S.	0.97	0.67	-1.22	17

Table 1. Dilatancy material parameters employed.

4 VERIFICATION ON MICRO-LEVEL

This strategy has been shown to accurately simulate measured masonry micro-shear behaviour [3]. Figure 4 (a) and (b) show the experimental set-up and FE models of the shear displacement and normal force controlled test.

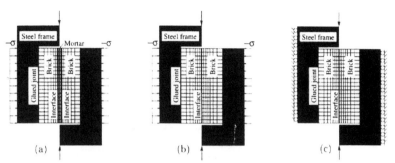

Figure 4. (a) Detailed and (b) simplified modelling of masonry micro-shear test. (c) Normal displacement constrained micro-shear test.

The material parameters employed here were obtained by averaging and regression of experimental results [2], [4] are shown in table 2.

	Brick				Mortar	Interface					
	E $\frac{N}{mm^2}$	ν	f_t $\frac{N}{mm^2}$	G_f^I $\frac{N}{mm}$	G $\frac{N}{mm^2}$	f_t $\frac{N}{mm^2}$	G_f^I $\frac{N}{mm}$	c_u $\frac{N}{mm^2}$	$\tan\phi_o$	$\tan\phi_r$	G_f^{II} $\frac{N}{mm}$
(a)	16700	0.28	2.0	0.08	2300	0.6	0.012	0.88	1.01	0.75	$0.06-0.13\sigma$
(b)	13400	0.20	2.0	0.06	1375	0.1	0.005	0.28	0.97	0.75	$0.02-0.03\sigma$

Table 2. Material parameters employed for (a) clay and (b) calcium silicate brick specimen.

Both detailed (figure 4a) and simplified (figure 4b) approaches were followed. Figure 5(a) and (b) show that good agreement with measured stress-displacement behaviour is obtained with both approaches.

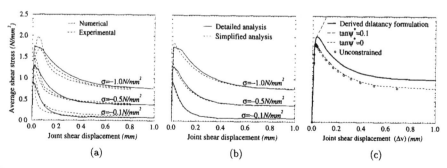

Figure 5. Comparison of (a) numerical with measured response and (b) detailed with simplified modelling response of micro-shearing clay brick specimens. (c) Normal displacement constrained response for various dilatancy options.

To demonstrate the importance of correct dilatancy modelling, the case of micro-shear with constrained normal displacement shown in figure 4(c) is analysed. Figure 5(c) shows that for even a small, but constant dilatancy angle no peak strength is found. The current variable dilatancy formulation captures the enhanced peak and residual shear strength coinciding with the normal stress at which the interface is smoothed and after which no further normal displacement across the interface occurs. Experiments to verify this predicted response are currently undertaken.

5   VERIFICATION ON MESO-LEVEL

Tensile response of small wall parts as obtained from displacement controlled tests of a calcium silicate specimen [5] similar to the Dutch specimen is employed here for verification of the model. Apart from global deformational behaviour only strength and stiffness were reported. Thus, being of similar material, the fracture energies and dilatation parameters for calcium silicate in table 2 were employed, rendering reasonable agreement of global stress-deformational response and the correct step-wise failure along head and bed joints, as shown in figure 6. Despite free upper and lower edges of the specimen, dilatational uplift along shear-slipping bed joints is resisted by the intact, undamaged masonry, causing wedging and accompanied residual strength. Also shown is the total strength degradation if zero dilatancy is assumed. Peak strength is governed by the head joints, followed by bed-joint shearing in the post-peak response. Therefore dilatancy influence is seen only in the latter phase.

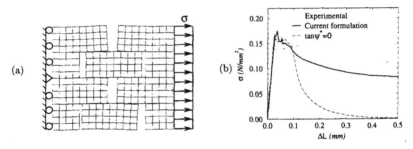

Figure 6. Verification on the meso-level. (a) Numerically found deformation and (b) numerical versus experimental average normal stress-displacement response of calcium silicate wall part.

6   PREDICTION OF WALL RESTRAINED SHRINKAGE RESPONSE

Having verified the modelling strategy on the micro- and meso-scale, large masonry walls subjected to base-restrained shrinking are analysed. A single leaf, non-bearing wall was chosen for this study. To restrict the model size only the central four units and their joints were discretised, enabling capturing of the observed primary crack shown in figure 7. The remaining wall is assumed to remain elastic with homogenised masonry elasticity parameters. A stiff, elastic concrete foundation of 200mm width was modelled. Slipping along the wall-base interface was not included for lack of experimental data, but also to demonstrate the worst case scenario.

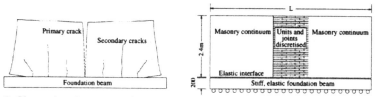

Figure 7. Observed crack pattern in base-restrained shrinking wall and model shown schematically.

The two masonry types with parameters listed in tables 1 and 2 were investigated, representing a typical clay brick of Dutch waal format ($210 \times 52 \times 100$mm) with 10mm 1:2:9 volume cement:lime:sand mix mortar joints (JOB) and a waal format calcium silicate unit with 10mm joints of $1:\frac{1}{2}:4\frac{1}{2}$ mortar (KZC).

Figures 8(a-d) show the crack propagation at different stages of shrinkage and figures 8(e) and (f) the primary crack width with varying shrinkage. Note that uniform shrinking in the walls is assumed. Also, the reduction in shrinking for increasing crack width means literally that the environment should enable the wall to adhere to this particular shrinking history for the shrinkage-crack width responses shown to realise. In reality, for increased shrinking the crack will open dynamically to a configuration of sufficient shrinkage resistance, as indicated for the KZC 6m wall.

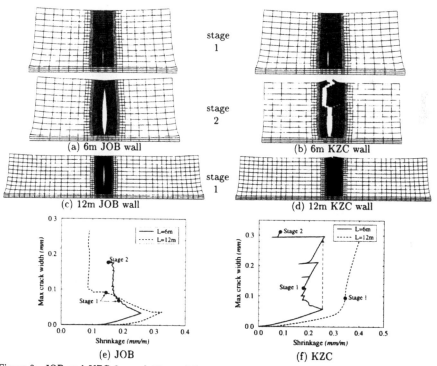

Figure 8. JOB and KZC 6m and 12m wall base-restrained shrinkage deformation at particular stages of the shrinkage-crack width development.

Vertical cracking is found in the clay brick wall as well as in the lower third of the 6m KZC wall, followed by step-wise cracks along head and bed joints in the latter. Limited discretisation possibly inhibits the step-wise cracking, enforcing further vertical cracking until the upper, unbounded part again fails in step-wise fashion. From table 2 it can be seen that the JOB joints are quite strong, forcing cracks to jump vertically between head joints through bricks. KZC has low adhesion strength, but due to the confinement near the base Coulomb-frictional strength of the bed joints prevent shear-slip, enforcing brick cracking. With less confinement, but also relatively low required uplift during shearing (table 1) bed joint failure occurs higher up in the wall. A detailed discussion of seemingly greater resistance to shrinking of longer walls is given elsewhere [6].

To investigate shrinkage behaviour further, a simplified strategy to make the analyses viable, is investigated. Because a nearly uniaxial stress-state exists in the region of the primary crack, the response of a periodic wall part under uniaxial tension is studied. By discrete modelling the total tensile deformational response of such a part is found. The nonlinear behaviour found is now attributed to an equivalent vertical crack, as shown in figure 9 for an example where step-wise cracking along the two head joints and a single bed joint occurs. The surrounding masonry is modelled continuously with elastic masonry properties. The stress-deformational comparison shows a discrepancy in crack width (w_c) at peak stress, which is ascribed to the crack width averaging process implied. This discrepancy can be seen to disappear soon after peak response.

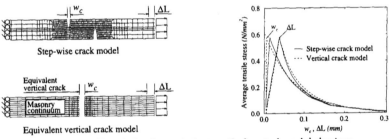

Figure 9. Translation for equivalent vertical central crack behaviour.

The influence of (1) boundary conditions simulating the wall part confinement, as well as (2) the energy dissipated by the non-failing joints and/or cracking bricks, which depends mainly on material parameter spatial variation, is of importance. It is shown elsewhere [6] that a *lower* and *upper* bound of wall part toughness can be found by investigating extreme cases of the above two factors, with as *average* the response for an assumed spatial variation in material parameters. The upper and lower bounds have equal peak strengths ($f_t = 1.14 N/mm^2$ for JOB and $f_t = 0.96 N/mm^2$ for KZC), but the upper bound has higher toughness due to wider spread micro-cracking. Figure 10 shows the bounds of wall shrinkage-crack width behaviour obtained with the equivalent vertical crack when the bounds of wall part response are used for the translation.

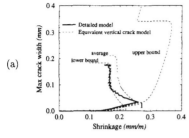 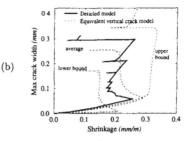

(a)   (b)

Figure 10. Verification of simplified modelling strategy. (a) JOB and (b) KZC base-restrained 6m wall shrinkage response for detailed and equivalent vertical crack approach.

Now employing the simplified vertical crack model walls of various lengths are analysed. Figure 11(a) shows the crack width-shrinkage responses of JOB walls for upper bound behaviour. By plotting the shrinkage values that will cause a particular crack width just not to be exceeded, here say 0.3mm, against the various wall lengths, the connection with practical design rules is made. These values of wall lengths can be considered as crack-free, or rather *unacceptable crack width*-free wall lengths for the applied shrinkage level. In figure 11(b) the latter, numerical values for $w_c \leq 0.3$mm are compared with strength-based analytical rules [7, 8, 9].

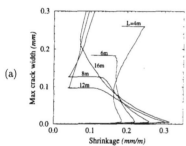 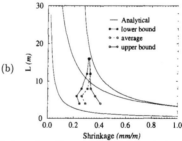

(a)   (b)

Figure 11. JOB (a) lower bound equivalent vertical crack wall response to base-restrained shrinkage. (b) Wall length containing a crack not exceeding 0.3mm width at the shrinkage strain shown.

7 CONCLUSIONS

A discrete modelling strategy employing nonlinear interface formulated in multisurface plasticity has been shown to simulate masonry behaviour on the micro- and meso-level. Realistic behaviour of large base-restrained walls are predicted with the strategy. This response is considered as reference behaviour for verification of simplified modelling strategies, as illustrated with an equivalent vertical crack approach for analysing the shrinking walls. It is shown how the results obtained in this rational manner can be translated into eventual design rules.

8 ACKNOWLEDGMENTS

This research is supported by the Netherlands Technology Foundation STW, applied science division of NWO under grant DCT 44.3406, and by CUR committee A33. The model described here was implemented into a pilot version of the DIANA finite element package, which was subsequently employed for the calculations.

REFERENCES

[1] Lourenço, P.B., *Computational strategies for masonry structures*. Dissert., Delft Univ. of Tech., Delft, The Netherlands, 1996.

[2] Pluijm, R.v.d., *Deformation controlled shear tests on masonry* (in Dutch). Rep. BI-92-104, TNO-Bouw, Delft, The Netherlands, 1992.

[3] van Zijl, G.P.A.G., *Masonry micro-shear behaviour along bed joints*. Rep. 03-21-22-0-01, Delft Univ. of Tech., Delft, The Netherlands, 1996.

[4] Pluijm, R.v.d. and Vermeltfoort, A.T., *Deformation controlled tension and compression tests on units, mortar and masonry* (in Dutch). Rep. B-91-0561, TNO-Bouw, Delft, The Netherlands, 1991.

[5] Backes, H.P, *The behaviour of masonry under tension in the direction of the bed joints* (German). Dissert., Aachen Univ. of Tech., Aachen, Germany, 1985.

[6] van Zijl, G.P.A.G. and Rots, J.G., Understanding masonry wall restrained shrinkage behaviour, to be published in: *Proc. 11th Int. Brick and Block Masonry Conference*, Shanghai, China 1997.

[7] Copeland, R.E. Shrinkage and temperature stresses in masonry, *ACI Journal*, **53**, pp.769-780, 1957.

[8] Schubert, P., About the crack-free length of non-loadbearing masonry walls (in German). *Mauerwerk-Kalendar*, pp.473-488, 1988.

[9] Hageman, J.G., *Study of shrinkage cracks* (in Dutch). Research Center for Calcium Silicate Industry, Rep. no. 189-1-0/189-2-0, The Netherlands, 1968.

EXPERIMENTALLY-BASED COMPUTATIONAL MODELLING OF MASONRY

V. Bosiljkov[1], R. Žarnić[1], B. Kralj[2] and G.N. Pande[2]
[1]Faculty of Civil Engineering and Geodesy, University of Ljubljana, Jamova 2, SI-61000 Ljubljana, Slovenia and [2]Department of Civil Engineering, University College of Swansea, Singleton Park, Swansea SA2 8PP, UK

ABSTRACT

The main scopes of this investigation were the experimental and numerical studies of the influences of different mortar types on the behaviour of masonry subjected to uniaxial compressive loading. The numerical model used was a micro-model based on the finite element method. The numerical analysis was used in order to analyse the influence of individual mechanical parameters on the deformation behaviour of compressively loaded wallette. Further on, the conditions where two typically different failure mechanisms of compressively loaded wallettes occurred during the compressive tests were determined.

1. INTRODUCTION

Masonry is a composite, heterogeneous, non-linear material consisting of two or more different constituent materials bonded together at an interface. The basic mechanical properties of such media are strongly influenced by the mechanical properties of it's constituents (brick and mortar), their volumetric ratio and their interface. A numerical model of such material can be developed using micro-modelling in such a way that each element contains only one material. The interface between the elements and the potential cracks within the constitutive materials can be modelled using link elements. This approach can be very accurate and can give a good insight into micro stresses in masonry. On the other hand, the micro-modellig approach has its limitation in a large number of elements required (CPU time) and large input of mechanical characteristics of constituents and their junctions. For these reasons this approach is more suitable for the modelling small masonry elements such as wallettes, prisms etc. or for the calibration of the mechanical parameters for macro-models for masonry, see [1].

Computer Methods in Structural Masonry – 4, edited by G.N. Pande, J. Middleton and B. Kralj.
Published in 1998 by E & FN Spon, 11 New Fetter Lane, London EC4P 4EE, UK. ISBN: 0 419 23540 X

2. SCOPE OF INVESTIGATION

The main scope of the investigation was to determine the behaviour of wallettes made from four different types of mortars subjected to compressive loading. Mortar with volumetric ratio of its components 1 : 1: 6 (cement : lime : sand) was taken as a basic type. Three other types of mortar were derived by modifying of the basic one. From experimental measurements and observations of the specimens during the testing two different types of failure mechanisms were found: one for wallettes made from the basic mortar (MIX-1) and the other for wallettes made from its modification (MIX-4). In the numerical part of the investigation conditions under which these two different types of failure mechanisms occur were determined. Also, parametric studies of influences of variation of mechanical parameters on the deformation of wallettes under compressive loading were undertaken.

3. EXPERIMENTAL WORK

The experimental part of the study consisted of two parts. First, an extensive research of over 20 different modifications of basic mortar (MIX-1) was made, from which three modifications were chosen and used in subsequent tests. All of the mortar specimens were cured and tested according to prEN 1015-11. The main criteria in choosing the modified mortars were their toughness and shrinkage. Toughness was determined on flexural specimens while shrinkage was determined on prisms 4x4x16 cm^3. Mix proportions for the chosen mortars were published elsewhere [2]. Their modifications as compared to the basic mortar (MIX-1) are given in Table 1.

Table 1. Modifications of the basic mortar mix

MIX	Description
MIX-1	basic mortar with volumetric ratio of cement : lime : sand = 1 : 1 : 6
MIX-2	polypropylenic fibre reinforced mortar with superplastisizer
MIX-3	polypropylenic fibre reinforced mortar where part of the binder was substituted by micro-silica
MIX-4	polypropylenic fibre reinforced with polymer modified mortar where part of the binder was substituted by micro-silica

The second part of the experimental investigation was testing the compressively loaded wallettes according to prEN 1052-1, including tests of their constituents according to prEN 772-1 (brick) and prEN 1015-11 (mortar). Bricks (250x120x65 cm^3) used in the specimens were extruded solid units, all from the same batch. All the bricks were prewetted before laying and the same mason made all the wallettes. The wallettes were cured under laboratory conditions according to prEN 1052-1 and were tested at the age of 60 days. For each MIX five specimens were built. Three of them were tested under monotonous loading and two of them under cyclic loading. Wallette instrumentation (Figure 1) included 9 LVDT's and an optical measurement of lateral deformation of mortar in the bed joint. LVDTs L1, L2, L6 and L7 were placed according to prEN 1052-1 for measuring vertical deformation of wallettes. LVDT's L5 and L9 were used for measuring local lateral deformations of wallettes in the mid-height of the specimens. Optional measuring devices for lateral deformation of the wallettes in the middle height of the specimens (global deformations) were L3 and L4. L8 was used to measure lateral deformation of the brick in the middle heights of the specimens.

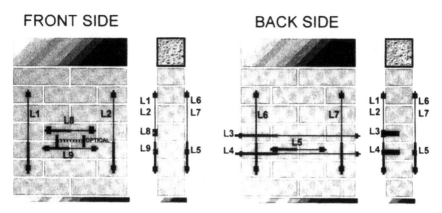

Figure 1: Wallette instrumentation

3.1. Results of the experiments

In the table below (Table 2) some major results of our testing are presented, where:

f_m .... is compressive strength of mortar acc. to prEN 1015-11

f_b .... is compressive strength of brick acc. to prEN 772-1

f_{eks} .... is characteristic compressive strength of wallettes acc. to prEN 1052-1

f_{EC-6} ..... is the calculated characteristic compressive strength of unreinforced masonry acc. to formula given by EC-6: $f_{EC-6} = K f_b^{0.65} f_m^{0.25}$

$E_{(30\%)}$ ... is secant modulus of elasticity of wallettes acc. to prEN 1052-1

Table 2: Results of testing

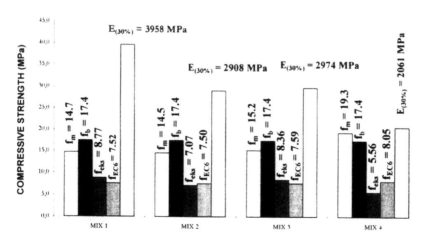

The obtained results were different from the expected one. For the mortar with the highest compressive strength (f_m) the compressive strength of the wallettes (f_{eks}) was the lowest which contradicts the expected strength as given by the EC-6 (f_{EC-6}). Also, the secant modulus of elasticity was much higher for the specimens made with unmodified mortar than for those made with mortar MIX-4.

3.2. Failure mechanisms

According to the experimental measurements and observations of the specimens during the testing, there were two different types of failure mechanisms: for wallettes made from basic mortar (MIX-1) and those made from MIX-4 (Figure 2).

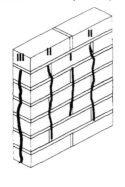 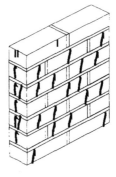

a)) failure mechanism for specimens built   *b) failure mechanism for specimens*
with basic mortar (MIX-1)   ,   *built with modified mortar (MIX-4)*

Figure 2. Failure mechanisms

For the specimens made with MIX-1, the crack pattern was as expected with the typical crack pattern I and II in the plane of the wallette and with the failure pattern (III) on the side of the wallette. Once the crack has started (usually in the unit) the mortar allowed the crack to grow through the height of the wallette. On the other hand, wallettes made with MIX-4 had their first cracks (I) usually in the bricks in the plane of the wallette. The first crack pattern consisted of randomly placed cracks within the plane of the wallette. The failure crack which occured on the side of the wallette (II), was as a previous one (I), very poorly connected through the height of the wallette. The typical results of measuring the failure of unmodified (MIX-1) and modified mortar (MIX-4) are presented in Figure 3.

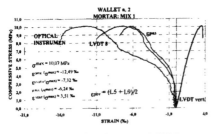 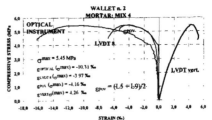

a) unmodified mortar MIX-1   *b) modified mortar MIX-4*

Figure 3. Local lateral and vertical deformations of the constituents and wallette

It can be seen that in the case for unmodified mortar (MIX-1), the wallette was acting as a composite material with strong interface between its constituents. In the transition zone from elastic region of lateral deformation to plastic region, we can see that all measurements (optical measurement of bed joint, L8, and average measurement (ε_{pov}) of L5 and L9) have the same shape following each other. On the other hand, the measurements on the wallette made with modified mortar (MIX-4) show sudden change in the transition zone from the elastic into the plastic range. Also the measurement of L8 and ε_{pov} were not continuous and did not follow each other. The main reasons for such behaviour (according to our assumptions) were poor bond strength and the polymer present in the mortar which prevents the evaporation of water from the mortar - still wet mortar (MIX-4). Beside its moisture condition, MIX-4 had (due to high percentage of fibres) high toughness.

4. NUMERICAL MODELLING

The finite element program developed according to [5][7] uses plane stress 8 noded rectangular elements. The constituents (brick and mortar) are represented by different elements, which allows each material's characteristics to be used in the program. Steel profile for loading the wallettes is represented by a linear element. The material model used for brick was linear elastic, while Mohr-Couloumb's ideal elasto-plasticity model was used for the mortar. The brick-mortar interface and the potential cracks within the brick and the mortar are represented by a series of link elements , Figure 4.

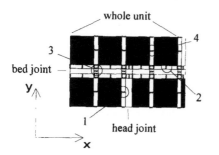

1. Brick - head joint
2. Brick - bed joint - brick
3. Mortar - mortar (bed joint)
4. Brick - brick

Figure 4. Type of link elements

Each link element is a fictitious element with a zero length and a stiffness which corresponds to its type. Using these elements cracking in the constituents and slippage along the brick mortar interfaces were simulated. A link element and its stiffness matrix are represented in Figure 5, where ($l=cos\theta$ and $m=sin\theta$).

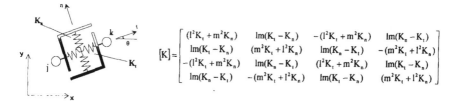

Figure 5. Link elements and their stiffness matrix

Each link element is defined by its: orientation angle (θ), normal and tangential stiffness, compressive and tensile strengths, and in the case of link element number 2 (el.N.2), coefficient of friction between bed joint and brick. The constitutive laws for link elements 1,3 and 4 are the same (elastic-plastic). The differences made for the link el.N.2 was an introduction of the yield threshold based on Coulomb friction law, for the tangential direction (Figure 6).

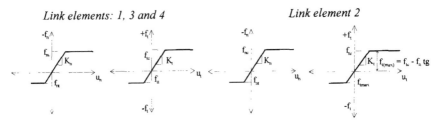

Figure 6. Constitutive laws for link elements

4.1. Material characteristics

In verifying the developed numerical model a <u>reference</u> numerical model based on the experimental results of various authors [3][4][5] was made. That reference model was later on used in parametric analysis.

Table 3. Material characteristics for reference numerical model

MATERIALS			
	BRICK	MORTAR	STEEL PROFILE
E	3.7E+07 kPa	2.00E+07 kPa	200 GPa
ν	0.10	0.25	
shear angle		36°	
compressive strength	17.64 MPa	2.5 MPa	
LINK ELEMENTS			
	1	2	3	4
k_n	8.30E+08 kPa	8.30E+08* kPa	2.00E+07 kPa	3.70E+07 kPa
k_t	8.30E+08 kPa	1.25E+05 kPa	2.00E+07 kPa	3.70E+07 kPa
f_{nt}	37 kPa	37 kPa	950 kPa	880 kPa
f_{nc}	9500* kPa	9500* kPa	9500* kPa	17640 kPa
f_{tt}	9500* kPa	1300 kPa	950 kPa	880 kPa
f_{tc}	9500* kPa	1300 kPa	9500* kPa	17640 kPa
φ		31°		
θ	90°	0°	0°	0°

(*) dummy value

4.2. Models

For the numerical analysis of <u>reference</u> model two models were made: with and without link elements. The model without link element assumes wallette as a set of plane stress elements with different material characteristics ideally bonded together. So, for that model any non-linearity can be only a result of the material non-linearity of the mortar. Otherwise, the model with link elements, which is an iterative one, includes two types of non-linearity:

- non-linearity due to progressive failure and

- material non-linearity of mortar.

Due to the fact that the our plane-stress model can follow crack pattern only within the plane of specimens, it was assumed that the model is acceptable for modelling experiment results up to 80% of their compressive strength (Figure 7). In the zone between 80% of compressive strength and the failure, the splitting crack on the side of the specimen is predominant, which leads the specimen to the failure.

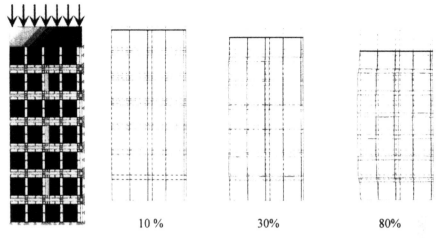

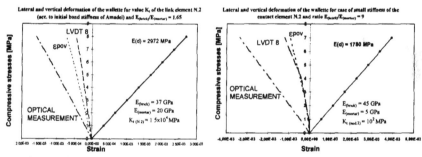

| 10 % | 30% | 80% |

Figure 7: Crack pattern for different level of loading

4.3. Verifying assumptions of different failure mechanisms

In the absence of the all mechanical parameters needed for verification of our model, some basic conclusions of our assumptions that we have made over the different mechanisms of failure can be made (Figure 8).

a) model with low ratio $E_{(brick)}/E_{(mortar)}$ and high K_t stiffness for el.N.2

b) model with big ratio $E_{(brick)}/E_{(mortar)}$ and low K_t stiffness for el.N.2

Figure 8

As it can be seen, for the case of big ratio $E_{(brick)}/E_{(mortar)}$ and low K_t stiffness for el.N.2, (which corresponds to still wet mortar in bed joint with poor bond) there are no similar shapes for the measurement point on ε_{pov} and LVDT 8. So in this case, when masonry behaves as an assemblage of poorly bonded constituents, it can be expected that brick can just slip over the bed joint and cause premature failure of the wallette.

4.4. Parametric analysis

The analysis of the numerical model as continuum and discontinuum showed minor differences in the modulus of elasticity of wallettes and significant difference in lateral deformation. By modifying Poisson's ratio for mortar we can significantly change the lateral deformation of wallette and cause minor changes for the modulus of elasticity of wallettes. Modifying stiffness K_t of link element N.2 we can also produce changes in lateral deformation of wallettes, as premature failure. The influence of the thickness of the bed joint and its modulus of elasticity on the modulus of the elasticity of wallette is given in the figures below (Figure 9).

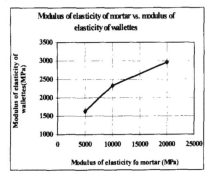 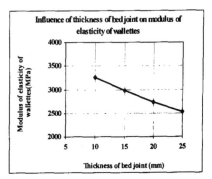

Figure 9

5. CONCLUSIONS

Despite the lack of our own experimental data of the constituents, we have found out the conditions for premature failure of our specimens made with modified mortar (MIX-4). The numerical model which was presented shows us better insight into the masonry material. With additional improvements which we have introduced through parametric analysis, this model will be used for the calibration of the homogenized approach for modelling masonry.

6. ACKNOWLEDGEMENT

The paper presents the results of the research work at the University of Ljubljana, financed by the European Community through the project COPERNICUS CIPA - CT 94 - 0174 - ATEM. Their support is gratefully acknowledged.

7. REFERENCES

1.   LOURENÇO, P.B. , "Computational strategies for masonry structures", *Ph.D. Thesis*, Delft University Press, Holland, February, 1996
2.   BOSILJKOV, V. , " On modelling mechanical properties of the masonry", *M.Sc. Thesis*, University of Ljubljana, Slovenia, October, 1996
3.   ATKINSON, R.H., AMADEI, B.P., SAEB, S. & STURE, S., "Response of masonry bed joints in direct shear", *Journal of Structural Engineering*, Vol.115, No.9, pp.2276-2296, September, 1992
4.   KHOO C.L. , "A Failure Criterion for Brickwork in Axial Compression", *Thesis presented to Univ. of Edinburgh*, at Edinburgh, Scotland, 1972
5.   SCHUBERT, P., "Eigenschaftswerte von Mauerwerk, Mauersteinen und Mauermörtel", *Mauerwerk-Kalendar 1996*, Verlag Ernest & Sohn, Berlin, 1996
6.   LOGAR, J & PULKO, B, "The elastoplastic analysis of sheet pile walls", *Razprave prvega posvetovanja Slovenskih geotehnikov*, Bled 93, Knjiga 1, pp. 115-124, 1993
7.   RIDDINGTON, J.R. & NAOM, N.F., "Finite element prediction of masonry compressive strength", *Computers & Structures*, Vol. 52, No. 1, pp.113-119, 1994

DRY BLOCK ASSEMBLY CONTINUUM MODELLING FOR THE IN-PLANE ANALYSIS OF SHEAR WALLS

G. Alpa Professor, **L. Gambarotta** Professor and
I. Monetto Graduate Research Assistant
Department of Structural and Geotechnical Engineering,
University of Genoa, Via Montallegro 1, 16145 Genova, Italy

1. ABSTRACT

A dry block masonry continuum model for the in-plane analysis of shear walls is considered. The effects of the block geometry and assembly on the inelastic strain mechanisms are taken into account by means of proper constitutive equations. Both the incipient inelastic strain mechanisms and the subsequent incremental ones, which depend on the active inelastic strain, are considered. The stress and strain states are proved to be the leading internal field variables governing the limiting surface evolution. The proposed constitutive model is applied to the analysis of masonry walls by means of a finite element procedure. The specialized algorithm, developed for the integration of the stress - strain relationships in the finite load step, is synthetically shown.

2. INTRODUCTION

The increasing need for a better understanding of the mechanical behaviour of old masonry buildings poses problems about the masonry modelling. Based on the observed weak tensile strength of masonry, a phenomenological approach to masonry structures has been often carried out where continuous media composed of a no-tension material are considered. Recently, with reference to masonry walls constituted by regular blocks, as typical in most historical and monumental buildings, more refined models have been proposed which are able to consider more realistic mechanisms of inelastic deformation. In the case of regular assembly of blocks and units, as a consequence of the periodic internal structure, a detailed analysis of possible motions and internal forces governing the equilibrium can be carried out by taking into account the influence of block geometry and assembly on the overall mechanical behaviour.

Although the analysis of block masonry can be carried out by considering discrete elements [1, 2], in the case of small block sizes compared with the dimensions of the whole structure the continuum approach turns out to be more suitable. In this way, displacement jumps occurring at joints and related to well defined opening and sliding

Keywords: Dry block masonry, In-plane response, Homogenization, Inelastic analysis.

Computer Methods in Structural Masonry – 4, edited by G.N. Pande, J. Middleton and B. Kralj.
Published in 1998 by E & FN Spon, 11 New Fetter Lane, London EC4P 4EE, UK. ISBN: 0 419 23540 X

mechanisms depending on the block dimensions and block assembling are taken into account and treated as inelastic strain contributions.

In this context, based on homogenization techniques, several models have been recently developed for brick masonry with mortar joints. In [3] de Buhan and de Felice focus on the limit analysis of brick masonry by applying a homogenization procedure to the case of multiple yield planes. On the other hand, a brittle damage model is proposed in [4, 5]. With reference to the dry block masonry, Alpa and Monetto [6] have derived the incipient inelastic strain mechanisms able to describe the dependence of the material response on the block assembly. Based on such a formulation, constitutive equations have been developed in [7] and subsequently generalized in [8], where a larger number of inelastic mechanisms depending on the active inelastic strain state has been considered.

In this paper, after a concise description of the constitutive model, the numerical procedure for the in-plane analysis of shear walls is described. The specialized algorithm which has been developed to integrate the rate constitutive equations in the finite load step and applied within an incremental iterative finite element analysis for dry block masonry walls is presented. Finally, the capability of the proposed method is provided by an example pointing out the importance of the multiple block mechanisms considered in the model.

3. A CONTINUUM MODEL FOR DRY BLOCK MASONRY

The considered masonry is a dry assembly of regular full-thickness rectangular blocks of width a and height b, resulting in a periodic internal structure with aligned bed joints and non-aligned head joints. If block size is small with respect to the wall size, a continuum approach can be used, as proposed by Alpa and Monetto in [6] with reference to the incipient inelastic strain mechanisms and subsequently generalized in [7, 8].

Blocks are assumed homogeneous, isotropic and perfectly elastic. Moreover, due to the absence of mortar, the contact between blocks is modelled as unilateral and linearly frictional. On the assumption of a Cauchy equivalent continuum, the constitutive equations are derived in terms of mean stress $\underline{\sigma}$ and mean strain $\underline{\varepsilon}$ [9] based on the equivalence between the repetitive elementary masonry portion $\mathcal{A}$ shown in figure 1 and the equivalent solid one. Separating the elastic and inelastic contributions to the mean strain, they turns out to be expressed as follows:

$$\underline{\varepsilon} = \underline{K}\underline{\sigma} + \underline{\varepsilon}^{i}, \tag{1}$$

where $\underline{K}$ is the compliance isotropic matrix of the compact masonry and $\underline{\varepsilon}^{i}$ is the contribution of inelastic mechanisms to the mean strain. The last term is due to frictional sliding and opening activated at block joints and is evaluated by means of a proper homogenization procedure considering elementary mechanisms of relative displacements between portions of the representative element $\mathcal{A}$.

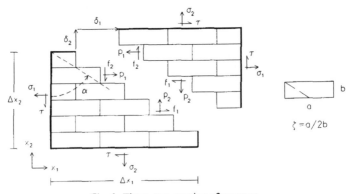

Fig. 1. Elementary portion of masonry.

With reference to figure 1, inelastic mechanisms of relative displacements between two parts of $\mathcal{A}$ are assumed to be activated at a separating path developing along the head and bed joints and defined by an angle α related to the block geometry parameter ζ (tan $\alpha = \pm\zeta$). Each part is in equilibrium under the external mean stresses and the internal contact forces, represented by the normal p_i (i= 1, 2) and tangential f_i components, which are assumed as uniformly distributed on the opposite block faces.

In the case the masonry is compact, the contact forces equal the external stress components. On the other hand, if relative displacements between blocks occur, the contact forces can be evaluated by equilibrium:

$$\begin{cases} \sigma_1 + \tau \tan\alpha - p_1 - f_1 \tan\alpha = 0, \\ \sigma_2 \tan\alpha + \tau - p_2 \tan\alpha + f_2 = 0, \end{cases} \tag{2}$$

combined with the unilateral contact conditions:

$$p_i \leq 0, \quad \delta_i \geq 0, \quad p_i\,\delta_i = 0, \quad (i = 1,2), \tag{3}$$

δ_i being the i-th component of the relative displacement between the two parts, and with the Coulomb frictional conditions:

$$\Phi_1 = |f_1| + \mu p_2 \leq 0, \quad \dot{\delta}_1 = \frac{f_1}{|f_1|}\dot{\lambda}_1, \quad \dot{\lambda}_1 \geq 0, \quad \Phi_1\dot{\lambda}_1 = 0, \quad \dot{\Phi}_1\dot{\lambda}_1 = 0, \tag{4}$$

$$\Phi_2 = |f_2| + \mu p_1 \leq 0, \quad \dot{\delta}_2 = \frac{f_2}{|f_2|}\dot{\lambda}_2, \quad \dot{\lambda}_2 \geq 0, \quad \Phi_2\dot{\lambda}_2 = 0, \quad \dot{\Phi}_2\dot{\lambda}_2 = 0, \tag{5}$$

where μ is the friction coefficient.

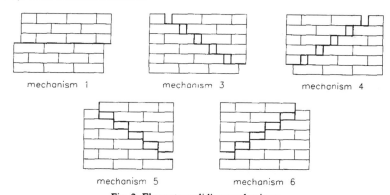

mechanism 1   mechanism 3   mechanism 4

mechanism 5   mechanism 6

Fig. 2. Elementary sliding mechanisms.

Varying the angle α, five couples of elementary sliding mechanisms can be identified, as shown in figure 2: pure sliding on the horizontal plane (mechanisms 1 and 2) and sliding along planes inclined at an angle $\pm\alpha$ on the bed joints (mechanisms 3, 3' and 4, 4') or on the head ones (mechanisms 5, 5' and 6, 6'). Excluding the pure sliding mechanisms 1 and 2, which differ in opposite inelastic angular dilation rates, for the sliding mechanisms along inclined planes a primary (opening) mechanism, involving a positive inelastic dilation rate, and a secondary (closing) mechanism, involving a negative inelastic dilation rate, are included in each couple. The superscript denotes the secondary mechanisms. It is worth noting that at the initial state only positive inelastic linear dilation rates, typical of primary mechanisms, are admitted; on the other hand, at subsequent states both primary and secondary mechanisms can be activated.

From relations (4) and (5) the limit condition defining the stress states activating the i-th elementary sliding mechanism can be written as follows:

$$\Phi_i = \underline{\psi}_i^T \, \underline{\sigma} = 0, \quad i = 1, 3, \ldots, 6 \tag{6}$$

where $\underline{\sigma}^T = \{\sigma_1 \ \sigma_2 \ \tau\}$ is the mean stress vector. In the stress component space the equation (6) corresponds to a plane with outside unit normal vector $\underline{\psi}_i$.
The inelastic strain contribution due to the i-th mechanism is given by the following non-associated flow rule:

$$\underline{\dot{\varepsilon}} * = \frac{f}{|f|} \, \underline{v}_i \, \dot{\lambda}_i, \quad f = \underline{v}_i^T \underline{\sigma}, \quad \dot{\lambda}_i \geq 0, \tag{7}$$

where the constant vector $\underline{v}_i$ defines the direction of the inelastic strain contribution; f represents the generalized tangential contact force and distinguishes a primary mechanism ($f > 0$) from the related secondary one ($f < 0$). The following vectors correspond to the elementary mechanisms shown in figure 2:

$$\underline{\psi}_{1,2}^T = \{0 \quad \mu \quad (f \, / \, |f|)\}, \qquad\qquad \underline{v}_{1,2}^T = \{0 \quad 0 \quad 1\}, \tag{8a}$$

$$\underline{\psi}_{3,3'}^T = \{(f \, / \, |f|) \, / \, \varsigma \quad \mu \quad (f \, / \, |f|) + \mu \, / \, \varsigma\}, \qquad \underline{v}_{3,3'}^T = \{1 \, / \, \varsigma \quad 0 \quad 1\}, \tag{8b}$$

$$\underline{\psi}_{4,4'}^T = \{(f \, / \, |f|) \, / \, \varsigma \quad \mu \quad -(f \, / \, |f|) - \mu \, / \, \varsigma\}, \qquad \underline{v}_{4,4'}^T = \{1 \, / \, \varsigma \quad 0 \quad -1\}, \tag{8c}$$

$$\underline{\psi}_{5,5'}^T = \{\mu \quad (f \, / \, |f|)\varsigma \quad (f \, / \, |f|) + \mu \, \varsigma\}, \qquad \underline{v}_{5,5'}^T = \{0 \quad \varsigma \quad 1\}, \tag{8d}$$

$$\underline{\psi}_{6,6'}^T = \{\mu \quad (f \, / \, |f|)\varsigma \quad -(f \, / \, |f|) - \mu \, \varsigma\}, \qquad \underline{v}_{6,6'}^T = \{0 \quad \varsigma \quad -1\}. \tag{8e}$$

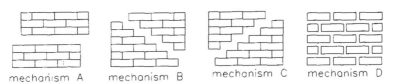

mechanism A   mechanism B   mechanism C   mechanism D

Fig. 3. Elementary pure opening mechanisms.

Furthermore, as a consequence of the unilateral contact, four elementary pure opening mechanisms can be identified, as shown in figure 3: opening along a horizontal plane (mechanism *A*), opening along planes inclined at an angle $\pm\alpha$ (mechanisms *B* and *C*) and the material disgregation (mechanism *D*). The following stress and inelastic strain states correspond to such mechanisms:

$$\underline{\sigma}_A^T = s \, \{1 \quad 0 \quad 0\}, \qquad \underline{\varepsilon}_{dA}^T = \{0 \quad \lambda_{d2} \quad \lambda_{d3}\}, \tag{9a}$$

$$\underline{\sigma}_B^T = s \, \{-\varsigma \quad -1 / \varsigma \quad 1\}, \quad \underline{\varepsilon}_{dB}^T = \{\lambda_{d1} / \varsigma \quad \lambda_{d2} \, \varsigma \quad \lambda_{d1} + \lambda_{d2}\}, \tag{9b}$$

$$\underline{\sigma}_C^T = s \, \{\varsigma \quad 1 / \varsigma \quad 1\}, \qquad \underline{\varepsilon}_{dC}^T = \{\lambda_{d1} / \varsigma \quad \lambda_{d2} \, \varsigma \quad -\lambda_{d1} - \lambda_{d2}\}, \tag{9c}$$

$$\underline{\sigma}_D^T = \{0 \quad 0 \quad 0\}, \qquad \underline{\varepsilon}_{dD}^T = \{\lambda_{d1} \quad \lambda_{d2} \quad \lambda_{d3}\}, \tag{9d}$$

where $s \leq 0$ and $\lambda_{di} \geq 0$ for i = 1, 2, 3.
The described approach allows to obtain the inelastic strain rate as follows:

$$\underline{\dot{\varepsilon}}^i = \sum_{i=1}^{6} \frac{f}{|f|} \, \underline{v}_i \, \dot{\lambda}_i + \underline{\dot{\varepsilon}}_d \,. \tag{10}$$

Finally, it must be noted that the model is characterized by only four parameters, i.e. the Young modulus E and the Poisson's ratio v of the block material, the friction coefficient μ and the ratio ζ defining the block geometry and assembly.

4. LIMIT ELASTIC DOMAINS OF THE BLOCK MASONRY

The stress states activating the inelastic opening and sliding mechanisms define a linear piecewise limiting surface in the stress space bounding the elastic domain (figure 4). As underlined in the previous section, each limit plane is related to an elementary sliding mechanism, whereas each edge is related to a combination of elementary sliding mechanisms or to an elementary pure opening mechanism. The vertex D corresponds to the configuration of disgregated masonry.

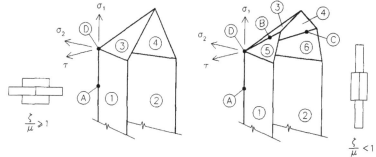

Fig. 4. Limiting surfaces in the stress space for an incipient inelastic strain state.

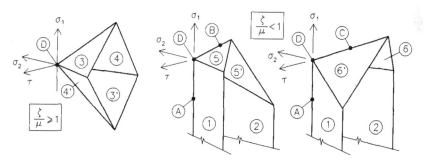

Fig. 5. Limiting surfaces in the stress space for a generic inelastic strain state.

The shape of the limiting surface depends on both the internal model parameters, by means of the ζ/μ ratio, and the active inelastic strain state. In the condition of incipient inelastic strain, as block cannot slide back, only primary mechanisms (named 3, 4, 5 and 6) can be activated (figure 4). On the other hand, at a generic inelastic strain state also secondary mechanisms (named 3', 4', 5' and 6') can become active. The modifications of the limit domain for a generic inelastic strain state are shown in figure 5.

To give a better explanation of this behaviour, let be considered the load history shown in figure 6 related to a masonry having $\zeta/\mu \geq 1$. From the condition of compact and non-loaded masonry, a vertical compression and a horizontal tension are applied, until the elastic state A is reached. Increasing the tangential stress component, A moves towards B on the limit plane related to the mechanism 3. The activated sliding on the bed joints along the plane inclined at an angle $\alpha{>}0$ induces an inelastic horizontal dilation $\varepsilon_1{*}$ which distinguishes B' from B. As a consequence, the limit domain suddenly changes because of the appearing of the secondary limit planes 3' and 4', which are only potential at the state B. If now the horizontal stress component is decreased, the masonry behaves as elastic from B' to C and a back sliding along the plane inclined at an angle $\alpha<0$ is activated. At the final state C' the masonry is compact again, even though an inelastic angular dilation remains active; its behaviour is elastic and the limiting surface is restored in its original shape. In figure 7 an analogous situation is shown with reference to a masonry having $\zeta/\mu<1$. At the final state C' also the inelastic angular dilation vanishes. Unlike the previous case, the activated closing mechanism 5' is the secondary one corresponding to the primary mechanism inducing the inelastic linear dilation $\varepsilon_2{}^*$.

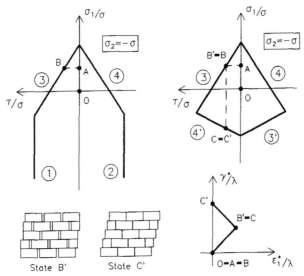

Fig. 6. Modification of the limit domain for $\zeta/\mu\geq1$.

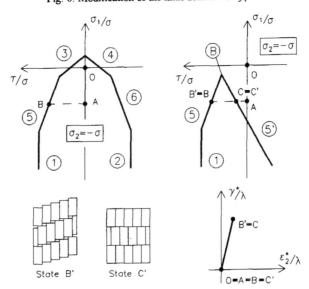

Fig. 7. Modification of the limit domain for $\zeta/\mu<1$.

It is worth noting that the proposed masonry model differs from the multiple elastic-perfectly plastic mechanisms model proposed by Koiter since it depends on both the stress and strain states. The unilateral contact between blocks, the coupling of unilateral and frictional mechanisms and the internal constraint on the linear dilations ($\varepsilon_1{}^i \geq 0$, $\varepsilon_2{}^i \geq 0$) result in a discontinuous evolution of the limiting surface. When a generic inelastic strain state is reached from the incipient one, potential limit planes are activated; whereas, if the initial configuration of compact masonry is restored, some active limit planes disappear. On the other hand, the proposed model must not be confused with an elasto-plastic model with kinematic hardening, where the continuous evolution of the limiting surface is related to the translation of always active limit planes.

5. ANALYSIS OF MASONRY WALLS UNDER IN-PLANE LOADS

The proposed constitutive model has been applied to the incremental iterative finite element analysis of masonry walls under in-plane loads. Based on a four nodes isoparametric finite elements discretization, the stress and strain increments at the n-th step are evaluated by means of an iterative procedure based on the modified Newton - Raphson method, which implies a specialized algorithm to integrate the rate constitutive equations in the finite load step. At a given equilibrium state, represented by the state variables $\underline{\sigma}$, $\underline{\varepsilon}^*$ and $\underline{\varepsilon}_d$, the procedure evaluates the stress and inelastic strain increments $\Delta\underline{\sigma}$, $\Delta\underline{\varepsilon}^*$ and $\Delta\underline{\varepsilon}_d$ induced by an assigned total strain increment $\Delta\underline{\varepsilon}$.

In the case of sliding mechanisms, the inelastic strain increment does not change direction in the load step, so that the corresponding stress increment $\Delta\underline{\sigma}$, describing the evolution of the stress point on the i-th limit plane, can be written as follows:

$$\Delta\underline{\sigma} = \underline{C}_i \, \Delta\underline{\varepsilon} = \underline{K}^{-1}\left(\underline{I} - \frac{\underline{v}_i \, \underline{\psi}_i^T \underline{K}^{-1}}{\underline{\psi}_i^T \underline{K}^{-1} \underline{v}_i} \right)\Delta\underline{\varepsilon}, \tag{11}$$

being $\underline{\psi}_i$ and $\underline{v}_i$ the constant vectors given by the relations (8). Analogously, the evolution of the stress point on an edge is governed by a stress increment $\Delta\underline{\sigma}$ directly proportional to $\Delta\underline{\varepsilon}$ by means of a constant matrix $\underline{C}_{ij}(\underline{v}_i, \underline{\psi}_i, \underline{v}_j, \underline{\psi}_j, \underline{K})$ involving the characteristic vectors of both the i-th and j-th mechanisms having limit planes intersecting at the considered edge. Finally, for pure opening mechanisms $\Delta\underline{\sigma}$, which has in this case a constant direction in the stress space, can be directly obtained combining the internal constraints on $\Delta\underline{\sigma}$ itself and $\Delta\underline{\varepsilon}_d$ given by relations (9).

Of course, the possibility that in the same increment $\Delta\underline{\varepsilon}$ the stress point can move from a limit plane to a neighbouring one or along an intersected edge is also considered. In this case the evolution of the stress point is followed subdividing the whole loading step into elementary ones. Once the vertex $\underline{\sigma} = \underline{0}$ is reached, no stress increment corresponds to the assigned $\Delta\underline{\varepsilon}$ which equals the inelastic strain increment $\Delta\underline{\varepsilon}_d$.

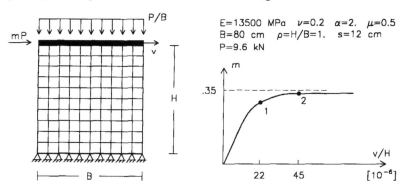

Fig. 8. Discrete model and structural response of the square masonry wall.

Finally, as structural application, the collapse analysis of a square masonry wall under horizontal forces superimposed on vertical ones has been carried out. The related discrete model and the mechanical response are shown in figure 8. From the maps of inelastic mechanisms at the states 1 and 2 (figure 9) it is evident that collapse is governed by multiple sliding mechanisms spreading over the wall.

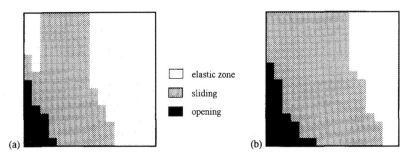

Fig. 9. Inelastic mechanisms of the square wall: (a) at the state 1 and (b) at the state 2.

6. CONCLUSIONS

The proposed constitutive model for the in-plane analysis of masonry walls is proved to be able to describe the main features characterizing the mechanical behaviour of dry block masonries: (i) the anisotropy in the inelastic response; (ii) the dependence on the block geometry and assembly; (iii) the dependence on the active inelastic strains also in the absence of physical causes of hardening. Especially, the combination of intrinsically different mechanisms, such as sliding and opening, results in a model which differs from an elastic perfectly plastic one as both the stress and strain states are the leading internal field variables governing the limiting surface evolution. Nevertheless, it must not be confused with an elasto - plastic model with kinematic hardening, as already underlined before.

Moreover, the model can be easily inserted into standard finite element non linear solution procedures for the continuum analysis of the collapse processes induced by monotonic or cyclic loads in masonry buildings and arches constituted by an assembly of regular blocks . With a good approximation it can be applied also in the case mortar joint are absent or characterized by a low cohesion.

7. REFERENCES

1. Baggio C. and Trovalusci P., 'Discrete models for jointed block masonry walls', VI North American Masonry Conference, Philadelphia, USA, 1993, vol. II, pp. 939-949.
2. Boothby T.E., 'Stability of masonry piers and arches including sliding', J. Engineering Mechanics - ASCE, 1994, vol. 120, pp. 304-319.
3. de Buhan P. and de Felice G., 'A homogenization approach to the ultimate strength of brick masonry', J. Mechanics and Physics of Solids, 1997, vol. 45, pp. 1085-1104.
4. Luciano R. and Sacco E., 'Homogenization technique and damage model for old masonry material', Int. J. Solids and Structures, 1997, vol. 34, pp. 3191-3208.
5. Luciano R. and Sacco E., 'A damage model for masonry structures', in press in European Journal of Mechanics and Solids.

6. Alpa G. and Monetto I., 'Microstructural model for dry block masonry walls with in-plane loading', J. Mechanics and Physics of Solids, 1994, vol. 42, pp. 1159-1175.
7. Gambarotta L. and Monetto I., 'Analisi incrementale di pareti in muratura a blocchi squadrati', VIII Convegno Italiano Meccanica Computazionale, Torino, Italy, 1994, pp. 37-43.
8. Alpa G., Gambarotta L. and Monetto I., 'Equazioni costitutive per murature a blocchi squadrati', Convegno Nazionale La Meccanica delle Murature tra Teoria e Progetto, (Editor Gambarotta L.), Messina, Italy, 1996, pp. 337-346.
9. Alpa G. and Gambarotta L., 'Mechanical models for frictional materials', International Conference New Developments in Structural Mechanics, Catania, Italy, 1990, pp. 69-81.

FE PREDICTION OF BRICKWORK PANEL SHEAR STRENGTH

J.R. Riddington Senior Lecturer and **P. Jukes** Visiting Research Fellow
School of Engineering, University of Sussex, Brighton BN1 9QT, UK

1. ABSTRACT

An iterative 2-D finite element program is described that was developed for the purpose of predicting the ultimate strength of brickwork panels when subjected to combined shear and vertical compressive loading. The program is intended to be used on panels where the bricks are relatively strong and failure lines develop primarily in the joints. When analysing a panel using this program, the bricks and mortar joints are modelled separately, but are enlarged so that there are fewer bricks and joints in the model, thus enabling the total number of elements required to be kept down to a reasonable level. The results from shear tests on twelve brickwork panels were used to assess the accuracy of the program. It was found that for eleven out of the twelve panels modelled, the predicted ultimate strength was within 10% of the test result. This finding was considered to verify the accuracy of the modelling method used. It was also found that for all of the panels analysed, mortar non-linearity had a minimal influence on ultimate strength; that the difference between failure initiation and ultimate failure tended to increase with the level of vertical compression and that for panels where the tensile and shear bond values were relatively high, there was very little difference between the initiation and ultimate loads.

2. INTRODUCTION

Because the in-plane shear strength of masonry panels depends on a large number of parameters, it is not practical to fully research the influence and interaction of these parameters on shear strength using full scale tests, as these are expensive and time consuming to conduct. Such a parameter study can however be undertaken using finite

Keywords : Shear strength, Walls, Finite elements, Masonry, Brickwork

Computer Methods in Structural Masonry – 4, edited by G.N. Pande, J. Middleton and B. Kralj.
Published in 1998 by E & FN Spon, 11 New Fetter Lane, London EC4P 4EE, UK. ISBN: 0 419 23540 X

element analyses, provided a program is available that can be shown to predict the shear strength of panels with a reasonable degree of accuracy. This paper describes a program that was developed for the purpose of undertaking such a parameter study on panels where failure lines develop primarily in the joints. It is therefore best suited for use in studying panels formed from relatively strong bricks. Currently the program is not capable of simulating failures that occur within bricks or blocks.

3. FINITE ELEMENT PROGRAM

Due to space restrictions, only a brief description of the program can be provided. Full details of the program are however available elsewhere[1]. The 2-D program was developed from a program[2] that had been produced to study triplet behaviour. In the current program, the bricks and mortar joints of a panel are modelled separately, but both are enlarged so that there are far fewer bricks and joints in the model, thus enabling the total number of elements required to be kept down to a reasonable level. Standard 4 node rectangular elements are used for both the brick and mortar components with the brick elements being assumed to behave elastically, whilst the stiffness of the mortar elements can be modified during the analysis' iterative procedure to simulate the non-linear elastic behaviour of the mortar. The shearing load is applied in steps until complete failure is indicated, with iterations being undertaken at each load step until a stable solution in terms of crack generation and mortar non-linearity is obtained.

In the model the bricks and mortar elements are connected together at one side of each bed and perpend joint by 2 node link elements of zero length and high stiffness whilst at the other side of the joints they are connected directly to each other. If during the analysis' iterative procedure a failure check indicates that a tension crack should form at a point on the interface, this is simulated by removing the link element at the point. If, however, a check indicates that compressive shear failure at a point should occur, this is simulated by reducing just the link's stiffness parallel to the interface to zero. The residual friction that would develop at the point is then simulated by applying equal and opposite forces to the brick and mortar nodes at the ends of the link.

The program allows the following forms of joint failure to be simulated :
1. Local shear failure on the interface in compression or tension, as defined by the failure envelope shown in Fig. 1, where τ_0' is the local bond shear strength, μ' is the local coefficient of internal friction and f_{tb}' is the local bond tensile strength. A similar form of failure criteria has been used by Rots and Lowenco[3].
2. Local bond tensile failure on the interface, which is assumed to occur when the stress across the interface is tensile and it exceeds the local bond tensile strength f_{tb}'.
3. Mortar tensile failure, which is assumed to occur when the principal tensile stress in the joint exceeds the tensile strength of the mortar f_{tm}.

When any of these forms of failure are detected, appropriate changes are made to the link elements. Failure checks at links are based on mortar element centroid stresses, since these have been found to give more consistent results compared with when link forces are used, with the stresses at a link normally being taken to be the average of the stresses in the two mortar elements attached to the link. At the corners of bricks, however, the mortar element at the junction of the bed and perpend joint has been found not to give reliable stress results, and so these elements are not included when calculating corner link stresses.

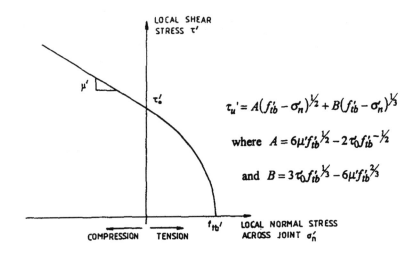

Fig. 1 Assumed local failure criteria for joint shear failure

4. PROGRAM VERIFICATION

The results from tests of 12 brickwork panels were used to determine the accuracy of the program. Fig. 2 shows the test arrangement used, whilst Fig. 3 shows the form of compressive loading applied to the top of the panels at failure; where σ_{c1} was the compressive stress applied by an external servo controlled hydraulic jack, which acted over the full length of the panel, and σ_{c2} was the compressive stress applied by tie rods, which acted on only half the length of the panel. The tie rods prevented the uplift of the panels.

All the panels were 1.94 m long and were constructed using UK dimension bricks (125 x 65 x 102.5 mm) and 10 mm mortar joints. The first set of 4 panels were formed from high strength (67.9 N/mm²) perforated clay bricks and 1:¼:3 (21.1 N/mm² cube strength) mortar and were 1.51 m high; the second set of 4 panels were formed from lower strength (18.3 N/mm²) solid-solid clay bricks and 1:5 masonry cement (6.8 N/mm²) mortar and were 1.57 m high; whilst the third set of 4 panels were formed from solid-frogged calcium silicate bricks (18.3 N/mm²) and 1:1:6 (6.6 N/mm²) mortar and were 1.16 m high. Further details of these tests can be found elsewhere[4].

The results of the tests are shown in Table 1, expressed in terms of the average shear stress τ_u and compressive stress σ_c acting on the panels at failure. Table 1 also shows the values of σ_{c1} and σ_{c2} at failure and these show that the distribution of stress acting across the top of the panels varied considerably from panel to panel.

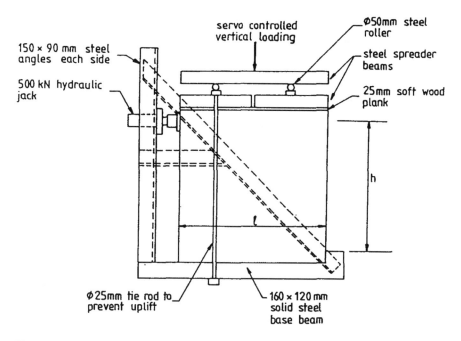

Fig. 2 Panel testing loading arrangement

$$\sigma_{c1} + \sigma_{c2} \qquad\qquad \sigma_{c1}$$

Fig. 3 Vertical stress distribution applied to the top of the wall panels

Panel No.	1.1	1.2	1.3	1.4	2.1	2.2	2.3	2.4	3.1	3.2	3.3	3.4
σ_{c1} N/mm^2	0.62	0.21	0.58	0.21	0.21	1.00	0.45	0.70	1.00	0.22	0.46	0.76
σ_{c2} N/mm^2	1.57	2.11	2.67	2.43	1.03	1.68	1.28	2.32	0.86	0.71	0.89	0.71
σ_c N/mm^2	1.41	1.26	1.92	1.42	0.72	1.84	1.09	1.86	1.43	0.58	0.91	1.11
τ_u N/mm^2	1.67	1.60	2.07	1.69	0.66	1.28	0.89	1.28	1.36	0.77	0.98	1.11

Table 1 Panel test results

Analyses of the panels were undertaken using the strength and stiffness values shown in Table 2. The τ_0' and μ' values were obtained from triplet tests undertaken with precompression with τ_0' being taken as 1.5 x average bond shear strength given by the tests. The f_{tb}' values were obtained from direct tension tests on couplets, with the load being applied via bolts passed through holes drilled in the bricks. f_{tb}' was taken as 1.9 x the average tensile stress applied to the specimen at failure. This factor accounts for the uneven stress distribution that develops in the joint with this loading arrangement. The mortar properties were obtained from compression and indirect tension tests on mortar cylinders formed using mortar extracted from joints after it had been in the joints for ten minutes. Other data used in the analyses, that is not shown in Table 2, such as the stress-strain curves used for the mortar, can be found elsewhere[1].

Panel set	1	2	3
Local bond shear strength τ_o' N/mm^2	1.79	1.07	0.68
Local coefficient of friction μ'	0.92	0.80	0.79
Local bond tensile strength f_{tb}' N/mm^2	1.30	0.25	0.20
Joint mortar tensile strength f_{tm}' N/mm^2	2.87	1.47	1.20
Brick elastic modulus N/mm^2	27800	10200	16000
Brick Poisson's ratio	0.15	0.15	0.15
Mortar elastic modulus N/mm^2	19400	8650	8440
Mortar Poisson's ratio	0.18	0.18	0.18

Table 2 Strength and stiffness values used in the analyses

For each analysis model produced, the steel spreader beams and the wood at the top of the panel were included, as was the steel section at the bottom, as is shown in Fig. 4. Possible uplift of panels from the steel base beam was also allowed for in the analyses, as was slippage of panels along this beam. Each panel was analysed by first applying the vertical compressive loading, as defined by the σ_{c1} and σ_{c2} values given in Table 1. The shear loading was then applied incrementally until complete failure of the panel was indicated.

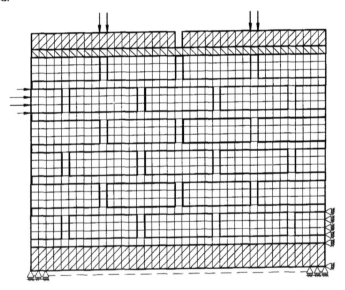

Fig. 4 Model and element grid when analysing panels using 1053 elements

As previously stated the panels were analysed using a reduced number of enlarged bricks and joints in order to reduce the number of elements needed for each analysis. In order to investigate the effect of this enlargement of the bricks, the 4 panels of set 3 were analysed using 4 different scaling levels. The results from these analyses are shown in Fig. 5, expressed in terms of the total number of elements used in the analyses, with the scaling and the element mesh adopted when 1053 elements were used, being shown in Fig. 4. The results indicated that the mesh arrangement and scaling level shown in Fig. 4 could be expected to produce results of reasonable accuracy. Consequently this arrangement was used for all the remaining analyses. It should be noted that changing the brick enlargement level did, in some instances, result in slight changes in the crack patterns predicted by the analyses. This is the reason why for panel 3.4, the failure load was slightly higher when 2835 elements were used compared with when 1551 elements were used.

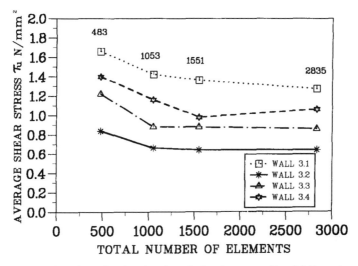

Fig.5 Influence of brick scaling level and mesh density on predicted failure stress

Comparisons between the measured strengths of panels and those predicted by the analyses are shown in Fig. 6 to 8. All the analysis results were obtained using the element grid shown in Fig. 4 and with mortar non-linearity taken into account. It was in fact found that for these analyses, mortar non-linearity had little effect on the ultimate load. It can be seen that for panel set 1, predicted strengths were all slightly below the measured strengths, with the greatest difference being 5.9 %; whilst for panel set 2, predicted strengths were all slightly above the measured strengths, with the greatest difference being 4.5 %. Larger differences occurred with set 3, where predicted strengths were both above and below measured strengths and where for one panel the difference exceeded 10 %, with for panel 3.2 the analysis predicting a strength 12.9 % below that measured. Overall, however, the results indicate that the program should normally be capable of predicting the shear strength of panels where failure occurs primarily within joints to within 10 %.

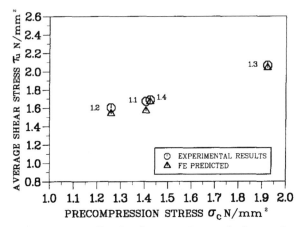

Fig. 6 Comparison between predicted and measured strengths for panel set 1

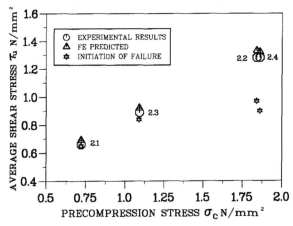

Fig. 7 Comparison between predicted and measured strengths for panel set 2

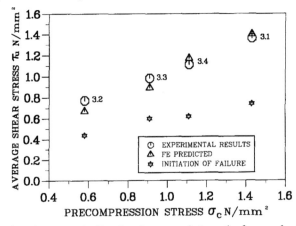

Fig. 8 Comparison between predicted and measured strengths for panel set 3

For test set 1, the analyses predicted that the panels would fail in a brittle manner, as did in fact occur in the tests. The failure initiation loads have not been shown in Fig. 6 because they were more than 95 % of the ultimate load in all cases. The brittle nature of these panels can be attributed to the high shear and tensile bond strengths of the joints. In the case of the other two panel sets, where bond strengths were much lower, failures tended not to be so brittle in nature and this was also predicted by the analyses. For these test sets the loads at which failures were initiated in the analyses are shown. These results indicate that the difference between the load that initiates failure and the maximum load will increase as the level of applied compressive loading increases, whilst it will reduce as the shear and tensile bond strengths of the joints increase.

5. CONCLUSIONS

The main conclusions that can be drawn from this work are :

1. The program that has been described could normally be expected to be able predict the in-plane shear strength of a brickwork panel to within 10 %, provided that the brick strength is such that failures would occur primarily in the joints and that accurate stiffness and strength data are known for the panel. The program can therefore be considered to be suitable for use in studying the influence of the various parameters that affect in-plane shear strength.
2. Mortar non-linearity had very little influence on the predicted ultimate load for all of the panels considered. It is likely that this will be true for most brickwork panels that are subjected to racking loading.
3. The results suggest that the difference between the load causing initiation and ultimate failure of a panel will tend to increase with the level of compressive load applied to the panel, whilst it will tend to reduce as the shear and tensile bond strengths of the joints increase. The results also suggest that panels with high bond strengths are likely to behave in a brittle manner.

6. REFERENCES

1. Jukes P., "An investigation into the shear strength of masonry joints", DPhil thesis, University of Sussex, 1997.
2. Jukes P. and Riddington J. R., "Ultimate strength prediction of masonry triplet samples", In Computer Methods in Structural Masonry - 3, Books & Journals International Ltd., 1995, pp 97-106.
3. Rots J. G. And Lowenco P. B., "Fracture simulations of masonry using non-linear interface elements", 6th North American Masonry Conference, Philadelphia, USA, 1993, pp 983-993.
4. Riddington J. R. And Jukes P., "A comparison between panel, joint and code shear strength", 10th International Brick/Block Masonry Conference, Calgary, Canada, 1994, pp 1481-1490.

MECHANICAL CHARACTERIZATION OF STONE MASONRY PANELS AND EFFECTIVENESS OF STRENGTHENING TECHNIQUES

S. Chiostrini Research Associate, **L. Galano** PhD and
A. Vignoli Associate Professor
*Department of Civil Engineering, University of Florence,
Via S. Marta 3, 50139 Florence, Italy*

1. ABSTRACT

The paper deals with an experimental and numerical research project concerning the behavior of stone masonry. In situ shear and compression tests were performed on wall - panels obtained from buildings in old Tuscany communes recently interested in an earthquake. Numerical F.E. simulations were performed to reach a better understanding of the material behavior and to check material parameter values to be employed for the modeling of whole buildings. Experimental setup and first results are presented, together with the scheme of the numerical simulations.

2. INTRODUCTION

The research project is aimed to the determination of the overall mechanical properties of structural masonry typical of old buildings in Lunigiana and Garfagnana (in the North-West part of Tuscany). These zones are mountain seismic regions with prevalence of old masonry structures with chaotic texture, so to pose particular problems in detection and restoration of the seismic vulnerability of building heritage. After severe earthquakes (the latter in 1920), in October 1995, these regions were interested in an earthquake of intensity 4.8 Richter, which caused generalized damaging in several buildings [1, 2, 3]. Due to this, Authorities decided to proceed to an investigation in order to assess the safety of buildings and detect some standardized interventions for reparation and strengthening.

Keywords : Masonry, Experimental Tests, Numerical Modeling

Computer Methods in Structural Masonry – 4, edited by G.N. Pande, J. Middleton and B. Kralj.
Published in 1998 by E & FN Spon, 11 New Fetter Lane, London EC4P 4EE, UK. ISBN: 0 419 23540 X

The major part of the buildings were constructed with mixed stone and brick masonry in the first decades of the century, showing a great variety of plant and elevation dispositions, thickness and texture of the walls. Hence, a large amount of variables should be taken into account in the development of a general approach to the safety detection and restoration strategy.

Several experimental tests concerning single components of masonry (stone units and mortar) and wall-panels were performed in the recent past by the same research group [4, 5]. These studies demonstrated the great variability of the mechanical characteristics of old stone masonry walls and the difficulties in predicting their actual behavior.

In this general framework, the Dept. of Civil Engineering of Florence is performing a series of experimental and numerical investigations to assess the strength of in situ masonry panels in their actual conditions and the effectiveness of common restoration techniques. A first part of the research comprehend: the isolation and testing of three panels (90 cm width and 190 cm height about) within the structural walls of different constructions; their reparation after first testing and then a second series of experimental tests performed on the same panels in the repaired conditions.

Experimental investigations comprehend both compression and shear-compression tests; the restoration technique consists in the construction of a sandwich with two reinforced jackets of about 5 cm in thickness on both faces of the panels, connected through the panel itself by ties.

The paper shows experimental setup and first results, together with some numerical simulations of the tests. Previous works (i. e. [6]) showed that the F.E. approach is the only reliable procedure to predict the location and severity of damage under seismic loading. To this aim the numerical analyses were performed to identify the most appropriate mechanical parameters to be employed within a F.E. approach in the definition of a masonry equivalent continuum.

3. TESTS SETUP

Two buildings were selected for the research purposes: the first is the old Town Hall in Pieve Fosciana (Garfagnana) (Fig. 1) and the second is a school in S. Anastasio (Garfagnana). Three test panels were obtained in the main shear walls of these buildings (two in Pieve Fosciana, named A and B, one in S. Anastasio, named C).

Cutting of the testing panels was performed with care in order of avoiding damages to the masonry texture; a concrete circular saw with 120 cm diameter was employed, driven by a row so as to cut the panels of the desired dimensions (Fig. 2).

Fig. 1. The Town Hall of Pieve Fosciana

Test panels were isolated by the rest of the walls through three cuts, so as to separate the two lateral sides and the upper base of the panel; the lower base was maintained attached to the original wall. Fig. 3 shows the typical texture of the walls in the building in Pieve Fosciana.

The preparation of the specimen was the same for both compression and shear-compression tests, that were set up to be performed in sequence, only employing different dispositions of the transducers.

Fig. 2. Circular saw used
to cut the testing panels

Fig. 3. Masonry texture
and cracks after the shear test

3.1 Compression tests

Compression tests were performed according to the setup of Fig. 4 (shear test) with a different instrumentation in order to measure horizontal and vertical strains instead of diagonal ones. The aim of the tests was the determination of the stress-strain law for the masonry and the value of Young modulus; application of the vertical stress was performed with the same technique employed for the shear tests, described below.

3.2 Shear tests

Shear tests were performed according to the general setup showed in Fig. 4, so as to obtain double bending conditions for two symmetric panels, the first above the horizontal testing action and the second below.

Fig. 4. Shear test setup

Fig. 5. Particular of the setting
of the horizontal actuator

The confining vertical stress was applied on the upper base of the panel by means of two hydraulic jacks forcing against a series of tie bars jointed to the base of the wall, in order of avoiding damaging in the part of the building above the specimen.

This test setup allows the application of the desired vertical stress during the test, independently by the actual vertical loads acting on the panel. Horizontal displacements of the upper side of the panel were restrained and measured. Horizontal testing load was applied by means of a third actuator and transferred to the panel using two steel bars (Fig. 5).

Instrumentation of the panel comprehended eight transducers applied on the two sides of the two semi-panels, to evaluate diagonal strains; other transducers were employed to measure upper base, middle span and lower base horizontal displacements. A load cell was employed to measure the reaction of the horizontal restrain of the upper base of the specimen.

Tests were performed applying firstly the vertical compression; once the desired vertical stress was reached, the oil circuit of the vertical jacks was closed, so as to maintain the same displacement of the two vertical jacks during the test. The horizontal testing action was then applied and increased until collapse.

4. RESTORATION TECHNIQUE

Fig. 6 shows a panel after the reparation with R.C. jackets applied on the two sides. Concrete with average compressive strength of 18 MPa and welded fabrics (ϕ 6, 15 × 15 cm) were used for the construction of the jackets. Tie bars were also employed to connect the jackets on the two sides (4 ties for m²).

This technique represents one of the most common restoration interventions, due to the relative simplicity of execution and the use of conventional materials. Effects of this strengthening technique were also examined in [7] by means of numerical simulations, also discussing the effectiveness of the jackets. The main drawbacks of this method are represented by high costs and strong modifications of the original structure.

Fig. 6. Jacketing of the testing panels

5. RESULTS

Fig. 7 shows the stress-strain curves obtained from compression tests performed on panel A, before (AC) and after (ARC) the reparation. In Table 1 values of the secant elastic modulus at the peak stress (E_s) and in the range 0.2-0.5 MPa ($E_{0.2-0.5}$) are reported, showing a good increment of the stiffness of the panels due to the restoration.

Compression test on panel C was performed until collapse due to the poor strength of this masonry texture.

Fig. 8 shows results of shear tests performed on panel B before (BT2) and after (BRT) the reparation. In Table 2 values of the referential shear strength τ_k and of elastic tangential modulus G are presented. In the Table, h indicate the height of the semi-panel, σ_v and τ_u the average vertical and shear stresses and μ the displacement ductility.

After the reparation τ_k values increased 1.75 (panel A) and 1.16 (panel B) times with respect to the original values.

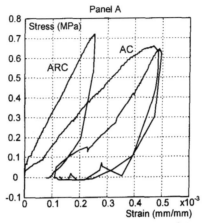

Fig. 7. Compression tests on panel A

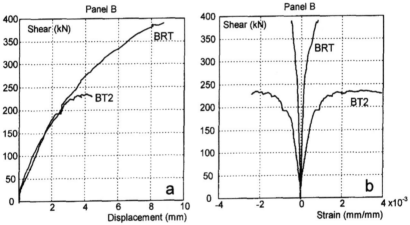

Fig. 8. Shear tests on panel B: original (BT2) and after reparation (BRT); (a) shear versus relative displacement; (b) shear versus bottom diagonal strains

6. NUMERICAL SIMULATIONS

Fig. 9 shows a general view of the F.E. model employed. Results from panel A in Pieve Fosciana were selected in order to perform numerical simulations of compression tests. The compression test was interrupted before collapse, in order to permit the subsequent execution of the shear test on the same specimen. Due to this, the test was utilized only for a preliminary identification of material parameters concerning the elastic and the plastic behavior, thereby excluding the modeling of the material in the cracked phase. Several simulations were performed to assess the influence of various material parameters, finally determining the values reported in Fig. 10.

Table 1. Compression tests results

Test	Panel	E_s (MPa)	$E_{0.2-0.5}$ (MPa)
AC - first cycle	A original	1345	1555
AC - second cycle	"	1591	1848
ARC - first cycle	A repaired	3489	3175
ARC - second cycle	"	2613	2693
BRC - first cycle	B repaired	5536	5329
BRC - second cycle	"	4934	5396
BRC - third cycle	"	5036	4886
CC - first cycle	C original	1921	1701(*)

(*)$E_{0.1-0.15}$

Table 2. Shear tests results

Test	A (cm^2)	h (cm)	σ_v (MPa)	τ_u (MPa)	τ_k (MPa)	G (MPa)	μ
AT	5765	96	0.38	0.38	0.24	179	2.95
ART	6473	97.5	0.35	0.57	0.42	/	/
BT1	4797	98	/	/	/	424	2.85
BT2	4797	98	0.43	0.49	0.32	/	/
BRT	5858	105	1.02	0.66	0.37	274	2.34

Results from the original panel B in Pieve Fosciana were selected to perform the numerical simulation of the shear test (BT2). The panel was initially subjected to a vertical load of about 230 kN and, in a second phase, to an horizontal testing load increasing until the collapse value of about 285 kN, which caused cracks in the lower part of the panel with 0.5÷2 mm width.

Several numerical models were developed to represent nonlinear behavior of masonry: elastic-plastic without cracking (Fig.12); and elastic-plastic with smeared cracking. Different schemes were employed for the modeling of the upper base horizontal bearing: "EL" refers to an elastic boundary element with fixed stiffness, "AN" refers to an anelastic bearing with imposed displacement corresponding to the value observed during the test.

In order to model the masonry in the cracked state, material parameters representing plastic behavior (c = 0.35 MPa, ϕ = 40°) were suitably reduced accordingly with Davis [8]:

$$c_k = \frac{c \cos\phi \cos\delta}{1 - \sin\phi \sin\delta},\qquad(1)$$

$$\tan\phi_k = \frac{\sin\phi \cos\delta}{1 - \sin\phi \sin\delta}.\qquad(2)$$

Uniaxial strength values required for the definition of strength dominion were set accordingly with experimental results, also taking into account the correct intersection of yield and strength surfaces (Drucker-Prager and Willam-Warnke), as explained in Fig. 11. Table 3 presents comparisons between experimental and numerical results in the collapse state, demonstrating the reliability of the numerical model.

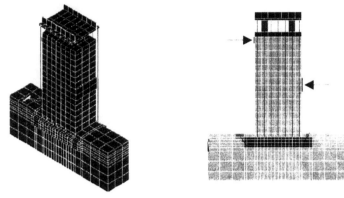

Fig. 9. Numerical F.E. model

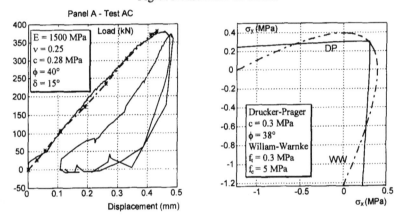

Panel A - Test AC

E = 1500 MPa
ν = 0.25
c = 0.28 MPa
φ = 40°
δ = 15°

Load (kN)

Displacement (mm)

Fig. 10. Simulation of the compression test
AC on the original panel A

σ$_y$ (MPa)

DP

Drucker-Prager
c = 0.3 MPa
φ = 38°
Willam-Warnke
f$_t$ = 0.3 MPa
f$_c$ = 5 MPa

WW

σ$_x$ (MPa)

Fig. 11. Plasticity and strength
dominions in plane stress

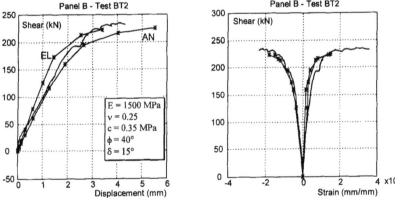

Panel B - Test BT2

Shear (kN)

EL

AN

E = 1500 MPa
ν = 0.25
c = 0.35 MPa
φ = 40°
δ = 15°

Displacement (mm)

Panel B - Test BT2

Shear (kN)

Strain (mm/mm)

Fig. 12. Comparison between shear test BT2 on panel B and numerical results

Table 3. Comparison between experimental and numerical results (test BT2)

		Test	Simulation AN Anelastic bear.	Simulation EL Elastic bear.
Horizontal load	(kN)	285.34	285.23	285.23
Vertical load	(kN)	238.88	240.00	240.00
Bearing reaction	(kN)	55.82	58.07	71.58
Vertical stress	(MPa)	0.498	0.500	0.500
Midspan (bottom) hor. displ.	(mm)	7.40 (2.98)	6.42 (2.60)	5.03 (2.18)
Relative hor. displ.	(mm)	4.42	3.82	2.85
Diagonal strains	($\times 10^3$)	- 2.42 (3.93)	- 2.30 (3.80)	- 2.00 (3.50)

7. CONCLUSIONS

The paper presents first results from an extensive research project performed to assess the safety characteristics of existing building heritage in two seismic regions of Tuscany. Both experimental and numerical results are presented; a new setup for the in situ shear-compression tests was employed in order to permit testing of panels independently from the vertical compression actually acting on the walls. The prosecution of the research will comprehend a larger number of tests and the investigation about the effectiveness of mortar injections as an alternative restoration technique.

ACKNOWLEDGEMENT: The paper presents the first results of the research project "Experimental investigation on the effectiveness of seismic strengthening techniques for historical masonry buildings in Lunigiana and Garfagnana", supported by "Regione Toscana".

8. REFERENCES

1. Postpischl D. ed. 1985, 'Catalogo dei terremoti italiani dall'anno 1000 al 1980', Graficoop, Bologna, (in Italian).
2. G.N.D.T., Regione Toscana, Progetto terremoto in Garfagnana e Lunigiana, Atti del Seminario 'Prima del disastro: la prevenzione', Firenze, Dicembre, 1992, (in Italian).
3. C.N.R. Istituto di ricerca sul rischio sismico, Regione Toscana, 'Pericolosità sismica e prime valutazioni di rischio in Toscana', a cura di V. Petrini, 1995, (in Italian).
4. Chiostrini S., Vignoli A., 'In-situ determination of the strength properties of masonry walls by destructive shear and compression tests', Masonry International, The British Masonry Society, Vol. 7, No. 3, 1994, pp. 87-96.
5. Chiostrini S., Vignoli A., 'An Experimental Research Program on the Behavior of Stone Masonry Structures', Journal of Testing and Evaluation, ASTM, Vol. 20, No. 3, May, 1992, pp. 190-206.
6. Karantoni F. V., Fardis M. N., 'Computed versus Observed Seismic Response and Damage of Masonry Buildings', Journal of the Structural Division, ASCE, Vol. 118, No. 7, July, 1992, pp. 1804-1821.
7. Karantoni F. V., Fardis M. N., 'Effectiveness of Seismic Strengthening Techniques for Masonry Buildings', Journal of the Structural Division, ASCE, Vol. 118, No. 7, July, 1992, pp. 1884-1902.
8. Davis E. H., 'Theories of plasticity and the failure of soil masses', in I.K. Lee (Editor), Soil mechanics: selected topics, Butterworth, London, 1968, pp. 341-380.

CYCLIC MODELLING OF REINFORCED MASONRY SHEAR WALLS FOR DYNAMIC ANALYSIS: INDICATIONS COMING FROM EXPERIMENTS

G. Magenes Research Associate and **S. Baietta** Graduate Student
Dipartimento di Meccanica Strutturale, Università di Pavia, Via Ferrata 1, 27100 Pavia, Italy

1. ABSTRACT

An empirical shear-displacement hysteretic law for reinforced masonry shear walls is calibrated on the basis of experiments. The capability of the model to reproduce the behaviour of walls failing in shear as well as walls failing in flexure is discussed. The law is suitable for use in simplified non-linear dynamic modelling. By means of an n - d.o.f. shear-type model, the seismic behaviour of multi-storey buildings can be simulated within the context of parametric studies, studying the influence of different predominant failure mechanisms.

2 INTRODUCTION

It is generally recognised, after the research and applications developed especially in the last two decades, that engineered reinforced masonry (r.m.) can be a suitable technique for the construction of medium-rise buildings in seismic areas. Nevertheless, a wide variety of construction techniques are available, in terms of brick or block material and shape, of layout of reinforcement, of bond, so that the resulting mechanical behaviour can be rather different from type to type of r.m. construction.

Although in general reinforced masonry has a better behaviour than unreinforced masonry, still it may exhibit significant strength and stiffness degradation under cyclic loading, depending mostly on the failure mechanism developed. In fact, r.m. buildings are wall structures, which can exhibit a variety of failure mechanisms (shear, flexure, mixed). Rather often, capacity design principles cannot be easily applied to enforce favourable (i.e. ductile) failure modes in r.m. walls, due to a series of limitations related to:

a) low shear ratios (especially true for low- and medium rise structures);
b) amount, positioning and anchorage of reinforcement;
c) difficulties in enhancing the compressive behaviour of masonry through confinement.

Keywords: Reinforced masonry, Shear walls, Cyclic modelling

Computer Methods in Structural Masonry – 4, edited by G.N. Pande, J. Middleton and B. Kralj.
Published in 1998 by E & FN Spon, 11 New Fetter Lane, London EC4P 4EE, UK. ISBN: 0 419 23540 X

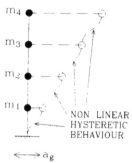

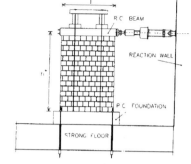

Fig. 1. Simplified 2-d model of a r.m. building for parametric dynamic analyses.

Fig. 2. Layout for quasi-static cyclic shear tests on r.m. walls.

Especially the factors indicated as b) may be strongly influenced by the type of construction technique. Because of the specific mechanical properties of r.m. systems and given the variety of construction techniques, each r.m. construction system requires a specific, experimentally based approach.

3. SCOPE AND OBJECTIVES

When the non-linear seismic behaviour of buildings is of concern, and when numerous parametric analyses must be performed to obtain general results, simplified modelling is often the most reasonable choice. A simplified approach which has been effectively used in the past [1,2,3] is to idealize a regular building as a multi-degree-of-freedom shear-type system (fig. 1). In such model, each interstorey shear-displacement relationship is given by one or more shear elements in parallel. In many cases only one single element per storey is used, characterized by a non-linear hysteretic behaviour which is supposed to encompass the global interstorey shear-displacement behaviour. The hysteretic law is derived from experimental observations (typically quasi-static cyclic tests on shear walls). In this paper the possibility of describing adequately the force-displacement cyclic behaviour of r.m. shear walls is discussed with reference to experiments. A hysteretic law is calibrated to be used in dynamic parametric analyses to estimate suitable values of "behaviour factor" for seismic design.

The type of masonry which is considered here is made of perforated clay bricks (45% void ratio), with low percentage of reinforcement. The horizontal reinforcement (deformed steel bars) is laid in bedjoints. The vertical bars are positioned in vertical flues created by gripholes, which are filled with the same mortar which is used for bedding, as each course is laid. This type of masonry is considered to be a suitable construction system both for its structural behaviour and for its improved habitability (i.e. thermal and acoustic insulation properties).

4. A HYSTERETIC LAW FOR R.M. SHEAR WALLS

A non-linear hysteretic law for shear-type masonry elements had been developed in the past years at the University of Pavia [1,4,5], based on cyclic diagonal compression

Table 1. Summary of the properties of the tested walls. Materials: steel type 500, masonry compression strength 10.8 MPa (mean)

Wall	h^{*} [m]	l [m]	h^{*}/l	σ_0 [MPa]	Spread reinforcement				Concentr. end reinf	Fail. mech.[2]
					vertical (dia. /spacing) [mm]	$\rho_v{}^1$ (%)	horizontal (dia. /spacing) [mm]	ρ_h (%)		
2	3 16	2.23	1.42	0.250	1∅10/ 480	0.055	2∅6 / 400	0.047	2+2∅18	M
3	3.16	2 23	1 42	0 278	1∅14/ 480	0.107	2∅6 / 200	0.094	2+2∅18	M
4c	3.16	2.23	1.42	0.285	1∅ 6/ 480	0.020	2∅6 / 400	0.047	1+1∅18	F
4s	3 16	2.23	1.42	0.472	1∅12/ 480	0.079	2∅6 / 400	0.047	no	F
9	1.76	1 61	0 65	0.238	1∅10/ 480	0.055	2∅6 / 400	0.047	2+2∅20	S
10	1.76	1.61	0.65	0.244	1∅14/ 480	0.107	2∅6 / 200	0.094	2+2∅20	S
11	1.76	1.61	0.65	0.235	1∅10/ 480	0.055	1∅6 / 400	0.024	2+2∅20	S

1 Geometric ratio calculated neglecting the concentrated end reinforcement.
2 Failure mechanism: F = flexure, S = shear; M = mixed flexure-shear

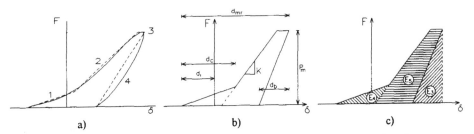

Fig. 3. Idealization of a force-displacement semi-cycle. Parameters of the semi-cycle.

tests on approximately square reinforced and unreinforced masonry elements. Given the type of test, the behaviour was almost completely dominated by shear strength and deformation. Recently, a series of tests on large panels made at the University of Pavia [6] within a co-ordinated research project on r.m. buildings [7] were designed to obtain information on the cyclic behaviour of walls characterized by different failure mechanisms, including shear and flexural mechanisms. The walls were tested as simple cantilevers (fig. 2), with a constant vertical load applied by two hydraulic jacks and a cyclic horizontal displacement history applied by means of an electro-mechanical jack. A thorough description of the relevant material properties and of the experiments can be found in [6]. The main properties of the tested walls are summarized in table 1. These quasi-static tests were taken as a reference in this paper to modify and calibrate the hysteretic law, so that both flexure- and shear-dominated behaviour could be modelled. Considering an experimental force-displacement curve taken from a cyclic shear test on a r.m. wall, each force semi-cycle is idealized by means of a multi-linear curve, made of four branches as shown in fig. 3a. The four branches correspond respectively to 1) crack closing, 2) approximately linear loading, 3) crack propagation 4) unloading. Given this idealization, a number of force-, displacement-, or energy-related parameters can be defined for the semi-cycle (fig. 3b and 3c). In addition, the following energy parameters are defined: $E_1 = E_3 + E_4 + E_5$ absorbed energy at peak displacement, $E_2 = E_1 - E_3 = E_4 + E_5$ dissipated energy, where E_4 is the "crack closing energy". The hysteretic law is based on a set of rules which establish a correlation among the parameters and allow to follow the four branches and to change the mechanical properties after each semi-cycle. The evolution of the parameters is governed by five relationships as follows.

Strength degradation:

$$\frac{P_{mx}}{P_{m0}} = 1 - a_p \cdot \frac{\sum E_{2x}}{E_R} \geq 0 \tag{1}$$

Stiffness degradation:

$$\frac{K_x}{K_0} = \frac{1}{\alpha_{K2}} \cdot \left(\frac{\sum E_{cx}}{E_R} \right)^{-\alpha_{K1}} \leq 1 \tag{2}$$

where: $\sum E_{cx} = \sum E_{sx} + f \cdot \sum E_{sy}$, $f = \dfrac{\alpha_{f1}}{\left(1 + \alpha_{f2} \cdot \dfrac{E_R}{\sum E_{sy}}\right) \cdot \left(1 + \alpha_{f3} \cdot \dfrac{\sum E_{sx}}{E_R}\right)} < 1$

Crack closing energy:

$$\frac{E_{4x}}{E_R} = \alpha_{E1} \cdot \frac{\sum E_{2y}}{E_R} + \alpha_{E2} \geq 0 \tag{3}$$

Crack closing displacement:

$$\frac{d_{cx}}{d_R} = \alpha_{dc1} \cdot \frac{\sum E_{2y}}{E_R} + \alpha_{dc2} \geq 0 \tag{4}$$

Unloading displacement:

$$\frac{d_{bx}}{d_R} = \alpha_{db1} \cdot \left(\frac{d_{mr}}{d_R} \right)^2 + \alpha_{db2} \cdot \left(\frac{d_{mr}}{d_R} \right) \quad \text{for} \quad \frac{d_{mr}}{d_R} \leq \left. \frac{d_{mr}}{d_R} \right|_V \tag{5}$$

$$\frac{d_{bx}}{d_R} = const. = \alpha_{db1} \cdot \left(\left. \frac{d_{mr}}{d_R} \right|_V \right)^2 + \alpha_{db2} \cdot \left(\left. \frac{d_{mr}}{d_R} \right|_V \right) \quad \text{for} \quad \frac{d_{mr}}{d_R} > \left. \frac{d_{mr}}{d_R} \right|_V$$

where the parameters P_{m0}, K_0 are the initial strength and stiffness of the wall, $d_R = P_{m0}/K_0$ and $E_R = P_{m0}^2/2K_0$. The subscripts x and y refer respectively to positive and negative semi-cycles. In eqn. 2 the weight function f is introduced to model the different effect of semi-cycles of the same or of the opposite sign on stiffness degradation, as found in tests with random loading histories [5]. In such a way, a weighed "conventional" dissipated energy E_c is introduced for each direction of loading. The present formulation of eqn. 2 is a simpler expression with respect to the original form [1].

5. CALIBRATION OF THE HYSTERETIC LAW ON TESTS

Figures 4 and 5 show the different experimental shear-displacement behaviour obtained with two representative cases taken from the tests on r.m. walls [6]. Wall 9 failed in shear with diagonal cracks without yielding of the flexural (vertical) reinforcement, while wall 4c failed in flexure without any significant diagonal crack and with moderate damage of the compressed corners. The different cyclic strength degradation displayed by the two mechanisms is apparent. In this study, it was found that the same correlation established by eqns. 1 to 5, that had been developed from diagonal compression tests, could be applied for walls failing in flexure or shear, with a proper best-fit calibration of the α coefficients for each wall. In the example of fig. 6 each experimental point is

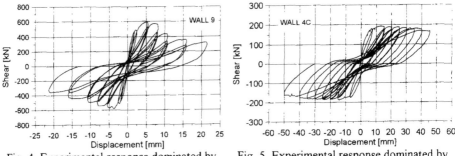

Fig. 4. Experimental response dominated by shear failure (wall 9).

Fig. 5. Experimental response dominated by flexural failure (wall 4c).

relevant to a single semi-cycle of the experimental curve of wall 4c, and the continuous lines are the representation of eqns. 1 to 5 after calibration of the α coefficients. Figures 7 and 8 show how the hysteretic model would reproduce the experimental static tests for the two different failure mechanisms.

Considering the results of the calibrations, the following conclusions could be drawn.

Strength and stiffness degradation. The two most important relationships for the hysteretic model are eqns. 1 and 2, describing cyclic strength and stiffness degradation. In fig. 9 it is shown that calibration of eqn. 1 (strength degradation) led in practice to the same result (i.e. same α coefficients) for walls with the same failure mechanism, but a strong difference was found between different failure mechanisms. Within the same mechanism, the strength degradation law seems to be in practice insensitive to other parameters such as the different shear reinforcement ratio for walls failing in pure shear (walls 9,10,11) or to axial load level or distribution of flexural reinforcement for walls failing in flexure (walls 4c and 4s).

The differences in the stiffness degradation law (fig. 10) were not as striking as for strength degradation, although it could be seen that flexural failure gives a slightly less degrading behaviour.

Energy dissipation. The hysteretic model can reproduce fairly well the overall energy dissipation obtained from experiments, both for shear and flexural failure. In figure 11 the cumulative dissipated energy in the experiments is compared with the cumulative energy dissipated by the corresponding numerical simulations of the quasi-static tests.

As a further element of interest, it is reported that if we accept to describe the "shape" of a single experimental hysteresis cycle with a single dimensionless scalar parameter such as the equivalent viscous damping $\xi_e = W_d/(2\pi W_e)$ which is given by the ratio of the dissipated energy over the stored elastic energy at peak displacement, no significant differences in the values of ξ_e were found between shear or flexural failure mechanism. The values of ξ_e were increasing with the displacement demand, ranging up to $\xi_e \cong 15$-20% for the cycles with the highest peak displacement.

Role of spread shear reinforcement. The role played by spread shear reinforcement in the cyclic behaviour or r.m. walls failing in shear (walls 9,10,11) was to affect mainly the absolute shear strength of the walls, i.e. a higher spread reinforcement ratio gives a higher shear strength [6] (the reinforcement ratios ranged approximately from 0.25‰ to 1‰). This role is not apparent when non-dimensional normalized quantities are looked at, as for eqns. 1-5 and for the parameter ξ_e. The non-dimensional strength and stiffness

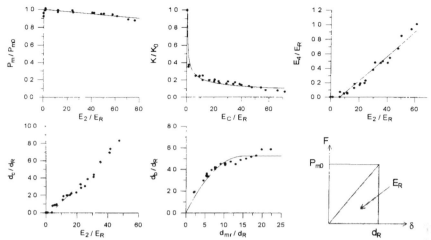

Fig. 6. Representation of eqns. 1-5 as calibrated on the experimental response of wall 4c (flexural failure).

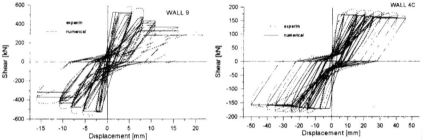

Fig. 7. Numerical simulation of the quasi-static test of wall 9.

Fig. 8. Numerical simulation of the quasi-static test of wall 4c.

degradation laws are not very sensitive to the spread reinforcement ratio, and the values of ξ_e were even found to be slightly decreasing with the increase in reinforcement ratio. However, the effect of a higher reinforcement ratio is directly reflected in an increase in absolute strength and consequently of absolute dissipated energy.

6. SOME APPLICATIONS IN DYNAMIC ANALYSES

The hysteretic model discussed above is being used at present to perform a parametric study to evaluate suitable values of the behaviour factor to be used for seismic design. Within this context, "behaviour factor" is defined as the ratio k between the elastic design response spectrum and the anelastic design spectrum corresponding to a specified limit state.

In the analyses, a regular r.m. building of n storeys is idealized as an n-d.o.f. shear-type system as in fig.1. The interstorey force-displacement hysteretic behaviour is modelled considering either shear-dominated or flexure-dominated behaviour. The seismic input is made by families of non-stationary spectrum-compatible accelerograms (i.e. compatible

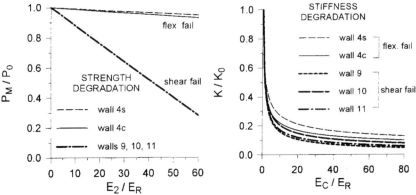

Fig 9. Graphical representation of eqn.1 calibrated for each tested wall.

Fig. 10. Graphical representation of eqn.2 calibrated for each tested wall.

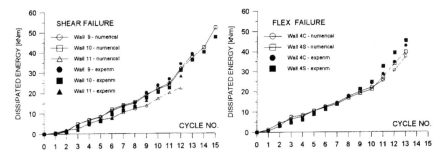

Fig. 11. Comparison between the energy dissipation of the experiments and of the numerical simulation of the experiment.

with given reference acceleration design spectra). The following issues concerning the behaviour factor are being investigated:
- the influence of the type of dominating failure mechanism,
- the influence of frequency content and duration of the seismic input,
varying the initial natural period (in the range 0.1-0.55 sec) and the number of storeys (3 to 5). Two possible limit states are considered: one associated to a maximum strength degradation with respect to the initial strength (20% or 30% degradation, corresponding to 80% or 70% residual strength), one associated to a maximum interstorey drift (1%). For each combination of parameters, a dynamic analysis is performed with in-time step-by-step integration of the equations of motion.
The first results of these parametric analyses with short duration accelerograms were presented in [8]. The role of the dominating failure mechanism was found to be of paramount importance especially in relation with the frequency content of the seismic input (figs 12 and 13) Accelerograms of longer duration are presently being considered.

7. CONCLUSIONS AND FUTURE DEVELOPMENTS

The hysteretic law presented in this paper was shown to be a flexible tool to reproduce the most important features (strength and stiffness degradation, energy dissipation) of the

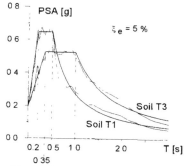

Fig. 12. Mean response spectra and reference design spectra of the accelerograms (short duration).

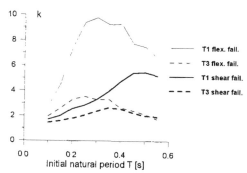

Fig. 13. Calculated behaviour factor *k* for different soil types and different failure mechanisms.

non-linear shear-displacement cyclic behaviour of reinforced masonry shear walls failing in shear of flexure. The law is presently being used for simplified non-linear modelling of the seismic response of r.m. buildings within a parametric study in which numerous analysis must be performed. In these analyses the role of the dominating failure mechanism in the seismic response can be easily investigated by setting suitable values of the parameters of the model, taking as a reference the calibration performed on the results of experimental testing.

A non-linear shear-type element based on the above hysteretic model is presently being implemented in a finite element code in view of three dimensional seismic analyses of r.m. buildings.

8. ACKNOWLEDGEMENTS

Research funded under the BRITE-EURAM BE-4001, Contract BREU-0575.

9. REFERENCES

1. Calvi G.M. and Macchi G., "Dynamic parametric simulation of masonry structures", Volume dedicated to the 60th Anniversary of Prof. T.P.Tassios, National Technical University of Athens, 1990, pp. 461-485.
2 Modena C., "Seismic behaviour of masonry structures: experimentally based modelling - Parts 1 and 2", *Masonry International*. Vol. 6, No. 2, 1992, pp. 57-68.
3. Tomazevic M., "Dynamic modelling of masonry buildings: storey mechanism as a simple alternative", *Earthquake Engineering and Structural Dynamics*, Vol. 15, 1987, pp. 731-749.
4. Cantù E., and Zanon P., "Combined cyclic testing procedures in diagonal compression on hollow clay block reinforced masonry", Proc. of the 6th I.B.Ma.C., Roma, 1982, pp.1007-1020.
5. Calvi G. M., Macchi G., and Zanon, P., "Random cyclic behaviour of reinforced masonry under shear action", Proc. of the 7th I.B.Ma.C., Melbourne, 1985, pp. 1249-1256.
6. Magenes G., Calvi G.M., and Gaia, F., "Shear tests on reinforced masonry walls", Scientific Report RS-03/96, Dipartimento di Meccanica Strutturale dell'Università di Pavia, 1996.
7. BRITE-EURAM Project "D.Re.Ma.B." - Industrial Development of Reinforced Masonry Buildings - Synthesis Report, September 1996.
8. Magenes G., Baietta, S., and Calvi G.M. "Una ricerca numerico-sperimentale sulla risposta sismica di edifici in muratura armata", Atti 8° Convegno Nazionale ANIDIS *L'Ingegneria Sismica in Italia*, Taormina, 21-24 settembre 1997, pp. 143-150 (in Italian).

NUMERICAL MODELING OF SHEAR BOND TESTS ON SMALL BRICK-MASONRY ASSEMBLAGES

G. Mirabella Roberti Research Associate, **L. Binda** Full Professor and
G. Cardani Architect
Department of Structural Engineering, Politecnico of Milan, Piazza L. da Vinci 32, 20133 Milan, Italy

1. ABSTRACT

An experimental and numerical investigation has been carried out in Milan for the characterisation of the materials of a two storey, full-scale, brick masonry building prototype set up at the University of Pavia. Among other tests, shear bond tests have been carried out on triplets. The displacements in the vertical and horizontal directions have been carefully measured on bricks and mortar joints under different normal stress levels.

For the numerical interpretation of the material behaviour, a Mohr-Coulomb type failure envelope, calibrated on the experimental results, was applied to the brick/mortar interfaces of a FE model in plane stress conditions. Contact elements with finite sliding capacities under large displacements have been used to model the test.

2. INTRODUCTION

An extensive experimental programme was promoted by C.N.R.-G.N.D.T. (the National Group for Defence from Earthquakes) for the chemical-physical and mechanical characterisation of the materials used to build a two storey, full-scale, brick masonry building prototype to be tested under seismic action [1]. One of the aims of this project was to compare the experimental results with the prediction obtained from numerical models based only on material test results [2].

Among other tests for characterising the mortar, the bricks and their assemblages [3], shear bond tests were performed on triplets in order to measure the bond strength between mortar joints and units, that is one of the most critical parameters for the implementation of analytical models [4]. Tests were performed according to RILEM recommendations (RILEM 127 MS.B.4) and the first draft of CEN standard, at different vertical loads, measuring on the specimens vertical and horizontal displacements, both absolute and differential.

3. EXPERIMENTAL SET-UP

The materials adopted for the prototype were: (i) solid soft mud bricks with high water absorption and rather low strength, of the type frequently used in practice for restoration, (ii) mortar composed by hydraulic lime and siliceous sand, prepared according to the Italian code [5]. The mean values of the elastic parameters, achieved from different experimental tests, are reported in table 1.

Bricks	Mortar	Ratio
$E_b = 2000 \ N/mm^2$	$E_m = 500 \ N/mm^2$	$E_b/E_m = 4.0$
$v_b = 0.1$	$v_m = 0.2$	$v_b/v_m = 0.5$

Table 1 - Elastic parameters of the materials used for the tests

Computer Methods in Structural Masonry – 4, edited by G.N. Pande, J. Middleton and B. Kralj.
Published in 1998 by E & FN Spon, 11 New Fetter Lane, London EC4P 4EE, UK. ISBN: 0 419 23540 X

Shear bond tests were carried out on stack bond prisms 190 mm high, 120 mm wide and 240 mm long, the bed-joints being 10 mm thick. The prisms were cured at 20 °C and 90% R.H. for 365 days. A thin gypsum layer was used as capping to ensure a good contact with the testing device. A testing machine suitable for biaxial tests was used, in which two hydraulic double effect jacks can be used simultaneously with a separate feedback control system for each actuator. The tests were carried out applying monotonic loads under displacement control, with a rate of displacement of 10μm/sec.

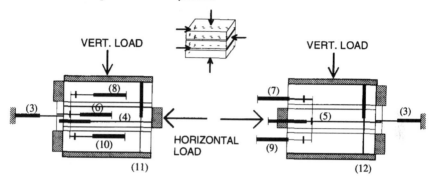

Figure 1 - Scheme of the shear bond test on triplets: position of LVDTs

As shown in Figure 1, two LVDTs measured the vertical displacements (n.11 and 12) and 8 LVDTs measured the absolute and relative horizontal displacements
Four different levels of pre-compression load, perpendicular to the bed-joints, were applied to the specimen, respectively corresponding to a mean stress of 0.12, 0.40, 0.80 and 1.25 N/mm². Three triplets were tested for each level of vertical load. The resulting load displacement curves are shown in Figure 2. In table 2, the values of the main parameters are reported for each test.

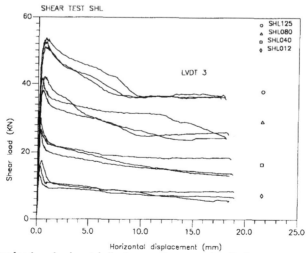

Figure 2 - Shear load vs. horizontal displacement curves; the displacement correspond to the LVDT (3) referred to the central brick.

NAME	σ_v [N/mm^2]	τ_p [N/mm^2]	τ_r 5/10 mm [N/mm^2]	T_p [KN]	AREA [cm^2]	MIX 2 mortar mixture	curing days
SHL012A		0.31	0.16/0.14	17.534	283.2	II	398
SHL012B	0.12	0.27	0.17/0.14	15.636	285.6	II	399
SHL012C		0.23	0.15/0.13	13.014	284.4	II	403
SHL040A		0.46	0.32/0.27	26.173	284.4	III	397
SHL040B	0.40	0.47	0.36/0.32	26.654	285.6	III	398
SHL040C		0.52	0.35/0.27	30.004	288	III	398
SHL080A		0.74	0.58/0.44	42.082	285.6	II	405
SHL080B	0.80	0.67	0.55/0.46	38.059	285.6	II	407
SHL080C		0.73	0.59/0.55	41.714	286.8	II	410
SHL125A		0.93	0.78/0.64	52.523	283.2	V	405
SHL125B	1.25	0.95	0.85/0.64	53.736	282	V	409
SHL125C		0.90	0.78/0.63	50.954	284.4	V	409

Table.2 -Experimental results of the shear tests on small masonry prisms (SHL).

Since the shear stress distribution along the brick-joint interfaces is not constant, the mean value has been calculated dividing the ultimate load by two times the area of the horizontal joint. A egression line was obtained interpolating the peak values τ_p of mean shear stress with respect to the correspondent average normal stress σ_v (see Figure 10), according to a Mohr-Coulomb criterion, in the form:

$$\tau_p = 0.23 + 0.57 \, \sigma_v$$

were: τ_p = shear strength; 0.23 = shear bond strength at initial compression equal to zero, (cohesion); 0.57 = friction coefficient between mortar and brick (corresponding to a friction angle of ~30°); σ_v = normal pre-compression stress.

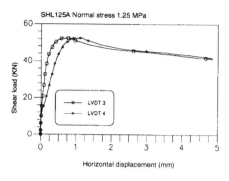

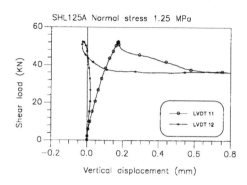

Figure 3 - Shear load vs. Horizontal displacements on the central brick (specimen 125A).

Figure 4 - Shear load vs. vertical displacements curves of specimen 125A

Comparing the absolute horizontal displacement of the central brick (LVDT n.3) with the relative displacement of the same (LVDT n.4), an horizontal contraction of the central brick is observed, see Figure 3.

The main limit of this test is the presence of a bending moment, which is inevitably applied to the specimen. Thus the specimen fails not only owing to shear stresses but also to flexural stress. The two vertical LVDTs n.11 and 12 with their opposite signs showed the presence of bending deformations, see Figure 4.

The failure mechanisms of the triplets showed cracks usually at the brick-mortar interface often together with a typical crack at 45° through the mortar at the end of the joint.

4. NON LINEAR FE ANALYSIS

A Mohr-Coulomb model of the failure limit was adopted for the brick-mortar interface, in order to investigate the state of stress induced on the triplets tested, under different normal stress levels.

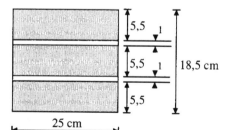

WIDTH	AREA
12 cm	300 cm^2

Figure 5 - Geometrical dimensions for the numerical model

A "micro-model" was chosen allowing to discretize the geometry of the specimen in brick and mortar joint elements, characterised by their own mechanical properties; 4 nodes, first order elements have been adopted in plane stress conditions for the FE model. The mesh is more refined in strategic points: on top and base of the specimen, near the loading and contrast plates, on the brick/mortar interface (see Figure 6).

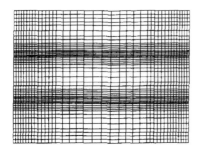

Figure 6 - Mesh adopted in stress analysis.

Two different series of nodes with the same initial position are defined at the interfaces, one for the joint and one for the brick, connected together with contact elements that allow for finite relative sliding and separating. Dummy springs connected to ground and running in the horizontal direction are able to remove instability. The boundary conditions vary during the steps of the numerical simulation, according to the real testing situation:

a) during the first phase of pre-compression, normal to the bed-joints, an asymmetry in the stress distribution is observed due to the lateral expansion restrained by the contrast plates.

b) during shear loading the nodes of the loaded face of the central brick changes configuration. They can undergo only a rigid horizontal displacement following the load plate (see Fig. 7).

The load history is represented in the plot of Figure 8.

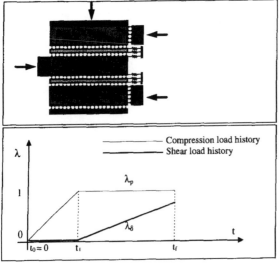

Figure 7 - Scheme of the constraints applied to the model.

Figure 8 - Loading multipliers during test simulation
λ_p pre-compression multiplier
λ_δ displacement multiplier

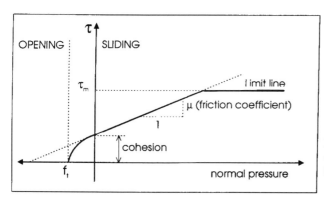

Figure 9 - The failure envelope assumed for the interfaces.

As far as the friction problem is concerned, a modified Mohr-Coulomb failure envelope was assumed in order to avoid the infinite growth of the shear stress on the brick/mortar interface with increasing normal pressure. Also the bond tensile failure limit is modified. A parabolic failure envelope is assumed in tension (see figure 9), where bond tensile strength was achieved from tests on small specimens [1].

5. NUMERICAL RESULTS

Since the materials are considered purely elastic, the phase of vertical pre-compression confirm the presence of tensile stress normal to vertical loads and show the effect of mortar expansion on the joint extremities. The shear stresses contour plot show a quite uniform distribution on the brick/mortar interface inside the specimen, but near the joint extremities a complicated stress concentration is registered.

The peak shear stress with different compression load levels calculated with numerical analysis are reported in Table 3 and plotted in Figure 10.

The results obtained are very similar to the experimental results particularly for the intermediate vertical loads. The differences increase with low or high compression levels because the failure limit adopted in the model (see Figure 13) is obtained from a regression line based on the experimental results.

NUMERICAL ANALYSIS					
σ_v [N/mm²]	P [kN]	T_p [kN]	Name of Specimens	T_p mean value [kN]	Difference (%)
0.12	3.53	16.00	SHL012	15.33	+4.3
0.40	11.77	26.90	SHL040	27.61	-2.6
0.80	23.54	39.84	SHL080	40.62	-1.9
1.25	36.79	52.02	SHL125	52.40	-0.7

Table 3 - Results of the numerical analysis in comparison with the experimental results.

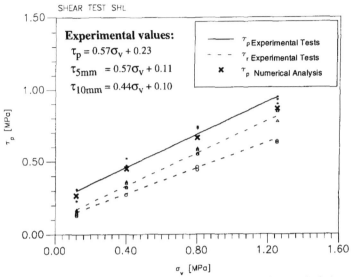

Figure 10 - σ/τ plot with regression lines of peak τ of the numerical analysis in comparison with the experimental results.

In Figure 11 the peak shear forces vs. horizontal displacements measured on the two faces of the central brick is reported. These displacements correspond to the measurements of the LVDTs 3 and 4. It can be seen that displacements u2, on the free side of the central brick, are lower than the u1 on the loaded side. This situation is caused by the shortening of the central brick due to shear load combined with the contrast effect of the two mortar joints.

The behaviour is elastic until the 60% of the ultimate strength, then joints start to separate and/or slip. With respect to the experimental results, in this plots "locking" and "softening" branches do not appear, because they are not taken into account in the definition of the numerical model.

The FE program (Abaqus ver.5.4) allows to check the behaviour for every loading or displacement increment and so for the opening or sliding of the interfaces (see figure 12).

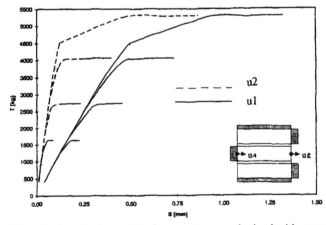

Figure 11 - Shear load vs. horizontal displacement curves obtained with numerical analysis.

For all the levels of normal load, the specimen behaviour is elastic up to the 60% of the ultimate load. Then cracks form, starting from the ends of the interface mortar joint/central brick: where the horizontal load is applied, a relative slip is observed, whereas on the opposite side tensile cracks are evident, due to the bending effect. This last effect decreases with the increase of the normal load. The interface mortar joint/ upper and lower bricks do not present any failure, until the expulsion of the central brick.

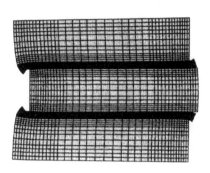

Figure 12 - FE analysis: deformed shape at 80 % of maximum load compared to the failure mechanism of a specimen.

Along the brick/mortar interface, the shear stress distribution is non-uniform: in fact, shear stresses are higher near the loading plate, decrease in the centre of the specimen and increase again near the other extremity, before a downfall due to the presence of tensile stress. Near the shear load peak, the shear stresses in the central part of the brick/mortar interface become more uniform, while towards the ends the influence of the machine platens is even stronger. In addition the shear strength distribution on the interface is dependent from the normal stresses, which are not uniformly distributed along the joint (see Fig. 13 and 14).

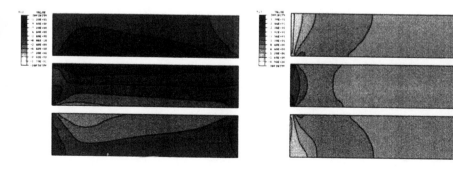

Fig. 13 - Shear stress contours at 80% of the peak load

Fig. 14 - Vertical stresses at the same load level

6. CONCLUSIONS

Although the shear bond test arrangement was modified in the last years by CEN, the experimental results from the triplet test give parameters which are representative of the shear bond behavior of the joints.

Althougth the experimental parameters are obtained as average values on the whole specimen, they can be successfully used in the numerical analysis as local parameters of the mortar-brick interface.

In fact, even if the stress distribution during the test is not uniform, nevertheless the shear stress along the interface becomes nearly constant at the ultimate load. The numerical model appears very useful for analyzing the stress distribution, both in the mortar and in the bricks, all along the test until failure.

7. ACKNOWLEDGEMENTS

The author wish to thank M. Cucchi and C. Tiraboschi for the experimental work, L. Ghirardelli and M. Tacchini for the elaboration of the results, L. Branduardi for the help in numerical analysis. CNR-GNDT support in the experimental research is gratefully acknowledged

8. REFERENCES

[1]   Binda L. et al., Measuring masonry material properties: detailed results from an extensive experimental research, part I and II, CNR-GNDT report 5.1, Milan 1996.

[2]   U.S.-Italy Workshop on Guidelines for Seismic Evaluation and Rehabilitation of Unreinforced Masonry Buildings, Pavia, 20 luglio 1994.

[3]   L. Binda, G. Mirabella Roberti, C. Tiraboschi, "Problemi di misura dei parametri meccanici della muratura e dei suoi componenti", Atti del Convegno Nazionale "La meccanica delle murature tra teoria e progetto", Messina, 18-20 settembre 1996.

[4]   Hofmann, P., Stöckl, S., Tests on the shear-bond behaviour in the bed joints of masonry, *Masonry International*, No.9, 1989.

[5]   D.M. 31.08.72, *Norme sui requisiti di accettazione e modalità di prova degli agglomerati cementizi e delle calci idrauliche*, Gazzetta Ufficiale n.287, 6.11.1972.

FINITE ELEMENT ANALYSIS OF MODEL SHEAR TESTS

V. Bosiljkov[1], B. Kralj[2], R. Žarnić[1], G.N. Pande[2] and J. Middleton[2]

[1]*Faculty of Civil Engineering and Geodesy, University of Ljubljana, Jamova 2, SI-1000 Ljubljana, Slovenia and [2]Department of Civil Engineering, University College of Swansea, Singleton Park, Swansea SA2 8PP, UK*

ABSTRACT

The possibility of using homogenized approach in modelling shear walls with and without openings was investigated. By using 3D homogenization method which were developed at the University of Swansea we made effort to test its ultimate options in modelling different crack patterns of single-wythe unreinforced masonry (URM) shear walls made from solid brick. The main scope of our investigation was to test, and according to its limitation to suggest further improvements of the numerical model which should be used for the modelling of our future experimental work on URM shear walls. In order to compare our numerical investigations, experimental as well as numerical work of other authors were used (see [1]).

1. INTRODUCTION

One of the first scientific investigations of masonry shear walls was performed in 1873 by Bauschinger, who carried out shear tests on coupled bricks [2]. Since then masonry shear walls have been investigated through numerous different varieties. As a general term *'masonry shear walls'* can be described not only in terms of their constituents (masonry units and mortars), but also in terms of their construction. In terms of its constituents it can be described as: solid or hollow, brick or block, and clay, stone, calcium-silicate or concrete shear wall. In terms of their construction type they can be: loadbearing or nonloadbearing, reinforced or unreinforced, single-wythe or multiwythe, solid or perforated, rectangular or flanged, and cantilevered or coupled shear wall. Shear wall behaviour is affected by the shapes of shear walls in plan, sizes and distribution of multiple openings, and boundary elements such as cross walls, returns, and columns [3].

2. SCOPE OF INVESTIGATION

The main scope of the investigation was numerical modelling of single-wythe unreinforced masonry (URM) shear walls made from solid clay brick. By using 3D homogenization technique for modelling masonry material we made effort to model different crack patterns for walls with and without openings which could be expected

Computer Methods in Structural Masonry – 4, edited by G.N. Pande, J. Middleton and B. Kralj.
Published in 1998 by E & FN Spon, 11 New Fetter Lane, London EC4P 4EE, UK. ISBN: 0 419 23540 X

and were observed in experiments made by Raijmakers and Vermeltfoort see [1]. During the investigation some parameters which can influence the results such as: modifying boundary conditions, mesh sensitivity and different sandwich materials were discussed. Further on, some rough comparisons with the experimental results of testing were carried out. According to our numerical results further improvements of the existing code were suggested.

3. ON NUMERICAL MODELING OF THE MASONRY

When starting with micro and continuing to macro modelling of masonry one can find homogenization approach the most rewarding one. On one hand it does not need large amount of input data which are refer to material characteristics of the constituents and their junction, and it does not produce large number of elements (CPU time), as it is needed for micro-modeling. On the other hand, by using homogenization approach we can still model large masonry structures avoiding costly experiments for biaxial testing for each new material or application of known material which is needed for the evaluation of the macro-models for masonry. By using homogenization approach of modeling masonry we can obtain an anisotropic macro-constitutive law from the micro-constitutive laws and geometry of the constituents, in such a way that the macro-constitutive law is not actually implemented. This would mean that a change in the geometry of the constituents would not imply different material models or costly experiments. That would allow us to exploit the existing data-bank of the material parameters of the constituents and the masonry without further experimental costs.

4. BASIC EQUATIONS AND ASSUMPTIONS OF THE NUMERICAL MODEL

The numerical model is based on two stage homogenization technique: the homogenization of masonry and the homogenization of cracked material. The homogenization of masonry includes substitution of masonry by a homogeneous material with orthotropic material properties (Eq. 2). Orthotropic material properties are derived according to two steps homogenization where bed joint and brick are homogenized first and than followed by head joint homogenization.

With this approach the overall behaviour of the homogenized material is equivalent to the composite. Homogenization is based on the equivalence of the deformation energies, and the homogenized properties of homogeneous material are given in analytical form. Symbols represented in equation (2) which are more introduced in more detail elsewhere (see [5][6]), are functions of dimensions of the bricks, Young's modulus and Poisson's ratio of the brick and mortar and the thickness of the mortar joint.

$$\bar{\varepsilon} = [\bar{C}]\bar{\sigma} \qquad \text{Eq. 1}$$

$$[\bar{C}] = \begin{bmatrix} \dfrac{1}{\bar{E}_x} & -\dfrac{\bar{\nu}_{xy}}{\bar{E}_x} & -\dfrac{\bar{\nu}_{xz}}{\bar{E}_x} & 0 & 0 & 0 \\ -\dfrac{\bar{\nu}_{yx}}{\bar{E}_y} & \dfrac{1}{\bar{E}_y} & -\dfrac{\bar{\nu}_{yz}}{\bar{E}_y} & 0 & 0 & 0 \\ -\dfrac{\bar{\nu}_{zx}}{\bar{E}_z} & -\dfrac{\bar{\nu}_{zy}}{\bar{E}_z} & \dfrac{1}{\bar{E}_z} & 0 & 0 & 0 \\ 0 & 0 & 0 & \dfrac{1}{\bar{G}_{xy}} & 0 & 0 \\ 0 & 0 & 0 & 0 & \dfrac{1}{\bar{G}_{yz}} & 0 \\ 0 & 0 & 0 & 0 & 0 & \dfrac{1}{\bar{G}_{zx}} \end{bmatrix} \qquad \text{Eq. 2}$$

The second stage of the homogenization includes the homogenization of cracked material, where failure criterion is based on maximum principal stresses and is checked for each constituent material (units, bed joint and head joint). Crack material is assumed as weak material perfectly bonded to intact masonry and homogenized with intact material.

The basic assumptions made in the derivations are:

1. Brick and mortar are perfectly bonded
2. Perpend (head) mortar joints are assumed to be continuous

The second assumption is necessary in the homogenization procedure, and it is shown elsewhere (see [6]) that the assumption of continuous head joints instead of staggered joints does not have crucial effect on the stress states of the constituent materials.

5. DISPOSITION OF SHEAR TESTS

For the application of our numerical model on URM shear walls, the tests carried out in the Netherlands (TU Eindhoven) [1] were used. They were chosen for two reasons, first they were well published and as a difference from the majority of tests on URM shear walls which were carried out under cycling racking load, this one was under monotonous racking load. On the other hand the size of the walls without openings is suitable for our future experimental work on URM shear walls and in the same time, it could be representative, e.g. as a spandrel for modeling multistory shear walls.

5.1. Disposition of shear tests on URM walls without openings

The arrangement of the investigated shear wall without openings is shown in Figure 1 -a. The tests were carried out for tree different level of precompression.

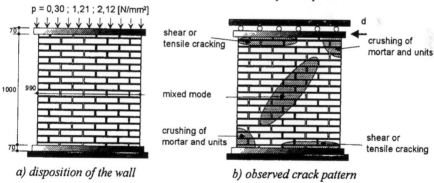

a) disposition of the wall   b) observed crack pattern

Figure 1: Walls without openings

In general, depending upon the nature of the test, crack pattern and failure of the URM shear walls depend on the combination of the applied loads (ratio of the racking to compressive load - R/C), wall geometry, and properties of the constituent materials. Some basic crack patterns and their explanation are shown in Figure 1 -b. Mixed mode - stepped crack pattern for high level of R/C load can be caused either by exceeding the bond and shear friction resistance of unit-mortar junction, or mortar tensile strength. For a low level of R/C load, mixed mode - diagonal crack pattern can be caused by exceeding unit and mortar tensile strength. Failure of the URM walls can happened either in a brittle manner following immediately the first crack, or more gradually upon increasing the applied racking load.

For the walls observed in the experiments, the first crack occurred near the heel of the wall, followed (depending by the R/C ratio) by the mixed mode crack and finally by crushing of the compressed toe.

5.2.   Disposition of shear tests on URM walls with openings

Shear walls with openings were investigated in the sense of testing the homogenization approach for the analysis of masonry assemblies within large masonry structures. The disposition and possible crack pattern are shown in Figure 2. For the same level of precompression two walls with openings were tested. The observed crack pattern was as follows; first with diagonal crack which arise from corners of the opening at four possible locations; than followed by tensile cracks situated by the outer side of two piers around the opening; and finally with compressive cracks as it is shown in Figure 2 -b which formed the failure mechanisms.

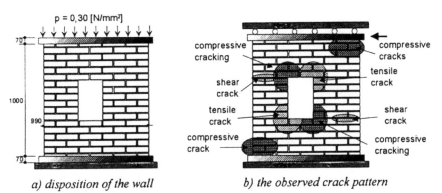

a) disposition of the wall   b) the observed crack pattern

Figure 2: Walls with openings

5.3.   Material properties of the constituents

All material properties of the constituents except [*)] and [**)] were taken from the experimental results - Table 1. Owing to the lack of the data for values [*)] and [**)], they were taken according to the first author's opinion.

Table 1

Material	Properties			
	E [GPa]	ν	f_c [MPa]	f_t [MPa]
Solid brick (210x100x52) mm³	16,7	0,15	27	2,5
Mortar in bed joint	8,7	0,15	10,5	2,5 *
Mortar in head joint	8,7	0,15	10,5	1,5 **
Characteristics of sandwich material used in numerical analysis				
Steel	210	0,3	240	240
Concrete	30	0,2	35	-
Masonry	14,5	0,2	-	-
Mortar	8,7	0,15	-	-

6. RESULTS OF INVESTIGATIONS

6.1. Walls without openings

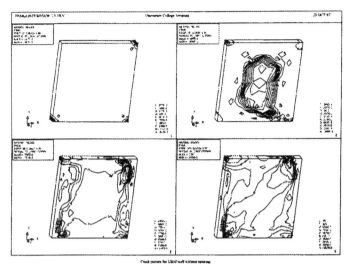

Figure 3: Developing of the crack pattern for different level of applied racking load

Figure 3 presents the development of crack pattern according to the level of the racking load. The first cracks would begin in the toe and the heel of the wall, which were of tensile nature. At approximately 15% of the applied racking load, mixed mode of crack pattern started to occur. Finally, typical failure diagonal crack is predominant at the ultimate racking load. In Figure 4 are represented the redistribution of the main compressive stresses in the homogenized material (1), brick (2), bed joint (3) and head joint (4) at the level of racking load when diagonal crack develops.

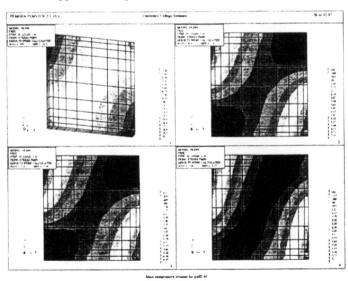

Figure 4: Main compressive stresses for homogenized material and constituents

6.2.   Walls with openings

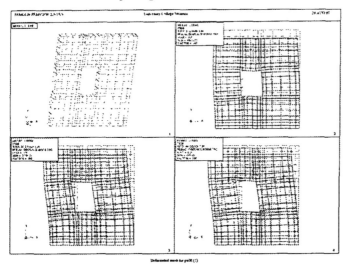

Figure 5: Deformed mesh

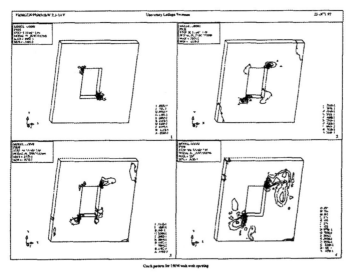

Figure 6: Development of crack pattern for different levels of applied racking load

Beside the existence of singularity points in the proposed model, our program showed good robustness. Deformed meshes of the wall with opening at different levels of racking load is presented in Figure 5. It is worth to mentioning that the appearance of the first cracks strongly depends on the level of precompression.

The development of the crack pattern was very well captured, except for the shear cracks in the piers around the opening (Figure 6). This was in some way expected due to the nature of the homogenized approach, which is not (at this stage) suitable for the modelling of such small masonry assemblies within large masonry panels.

Further investigation of the influence of different sandwich material on the crack pattern has shown that the stiffer sandwich material are, more accurate crack pattern can be obtained. On testing mesh sensitivity (both for walls with and without opening), we did not observe any changes in the obtained crack patterns.

6.3. Comparing experimental and numerical results

Although our primary task was to model different crack-patterns of URM shear walls with and without openings, some comparisons with experimental results and numerical modelling of other authors (micro-modeling) [1] can be seen in Figure 7 and 8.

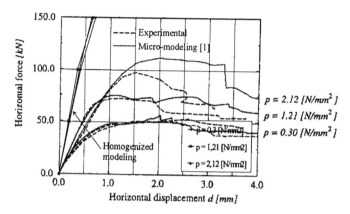

Figure 7: Comparisons of the experimental and numerical results for walls without an openings

In general, the homogenized approach gave us stiffer behaviour than it was observed in the experiments (Figure 7 and Figure 8). The reasons for this could be either the absence of the adequate experimentally determined material characteristics for our numerical model (Table 1) and/or, lack of parametric testing of the coefficients which determined strain-softening behaviour of the crack infill material. For the wall without an opening (Figure 7) similar behaviour under different level of precompressions, as it was observed both in experiments and with micro-modeling [1] was achieved - the highest shear stiffness was observed at the middle level of precompression.

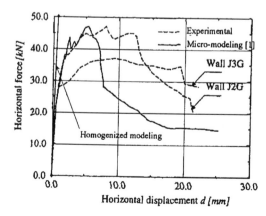

Figure 8: Comparisons of the experimental and numerical results for walls with an openings

7. CONCLUSIONS

As far as different crack patterns for URM shear walls are concerned, the following conclusions can be made:

- It is possible to get acceptable crack pattern for the URM shear walls, problems arise only in smaller masonry assemblies.
- There were no significant differences in the observed crack pattern by using either coarse (120 elements.), or fine meshes (800 elements.).
- Different sandwich materials can significantly influence the crack pattern on the URM shear walls.

Although our 3D techniques already showed good agreement with experimental results under different loading condition of masonry (compressive [4], lateral [5], in plane tensile loading [6]), some further improvements should be made. By introducing elasto-plasticity in the model, we should gain better modelling both for compressive and shear URM walls. By introducing shear band we should be able to get much softer behaviour of the shear walls and in the same time to bridge the gap of modeling smaller masonry assemblies. As monotonous racking load is rather rare in the engineering and experimental practice, it would be recommendable to extend the existing code on cyclic loads as well.

8. ACKNOWLEDGEMENT

The paper presents the results of the research work at the University of Swansea and University of Ljubljana, financed by the European Community through the project COPERNICUS CIPA - CT 94 - 0174 - ATEM. Their support is gratefully acknowledged.

9. REFERENCES

1. LOURENÇO, P.B., "Computational strategies for masonry structures", *Ph.D. Thesis*, Delft University Press, Holland, February, 1996

2. DIALER, C., "Bruch. und Verformungsverhalten von schubbeanspruchten Mauerwerksscheiben", *Mauerwerk-Kalendar 1992*, Verlag Ernest & Sohn, Berlin, 1992

3. DRYSDALE, R.G., HAMID, A.A. & BAKER, L.R., *"Masonry Structures - Behavior and Design"*, Prentice-Hall. Inc. A Simon & Schuster Company Englewood Cliffs, New Jersey, 1994

4. PANDE, G.N., KRALJ, B. & MIDDLETON, J. "Analysis of the compressive strength of masonry given by the equation $f_k = K (f_b)^a (f_m)^b$", *The Structural Engineer*, Vol. 72, No. 1, 1994

5. PANDE, G.N., MIDDLETON, J., LEE, J.S. & KRALJ, B., "Numerical simulation of cracking and collapse of masonry panels subject to lateral loading", *Proc. 10th International Brick and Block Masonry Conference*, Calgary, Canada, 1994

6. LEE, J.S., PANDE, G.N. & KRALJ, B. "A comparative study on the approximate analysis of masonry structures", *Materials and Structures*, to be published

NUMERICAL MODELLING OF MASONRY STRUCTURES REINFORCED OR REPAIRED

D.F. D'Ayala Lecturer
Department of Architecture and Civil Engineering, University of Bath, Claverton Down, Bath BA2 7AY, UK

1. ABSTRACT

An extensive review of the experimental literature on historic masonry structures led to the setting of a database of geometric and mechanical parameters. Data regression's curves have been derived, which enable the definition of the mechanical properties of the masonry when some of the mechanical and geometric parameters of the components are known. The database is linked to a f.e. non-linear analysis procedure which enables to simulate standard test and analysis of whole structures. The paper compares the results obtained with this procedure to the experimental results on repair of spandrel walls subjected to uneven settlement obtained at Bath University in collaboration with Karlsruhe University. Both unreinforced and reinforced spandrels are simulated.

2. INTRODUCTION

In assessing the structural behaviour of historic masonry buildings one of the main problems encountered is the collection of data concerning the mechanical features [1] . The preservation of the integrity of the building and its non homogeneity often work against each other in providing reliable in situ test results. While it is reasonably simple to extract single bricks and perform tests on a number of samples to obtain significant mean values, to extract undisturbed mortar samples is a much more difficult task. Moreover, tests on extracted masonry wallets might be limited by the historic importance, artistic value and structural soundness of the building considered.

One major problem for the analysis of masonry structures with numerical tools is the need of homogenisation due to the high computational costs associated with a direct simulation of the components when the analysis concerns real tridimensional structures. Different homogenising techniques have been proposed by Pande[2], Urbanski[3] etc., all relying on the proper identification of sometimes several constituents' parameters.

Keywords: f.e. non linear analysis, existing masonry, reinforcement

Computer Methods in Structural Masonry – 4, edited by G.N. Pande, J. Middleton and B. Kralj.
Published in 1998 by E & FN Spon, 11 New Fetter Lane, London EC4P 4EE, UK. ISBN: 0 419 23540 X

Thereafter a numerical procedure is defined, directly linked to a database which, given the topological and historic data of the masonry, enables a reliable estimate of the mechanical parameters to be introduced in a standard f.e. non linear program, without performing extensive destructive test. The upgradable database collects the mechanical data from tests on historic masonry of various periods and location available in the literature. A statistical analysis of the data is performed and a number of interpolating regression curves are drawn, which relate the geometric and mechanical parameters of the units, mortar and resulting masonry, or some of their non-dimentional ratios The values of the masonry parameters obtained from the regression curves are used as input in the non linear structural analysis procedure.

3. THE DATABASE AND REGRESSION ANALYSIS

The database is built in Microsoft Access for Windows95©, choosing among papers providing complete information about masonry, units and mortar in terms of geometric and physic characteristics, modality of tests, number of samples, and mechanical features. Papers dealing with either historic or new masonry are considered, the distinction between the two classes being related to the aim of the study. Thus, for some sets of data, the mortar or masonry samples may not necessarily be old, but reproduce historic masonry in laboratory conditions. For historic are intended masonry structures of 50 or more years of age built with non engineered techniques. The age and location of the samples are entered explicitly in the database allowing for comparisons of results of tests carried out in the same geographical area (similar component materials and craft techniques) or in different area but referring to the same age of construction.

The sources reviewed include World Brick Block Masonry Conferences and Earthquake Engineering Conferences, International Journals specialised in the field, International Conferences on Conservation and a number of Regional Conferences and specific reports. Of the approximately 700 papers reviewed to date only about one tenth had complete set of data suitable for the database. Only fired clay bricks are considered as masonry units. While there is scope for a similar database for stone masonry, this was not included in the present study. It is interesting to note that rather rarely papers relating to specific buildings provide complete data on the constitutive materials.

The tests considered are, uni-axial or tri-axial compression test, rupture test, shear-bond test, direct tension test, on the units, the mortar and the masonry, performed accordingly to the relevant ISO or CEN recommendation. The majority of the data are used to establish correlation between parameters of the same material, i.e. mortar or unit, while the compression test presents correlated data sufficient to allow a statistical analysis of the influence of the components' parameters on the masonry. The level of occurrence of the compressive strength is 80% for the mortar , 97% for the brick, 71% for the resulting masonry. The occurrence for the values of elastic module is 29% 59% 42% respectively.

A first group of regressions relates the compressive strength of the masonry to the compressive strength of the units and the mortar and to the heights of the units, of the bed joints, of the masonry sample, and their ratios. A second group relates the elastic modulus E of the masonry to its compressive strength; to the elastic modulus of the units or of the mortar; and to the heights of the units or the bed joints, or their ratios.

The first set of regressions shows a greater influence on the compressive strength of the masonry by the compressive strength of the mortar bed joint than by the strength of the unit. The similar distribution of the data and shape of the resulting curves suggests to combine the influence of the two parameters in a formula as follow (Fig. 1):

$$f_{cw} = 0.538 f_{cm} + 0.241 f_{cb} \qquad (1)$$

The high sensibility of the masonry strength to the height of the mortar bed joint was also highlighted and a very good correlation (r = 0.78) was obtained relating the masonry strength normalised with respect of the height of the sample, to the sum of the strengths of the two component materials, each normalised to its height (Fig. 2):

$$\frac{f_{cw}}{h_w} = 0.0216 \cdot \left(\frac{f_{cb}}{h_b} + \frac{f_{cm}}{h_m} \right) \qquad (2)$$

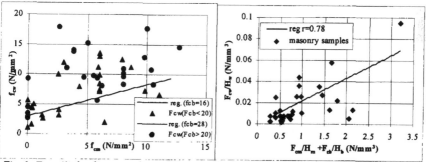

Fig. 1: Eqn. (1) for discrete values of F_{cb}.   Fig.2: Eqn. (2) regression coefficient r=0.78

The regressions define the direct proportionality between the elastic modulus of the unit, mortar, masonry and their related compressive strength (Fig. 3). The distribution of points over a wide range and the high values of the correlation coefficient in the three cases (0.935, 0.655, 0.937, for mortar, brick, masonry) make these curves highly reliable. Such a good correlation enables to propose, in absence of direct experimental tests, the use of those curves to define the elastic modulus when the strength of the material is known. It is worth noting that the slope coefficient for the masonry curve is c=417, more than half the value suggested by the Eurocode6[4] for the short term secant modulus in new constructions (c= 1000). However considering the maximum effect associated with creep, the long term value of the slope coefficient would reduce to c=660, which is still 1.5 times greater than the one obtained with the regression.

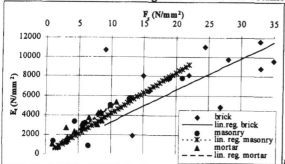

Fig. 3 Regression of E_i (f_c)

4. FINITE ELEMENT PROCEDURE

To date, finite element analysis of masonry structures within the typical design office environment has been hampered by the lack of simple yet reliable constitutive models to define appropriate mechanical properties for the material. This is particularly true of existing masonry buildings, because the requirement to assess their material properties entails extensive destructive testing. The commonly used method of circumventing this problem has been to extrapolate the constitutive laws of concrete to masonry. This however completely overlooks the orderly composed nature of masonry and the fact that the weakness of the material is concentrated at specific locations, i.e. the bed joints.

The strength of this approach[5] lies in the parallel use of the database and a preliminary finite element analysis to generate equivalent homogenous material properties for the masonry based on those of the constituent bricks and mortar. The preliminary finite element analysis simulates numerically the standard compression and bond test by modelling in two or three dimensions the individual bricks and mortar layers separately with their own failure criteria and constitutive law. These are derived from the database when the location and age of the masonry structure and composition of the mortar are known, together with the geometry of the units, thickness of the mortar and their compressive strength. The numerical simulation allows a wide variety of stress conditions to be considered, enabling the definition of a proper constitutive law and failure domain for the composite material. The values of the relevant parameters obtained in this way are verified by comparison with the database regression curves.

Once the masonry properties are in such way defined, a coarser mesh can be used to model larger portions of the structure up to the entire building. Each plate-shell element thus simulates the masonry as an homogeneous ortotropic material, using a smeared crack model and a modified Mohr-Coulomb failure criterion for the post-elastic behaviour. The constitutive law is approximated by a tri-linear curve. The algorithm is made up of
- a pre-processor which prepares the f.e. model, stores the material data, and defines the elastic limit of the analysis under the given load condition. It also define the number of increments and an initial incremental step for the load. On the basis of the material properties given, constitutive law and failure domains for each material are also defined.
- the main processor which solves the f.e. problem for a given incremental step is provided by the Algor Supersap© commercial package.
- the post processor which checks the state of stress at the centre of each element with reference to the failure domain, accordingly modifying the values of the stiffness parameters and operates the redistribution of the internal state of stress to the adjacent elements by equivalent nodal forces, when an element fails.

A four nodes or 3 nodes plate-shell element with 5 d.o.f. per node is used, allowing for the simulation of complex three-dimensional problems. Two failure criterion are used to allow for the non-associative nature of the materials. The first failure criterion, of the Rankine type, is defined in the principal stress plane and is used to define the state of stress internal to the single element of mortar or brick (Fig. 4). A tension cut off is included to take into account the reduced strength under bi axial tension. The bond between mortar and unit and the shear behaviour of the units are defined by the Morh-Coulomb type criterion shown in Fig. 5 with respect to the stress σ normal to the plane of the joint. The state of stress on the edges of the elements are verified with respect to this failure domain to define shear failure of the joint and direction of slip. The same failure criteria are used after homogenisation for the macromodelling, using the parameters for the masonry obtained by the simulation of the compression and shear bond test. A linear elastic tensile behaviour is assumed until the cracking surface is reached. The material becomes then orthotropic with characteristics defined for the direction parallel and normal to the crack. No variation of direction of the crack has been considered within the single element of mortar or brick. The reinforcement is simulated by bar elements connected at each node of the plate shell elements forming the mortar bed joints. During the elastic phase the bond with the mortar is considered full. The behaviour after cracking is simulated by a uni-axial elasto plastic model applied in the bar axial direction only, the stress released at the crack being applied at the common node on the bar.

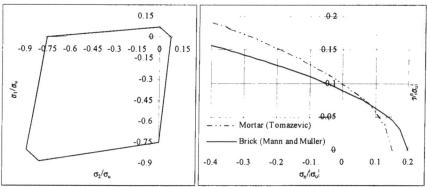

Fig. 4: Rankine type failure domain   *Fig.5 Mohr Coulomb type failure domain*

5. APPLICATION TO EXPERIMENTAL WORK

The finite element procedure described above has been applied to simulate the tests conducted by Cook[6] in collaboration with Karlsruhe University, on 14 specimen of full scale spandrel walls, which modelled the process of 'crack damage-repair-continued movement up to further damage'. The criterion of success was judged by the degree to which the damaged element, after a repair consisting of reinforcement of two bed joints, could tolerate further bending movement without re-aggravation of the original damage.

These tests were confined to spandrel walls of simple leaf brickwork, while the connection to piers was simulated by equivalent pre-stressing. Two types of masonry obtained with different geometry of unit and different mixture of mortars were tested. The overall geometry of the spandrel walls was however kept constant and in particular the *h/l* ratio . According to the regression analysis performed on the database the height of the bed joint has great influence on the strength of the masonry, and it was kept constant, while varying the height of the unit.

The information on the material parameters reduced to the compressive strength of mortar measured on cubes and brick for both cases; one partial compression test per set, performed directly on the spandrel wall, and aimed at defining the masonry initial tangential modulus. This value was kept for reference but not introduced initially as a datum in the Fe. analysis. Also a number of five bricks stack tests had been performed, to provide with a value of strength for the masonry. There was no indication of bond or tensile strength for any of the materials.

The finite element model simulated half of the spandrel tested in practice due to theoretical condition of symmetry. It can be however noticed from the crack pattern of the specimen that such condition were not met in reality(Fig. 6). Lacking however information on the causes and quantification of the occurred asymmetry it was not possible to include it in the numerical model.

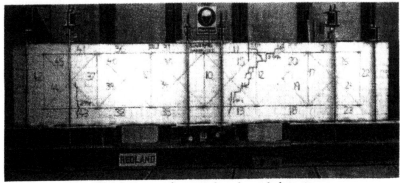

Fig. 6 Crack pattern of a unreinforced spandrel specimen

On the basis of the data available the initial tangent elastic modulus for mortar and bricks was derived from the database, together with their tensile and shear strengths. The finite element simulation of five brick stacks and masonry wallets based on those data showed a good agreement with the available experimental data. Therefore the chosen values were also used to simulate the shear bond test on triplets. Although there was no relevant experimental data these were equally carried out to compare the two resultant failure domains for the two types of masonry. As shown in Fig. 8 the Karlsruhe one is smaller in agreement with the lower strength of the components.

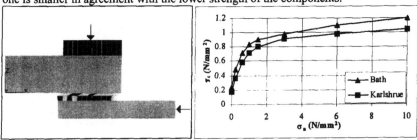

Fig. 7 Failure mechanism for triplet   *Fig.8: Failure domain as obtained by triplet test*

A different way to obtain the initial tangent modulus of the two components is to use the results of the spandrel test, applying some rather simple theoretical model and the standard homogenisation procedure in reverse order. Using the equation for deep beam with fix ends and concentrated load (Timoshenko), the modulus at cracking could be sought out, when a given value for the ratio E/G is established (chosen as 2.5 according to Hendry[7]). This yields the values collected in Tab. 1, for the two sets of tests.

Test set	Mortar			Brick			Masonry		
	$E_{//}$	$E_{reg.}$	$E_{eq. 4}$	$E_{//}$	$E_{half\,brick}$	$E_{reg.}$	E_{test}	$E_{reg.}$	$E_{Timoshenko}$
Bath	6290	5000	4950	15697		10000	8300	6000	7079
Karlsruhe		2000	2754		5950	6500	5283	5000	5176

From these values, using the homogenisation equation as follows:

$$E_w = \frac{E_j(h_b + h_j)(l_j E_j + l_b E_b)}{h_j(l_j E_j + l_b E_b) + E_j h_b(l_j + l_b)} \cdots \tag{4}$$

the value of tangent modulus of mortar could be calculated for a given estimate of the brick tangent modulus (see tab. 1).

The experimental tests, for which the only variable parameter was the number of brick layers and the geometric ratio *h/l*, showed quite scattered results, not only in terms of first crack load but also in term of stiffness. The parameter that seemed to maintain fairly constant in most of the tests was the cracking bending strain. Moreover the two sets of tests did not show considerably different crack loads or initial stiffness for similar ratios of *h/l*, notwithstanding the consistent difference in the materials used.

The author believes that a main reason for this is the relevant part that could have been played by friction between the upper plate and the specimen, effect which will initially show as an increase in stiffness and in cracking load . This effect, although has not been measured, given the rather high precompression applied at the spandrel's ends, might have been of the same order of magnitude in terms of stiffness than the one produce by the materials property. This would also explain the rather sudden drop in load capacity following the initial cracking.

Two different strategies were followed to numerically simulate the test. One used the value of Et for mortar and bricks obtained as mentioned above. This would yield lower values of first cracking load than the one obtained by the experimental test. The difference in stiffness would be converted in an equivalent spring system applied horizontally at the level of the plate with a limiting force equal to the one provided by friction at the plate, having assumed a friction angle of 30°. In this way the values obtained compared very well with the test results, for both test as shown in Figs. 9 and 10. However in absence of more detailed experimental evidence would be wrong to assume that this was the only cause for discrepancy of results.

Fig. 9 Crack pattern Bath numerical model   Fig. 10 Crack pattern Karlsruhe numerical model

The introduction of the reinforcement was analysed by simulating one of the spandrel specimen in which reinforcement was introduced at an initial state, before any loading occurred. As expected this model showed the same initial elastic behaviour as the unreinforced one until first cracking occurred. At this point the reinforcement appeared to be quite effective in increasing the load capacity for an increasing deformation up to 30% more than the initial elastic limit. The experimental and numerical results for all test are shown in Fig. 11, and the final crack pattern for the reinforced model in Fig. 12.

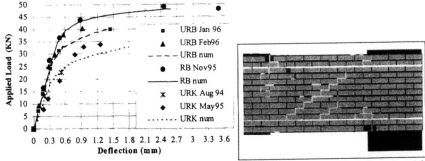

Fig. 11 Comparison of numerical and test results   Fig. 12 Crack pattern of reinforced model

Finally an homogenised mesh was analysed on the basis of the parameters derived from the smaller test for the Bath case. The reduction in computational effort is considerable having reduced to 1/3 the D.o.E. and even further the bandwidth. The results both in terms of load capacity/deformation and in crack pattern reproduce well the experimental results. However is not possible to localise the single failure to the bed joint or the brick.

6. CONCLUSIONS

The present study shows that, based on a range of experimental results obtained from different authors, it is possible to derive a number of regression curves and evaluate the strength and elastic modulus of the masonry from the ones of the components. Although several parameters of the experimental work simulated and relevant to the f.e. analysis had not been explicitly measured, the use of the regressions curves and homogenisation techniques allowed to obtain satisfactory results with the numerical simulation. This was able to clarify some aspects of the experimental set that can be improved. This technique will be used to extended the analysis to real structures interested by settlements .

7. REFERENCES

[1] Page A.W., Influence of material properties on the behaviour of brick masonry shear wall, *Proceedings of the VIII IBMAC*, Dublin 1988

[2] Lee, J.S., Pande G.N., Middleton J., Kralj B., Numerical Modelling of Brick masonry panels subjected to lateral loading. *Computers and Structures*, vol 61 n. 4, pp.735-745, 1996

[3] Urbanski A., Szarlinski J., Kordecki Z., Finite element modelling of masonry walls and columns by the homogenisation approach, *Computer Methods in structural masonry -3* 1995

[4] European Committee for Standardisation, ENV 1996-1-1, *Eurocode 6: Design of Masonry structures - Part 1-1*, Brussels 1995.

[5] D'Ayala D., Carriero A., A Numerical tool to simulate the non-linear behaviour of masonry structures, *Proceedings CMEM 95*, Capri Italy.

[6] Cook D.A., Ring S., Fichtner W., The effective use of masonry reinforcement for crack repair, *Proceedings Fourth International Masonry Conference*, vol. 2 pp.442-450, 1995

[7] A.W. Hendry (edit by), *Reinforced and Prestressed Masonry*, Longman 1991

MEASUREMENT OF TENSILE STRAIN SOFTENING AND ITS INFLUENCE ON BOND TENSILE TEST RESULTS

P. Jukes Visiting Research Fellow, **M.C. Bouzeghoub** Visiting Research Fellow and **J.R. Riddington** Senior Lecturer
School of Engineering, University of Sussex, Brighton BN1 9QT, UK

1. ABSTRACT

The method adopted to modify an existing masonry finite element program so that it could be used to investigate the effects of tensile strain softening, should it occur to any significant degree, is described. Results are presented from analyses using the modified program, of test specimens that have been used to measure tensile strain softening data, direct tensile bond strength and flexural tensile bond strength. It is concluded that it is difficult to obtain reliable strain softening data, that the effect of any strain softening on the ultimate strength of direct tensile bond strength samples will not be great, but that any strain softening could significantly increase the strength of flexural bond strength test specimens, which would lead to an overestimation in the value of bond strength calculated.

2. INTRODUCTION

Strain softening is a well known and established phenomenon that has normally been associated with concrete. Recently[1-3] it has been suggested that strain softening can also significantly influence the behaviour of masonry joints and that masonry FE programs should therefore allow for it. This paper reports the results from some preliminary work related to tensile strain softening that has been undertaken at the University of Sussex. As well as describing the method used to incorporate tensile strain softening into an established finite element program, the paper also considers the problems associated with obtaining the required data and the effects strain softening may have on joints subjected to different distributions of tensile stress.

Keywords : Strain softening, Tension, Masonry, Brickwork, Finite elements

Computer Methods in Structural Masonry – 4. edited by G.N. Pande, J. Middleton and B. Kralj.
Published in 1998 by E & FN Spon, 11 New Fetter Lane, London EC4P 4EE, UK. ISBN: 0 419 23540 X

3. TENSILE STRAIN SOFTENING

Tensile strain softening assumes that when a mortar joint fails in tension at a point, the failure is not brittle but plastic, with the tensile stress reducing progressively as the crack opens. Fig. 1 shows the form of relationship that has been assumed by Pluijm[1,3] and Rots and Lourenco[2], where f_{tb} is the tensile bond strength of the joint. The area under the curve defines the energy required to fully form a unit area of crack and this is known as the fracture energy G_f.

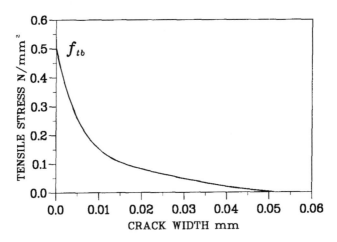

Fig. 1   Tensile strain softening relationship when $G_f = 5$ N/m and $f_{tb} = 0.5$ N/mm^2

The curve is described by Eq.1, which was developed by Hordijk and Reinhardt[4] for plain concrete.

$$\frac{\sigma}{f_{tb}} = \{1 + (c_1 \frac{w}{w_c})^3\}.e^{-c_2 \frac{w}{w_c}} - \frac{w}{w_c}.(1 + c_1^3).e^{-c_2} \tag{1}$$

where σ = tensile stress, $c_1 = 3$, $c_2 = 6.93$, w = crack width and

$$w_c = 5.14 \frac{G_f}{f_{tb}} \tag{2}$$

For the current work a tensile strain softening relationship of the form shown in Fig. 1 was assumed, with different fracture energy values G_f being used. Tensile strain softening data has to be measured experimentally and this is not at all easy. This topic will be considered later.

4. FINITE ELEMENT PROGRAM

Tensile strain softening was implemented into a 2-D finite element program that has been described elsewhere[5]. Although this program was developed specifically for use in studying triplet test specimens, it can also be used for analysing other forms of structure, as will be seen later. The program is based on standard four node rectangular elements, with the bricks or blocks and mortar being represented by separate groups of elements. At one side of the mortar joint, the connection between the two materials is made using link elements of zero length, that can be modified or removed according to the type of failure developing at a particular point. Prior to being modified the program was capable of simulating bond tensile failure, mortar tensile failure and shear failure along the interface, as well as mortar non-linearity. For the current work the program was changed so that it could also accommodate tensile strain softening.

The method adopted to simulate tensile strain softening was to replace any interface link where the direct stress across the joint exceeds the local bond tensile strength f_{tb}, by equal and opposite residual forces applied directly to the two nodes originally connected by the link. The value of the residual forces applied to the node pair is then adjusted using an iterative process, until the force value and the gap that develops between the node pair, complies with a force-displacement relationship that represents the strain softening. This force-displacement relationship is obtained by multiplying the stress in the Fig. 1 stress-displacement relationship by the interface area connected by the link being removed. It was found to be satisfactory to set the residual force value for the first iteration after the removal of a link to the maximum force given by the force-displacement relationship.

5. MEASUREMENT OF TENSILE STRAIN SOFTENING DATA

As stated previously, it is not easy to measure tensile strain softening data for masonry joints. The reason for this is that the displacements that occur as an interface crack is formed are extremely small, which makes it both difficult to apply the loading in a controlled enough manner, as well as difficult to measure the displacements. The only tests to obtain tensile strain data that have been conducted that are known to the authors, were undertaken by Pluijm[1] using a test arrangement of the form shown in Fig. 2. He found it very difficult to obtain points in the immediate post peak, steeply descending section of the stress-displacement curve and he found fracture energy values that varied considerably from one sample to the next. Strain softening data cannot therefore ever be expected to be obtained routinely, as might be expected of other masonry property data. Hence there could be a significant problem should it be shown that accurate strain softening data is needed to obtain accurate analysis results, since every brick/mortar and

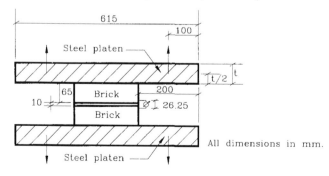

Fig. 2 Loading arrangement used to measure tensile strain softening data

block/mortar combination could be expected to have a different strain softening curve. Adjusting the curve used in a finite element analysis until the analysis results obtained fit test results cannot be considered to be a valid method for determining a material property, since it might simply serve to hide other problems with the analysis method, rather than to give the true material property.

The form of loading arrangement used by Pluijm, shown in Fig. 2, was analysed using the program in order to determine what influence the stiffness of the loading platens might have on the strain softening curve produced by this form of test. UK dimension bricks were used in the analyses (215×102.5×65 mm), but the stiffness and strength properties were values given by Pluijm[1]. These values are shown in Table 1. The strain softening curve was taken as that given by Eq. 1 & 2 when the fracture energy G_f is 5 N/m. The curve is shown in Fig.1. The bricks were assumed to be rigidly fixed to the platens and the element grid used in the analyses for the bricks and mortar was that shown in Fig. 3. It should be noted that the loads shown in Fig. 3 relate to the analyses of direct tension bond strength test specimens, that are considered in Section 6. Convergence tests using grids of different density indicated that the grid density shown in Fig. 3 was adequate. The platen width was taken to be the same as the brick width.

Brick elastic modulus N/mm^2	18000
Brick Poisson's ratio	0.2
Mortar elastic modulus N/mm^2	5000
Mortar Poisson's ratio	0.2
Steel elastic modulus N/mm^2	210000
Steel Poisson's ratio	0.3
Joint bond tensile strength N/mm^2	0.5

Table 1 Stiffness and strength properties used in the analyses

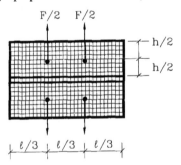

Fig.3 Direct tension bond strength test loading arrangement showing finite element grid

The results from the analyses are shown in Fig. 4, where the deflections across the joint at its outside edge over a gauge length of 26.25 mm, are plotted against the average applied tensile stress applied to the specimen as the applied deflection is increased, for a range of platen thicknesses. It should be noted that Pluijm's strain softening curve was obtained by taking measurements across the joint at its edge over this gauge length. If the loading arrangement produces a completely uniform stress distribution in the specimen, then the joint deflection will exactly match the strain softening curve assumed in the analysis, after allowance is made for the elastic deformations of the brick and mortar within the gauge length. In the case of the 150 mm and 200 mm thick platens there is a very close match, but this is far from true when the platen thickness is below 150 mm, as shown by the results obtained when the platen thicknesses were 100 mm and 50 mm. In these latter cases the stress distributions along the joint are far from uniform. These results indicate that the type loading arrangement used by Pluijm will only be capable of producing accurate measurements for the strain softening curve if the platens are extremely stiff. It is not known what stiffness platens were used by Pluijm.

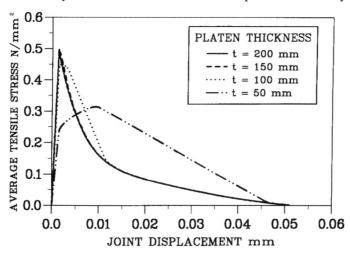

Fig. 4 Influence of platen thickness on joint displacement

Analyses were also undertaken using a range of fracture energy G_f values for the case when there is a very stiff, 200 mm thick, platen and the results are shown in Fig 5.

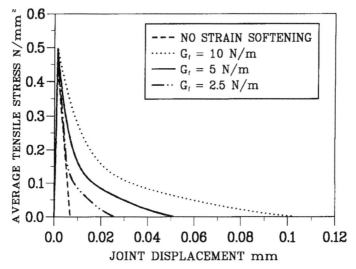

Fig. 5 Influence of fracture energy value on joint displacement

Although the stress-displacement results are significantly altered by the degree of strain softening, the ultimate load values obtained were little changed, with the ultimate load obtained with $G_f = 10$ N/m being only 1.5 % above the value when there was no strain softening ($G_f = 0$). If the loading system had produced a completely uniform stress distribution along the joint, then there would have been no difference in the ultimate strength values. If in addition and there was no strain softening and no bending of the loading platens, then the descending section of the stress-displacement curve would have been vertical. It should be noted that when there is no strain softening, the inclination of the descending line reduces as the flexibility of the platens increases and the stress distribution becomes less uniform. As a consequence, tests conducted with flexible platens will tend to produce results that will suggest strain softening, even if very little exists. Such results will in any case overestimate the amount of strain softening, even if it does exist to a significant extent.

6. EFFECT OF STRAIN STIFFENING ON BOND STRENGTH SPECIMENS

The possible effects of tensile strain softening on the results given by specimens that have been used at the University of Sussex to measure direct and flexural tensile bond strength were investigated using the program. Direct bond strength has been measured using couplets where the load is applied via bolts that pass through holes drilled through the bricks. Fig. 3 shows the form of specimen tested, the loading application points and the element grid used, which was the same as that used previously. When using this form of test, results have been factored so as to account for the non-linear stress distribution that develops along the joint as a result of the loading arrangement used, based on the assumption that a brittle form of failure occurs.

Flexural bond strength has been measured at the University of Sussex by undertaking 4 point bending tests on 9 brick high stacks. Fig. 6 shows the form of specimen tested together with the loading and support points. As is normal practice with this form of test, the bond strength has been calculated from the test results using simple bending theory, assuming a brittle form of failure. The element grid used for these analyses had 20 elements along the length of each brick, 4 over their height and 1 across the mortar joint. Convergence tests showed this density of grid to be adequate. Due to symmetry only one half of the beam was analysed.

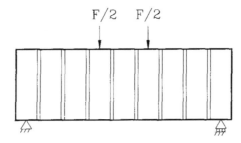

Fig.6 Flexural bond strength test loading arrangement

For the analyses of both types of specimen, it was assumed that they were formed from UK dimension bricks and that the brick and mortar properties were the same as those used in the analyses described in the previous section, as shown in Table 1. For each specimen form, analyses were undertaken with different degrees of strain softening being applied, as defined by the fracture energy G_f value used, as well as one analysis being undertaken without any strain softening ($G_f = 0$).

The results obtained from the direct bond test analyses are shown in Table 2. In these analyses the local bond strength was taken as 0.5 N/mm^2, and this resulted in failure being initiated at a value of average applied tensile stress of 0.282 N/mm^2. This is a consequence of the non-uniform stress distribution that develops in the joint when this loading arrangement is used. At the University of Sussex this non-uniformity has been allowed for by multiplying the average applied tensile stress at failure by a factor. This factor varies slightly with the ratio of the brick to mortar elastic moduli and in the present case a factor of 1.8 would be appropriate. It can be seen that, as might be expected, the average stress at final failure increases with the level of strain softening used in the analyses. The results indicate that if a significant amount of strain softening does occur, it could affect the results given by this type of test by around 10%.

G_f N/m	Av. stress at failure N/mm^2	% increase in ultimate load
0	0.282	0
1	0.283	0.4
1.25	0.285	1.1
2.5	0.297	5.5
3.33	0.302	7.1
5	0.307	8.9
10	0.315	11.7

Table 2 Influence of strain softening on results given by direct tension test

The results from the flexural test analyses are shown in Table 3. For the table the applied load at failure has been converted to the maximum stress that would have existed in the beam if it had behaved elastically and simple bending theory could be applied. In fact the maximum stress in the beam at failure in all cases is 0.5 N/mm^2, the assumed bond tensile strength f_{tb}. The results indicate that if strain softening does exist at a significant level, it could have a very considerable effect on strength if the stress distribution is highly non-uniform, as happens in a flexural test. The results also indicate that strain softening could result in the bond tensile strength being overestimated by as much as 40 %.

G_f N/m	Bending theory stress N/mm^2	% increase in ultimate load
0	0.501	0
1	0.507	1.2
1.25	0.513	2.4
2.5	0.547	9.2
3.33	0.576	14.9
5	0.615	22.8
10	0.701	39.9

Table 3 Influence of strain softening on results given by flexural tests

6. CONCLUSIONS

The main conclusions that can be drawn from this work are :

1. It is very difficult to obtain reliable strain softening data for masonry joints since the displacements involved as the crack opens are very small and the apparatus used needs to be extremely stiff as well as it producing a uniform stress distribution in the joint. Any non-uniformity in the stress distribution will tend to produce results that will suggest strain softening even if very little exists and which will in any case tend to overestimate the amount of strain softening.
2. When the stress distribution along a joint is not too uneven, as exists in a direct tension test, the influence of any strain softening on ultimate load will not be great.
3. When the stress distribution along a joint is highly non-uniform, as exists in a flexural test, the influence of any strain softening could be very significant and it could lead to an overestimation in the value of bond strength calculated.

7. REFERENCES

1. Pluijm R. Van Der, "Material properties of masonry and its components under tension and shear", 6th Canadian Masonry Symposium, University of Saskatoon, Canada, 1992, pp 675-686.
2. Rots J. G. and Lourenco P. B., "Fracture simulations of masonry using non-linear interface elements", 6th North American Masonry Conference, Philadelphia, USA, 1993, pp 983-993.
3. Pluijm R. Van Der, "Numerical evaluation of bond tests on masonry", Masonry International, Vol 9, No 1, 1995, pp 16-24.
4. Hordijk D. A. and Reinhardt H. W., "Testing and modelling of plain concrete under mode I loading", In Micromechanics of Failure of Quasi-Brittle Materials, Elsevier Applied Science, London, 1990, pp 559-568.
5. Jukes P. and Riddington J. R., "Ultimate strength prediction of masonry triplet samples", In Computer Methods in Structural Masonry - 3, Books & Journals International Ltd., 1995, pp 97-106.

INFLUENCE OF MODIFIED MORTARS ON MECHANICAL PROPERTIES OF MASONRY

R. Žarnić Assistant Professor, **V. Bokan-Bosiljkov** Teaching Assistant,
V. Bosiljkov PhD Candidate and **B. Dujič** Research Assistant
*Faculty of Civil Engineering and Geodesy, University of Ljubljana,
Jamova 2, SI-1000 Ljubljana, Slovenia*

ABSTRACT

A set of tests was accomplished in order to get better insight into the basic mechanical properties of masonry made of normal solid clay bricks and different type of mortars. The basic mortars were lime mortar, lime-cement mortar and cement mortar. For the study of grade effect differently graded sand was used in mortars. Polypropylene fibres, glass fibre meshes and several types of cement additives modified the basic mortars. Three types of tests were performed. Masonry wallettes were tested both in compression and diagonally in tension. Horizontal mortar joints were tested by bond wrench test. A significant influence of sand grade on bond strength was observed. It was found that the most effective way to increase shear strength is to add glass fibre mesh to mortar joints made of fine graded sand.

1. INTRODUCTION

The development of structural masonry is strongly supported by efficient computational models that enable better understanding of masonry behaviour. However, the essential part of the development is experimental studies of masonry. They are especially needed when new or modified materials are introduced. There are many demands generated from practice for the improvement and development of diverse masonry. Some of them tend towards the development of mortar that would enable easy recycling of bricks. Others tend towards the improved quality of brick to mortar bond and increased shear and tensile strength as well as ductility of masonry. The decision on the masonry type should be governed by the expected loading in the life cycle of the structure. A special attention should be paid to earthquake loading.

The variety of additives that are available at the market generates ideas how to improve mortars that are traditionally composed of mineral binders and sand. The lack of data on the influence of modifications of mortars may lead to misuse of additives.

Computer Methods in Structural Masonry – 4, edited by G.N. Pande, J. Middleton and B. Kralj.
Published in 1998 by E & FN Spon, 11 New Fetter Lane, London EC4P 4EE, UK. ISBN: 0 419 23540 X

It is often the case when polymer materials are introduced into inorganic cement binder in order to improve adhesion and tensile strength of cement mortar as well as its resistance to chemical and physical attack. Short polymer fibres are sometime used to improve mortar toughness and flexibility, but their prime role is to reduce cracks induced by the process of mortar hardening. Another approach to improve masonry is to introduce non-metal reinforcement embedded in mortar.

An extensive program of testing the influence of different mortar compositions has been started in the mainframe of an international collaborative project. In the current phase of investigations the influence of differently graded sand, polymer suspensions, polypropylene fibres and glass-fibre meshes on masonry are studied. The obtained results of the first series of experimental investigations are briefly presented in the sequel.

2. EXPERIMENTAL PROGRAMME

2.1.   The range of tests

The tests were divided into three main phases. In the first phase, the preliminary compressive tests of masonry wallettes, were carried out according to the European preliminary standard prEN 1052-1. The test wallettes were made of three different types of modified coarse sand mortar and of the basic, unmodified mortar mix. The basic mix was composed of cement, lime and sand in volume proportion 1:1:6. The modification of the basic mortar was obtained by adding polypropylene fibres, superplastisizer, polymeric dispersion and/or micro-silica. The preliminary tests are not subject of this paper because they were elaborated elsewhere [1], [2].

In the second phase of the test programme the horizontal mortar joints were tested by means of bond wrench test, which is standardised in USA (ASTM C1072-86) and Australia (AS 3700-1988). The test specimens in the form of mortar prisms were made of six types of mortar. The mortars were composed in volume proportions of the following materials:

- cement : sand in proportion of 1:3,
- cement : sand in proportion of 1:3, modified with polymeric binder (5% of polymeric particles with respect to the weight of cement),
- cement : lime : sand in proportion of 1:1:6,
- cement : lime : sand in proportion of 1:1:6 with addition of 3kg of polypropylene fibres to the 1m^3 of mortar,
- cement : lime : sand in proportion of 1:1:6 reinforced by glass-fibres mesh, and
- lime : sand mortar in proportion of 1:3.

Two grades of sand were used to prepare the specimens. The first type of sand was used as purchased. Its nominal maximum grain was 4 mm, but it was found that 14% of the sand had larger grains. Therefore, it was decided to examine the influence of oversized grains on mortar to brick bond. The second batch of mortars was prepared by sand sieved through a 4 mm sieve. The results of sieve analysis of both batches of sand are shown in Figure 1. In the second phase of testing also the pre-wetting of bricks prior to bricklaying was included as a variable. The tests were performed after specimens were aged for 120 and 30 days for coarser and finer sand mortar, respectively.

In the third phase of the test programme the masonry wallettes designed according to prEN 1052-1 were tested diagonally to obtain the data on their tensile strength. The test specimens were made of six types of coarser sand mortar composed in the same way as listed above. Specimens were tested at their age of 120 days.

2.2. Composition of specimens

The bond wrench test specimens were composed in form of three brick high masonry prisms. The diagonal tensile test specimens were made in form of 525 mm high and 525 mm long wallettes with a thickness of 120 mm (Figure 2). The thickness of bed joints was between 10 and 12 mm. The bricks used for specimens had standard dimensions according to the former Yugoslav codes: 250 mm in length, 120 mm in width and 65 mm in height. They were extruded, solid without any holes and taken from the same batch. All bricks were pre-wetted before bricklaying. The first pre-wetting procedure, used in the second as well as in the third phase of testing, was a combination of water immersion for 60 minutes and storage under PVC foil for 24 hours. The second pre-wetting procedure, used only in the second phase of testing, was water immersion of bricks for 30 minutes and their drying in the laboratory ambient conditions for 30 minutes.

The mortar was batched by weight to assure consistent mix properties. After the water content had been adjusted in order to obtain the desired mortar consistency measured by flow table (110±10 mm for coarser sand mortar and 150±5 mm for finer sand mortar), no re-tempering of the mortar was permitted. From each mortar batch at least 3 mortar prisms (4x4x16 cm^3) were prepared for flexural and compressive tests.

The masonry prisms and the wallettes were for the first 24 hours cured under PVC foil and after that exposed to laboratory ambient conditions.

2.3. Properties of mortar materials and mortars

Cementitious materials. The Portland cement PC 30dz 45S according to the former Yugoslav codes and hydrated lime in the dry form was used in all types of mortars.

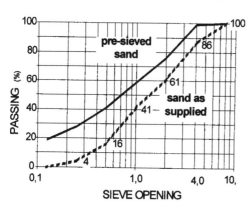

Figure 1: Gradation of mortar sands

Sand. Sand with maximum particle size equal to 4 mm was supplied by the sand processing plant and has been directly used for coarser sand mortars. Relatively low strengths were obtained both during bond wrench and diagonal tensile tests. Therefore the effect of sand gradation was examined by additional bond wrench tests. The sand used in these tests was then pre-sieved through a 4mm sieve. Figure 1 shows the particles size distribution of the sand before and after the pre-sieving.

Fibres, mesh and polymer. The mortar was micro reinforced by commercial polypropylene fibres "Krenit" with a fibre length of 12 mm, cross-section size 35x250-600 µm, ultimate stress 340-500 MPa, elastic modulus 8.5-12.5 GPa and ultimate strain of 8-10%. Commercial glass-fibres mesh with square-shaped openings 5x5 mm was used as mortar bed joints macro-reinforcement. In case of the cement:sand mortar the water polymer suspension with commercial name "Elastosil 34" was used as modifier of the cement binder.

Properties of mortars. The average compressive and flexural strengths of mortars used for specimens are shown in Table 1. In case of finer sand mortars both compressive and flexural tests were carried out, and for coarser sand mortars only flexural strengths were obtained. Apart from the cement:sand mortar, the flexural strengths of all other coarser sand mortars obtained at 120 days were between 40 and 80% higher than strengths measured in the cases of 30 day old finer sand mortar specimens. The polypropylene fibres added to cement:lime:sand mortar did not affect significantly neither the compressive nor the flexural strength of mortar. On the other hand, the modification of the cement:sand mortar with the polymer reduces the 30 day compressive strength of the mortar by about 22% and increases its 120 day flexural strength by about 63%.

Table 1: The average compressive and flexural strengths of masonry mortars

MORTAR MIX	C:S = 1:3	C:S = 1:3 + polymer	C:L:S = 1:1:6	C:L:S = 1:1:6 + PP fibres	C:L:S = 1:1:6 + mesh	L:S = 1:3
Compressive strength at 30 days [MPa]	26,97	22,13	9,43	8,47	10,98	0,29
Flexural strength at 120 days [MPa]	6,26	10,19	3,95	-	4,83	0,56
Flexural strength at 30 days [MPa]	7,20	6,87	2,15	2,22	3,41	-

2.4.   Testing of specimens

The bond wrench tests were carried out with two different lengths of the level arm. At the lime:sand mortar the length of the level arm was 540 mm and at the other five mortar types it was 1080 mm. The eccentric axial load was increased uniformly with the velocity of 17 N/min. For each mortar type five masonry prisms, 10 bed joints in total, were tested.

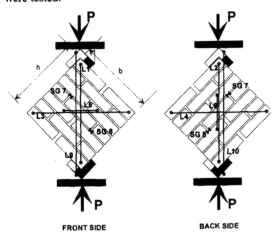

FRONT SIDE   BACK SIDE

Figure 2: Wallette instrumentation

Wallettes were diagonally tested in the servo-hydraulic testing machine Instron with the actuator capacity of 250 kN. Each specimen was loaded by conducting the actuator displacement with a velocity of 8mm/hour down to the negligible post-peak load-bearing capacity of the wallette. The wallettes were instrumented with 6 LVDT's for the measurement of local and global deformations in two perpendicular directions (Figure 2). The base length of two shorter LVDT's was 200 mm, and the base length of longer LVDT's was 650 mm.

The local deformations of bed joints were measured by two strain gauges (SG7 and SG8 on Figure 2) that were build in the two bed joints during the bricklaying.

3. TEST RESULTS AND DISSCUSSION

3.1. Bond wrench tests of masonry prisms

The results of bond wrench tests of the masonry prisms made from the six mortar types and from bricks cured in two different ways are given in Table 2. The difference between the strength of specimens made of coarse and those made of fine sand is well seen dispute the difference in age of the specimens. The beneficial effect of the second pre-wetting procedure in comparison to the first one is also significant. The simpler and shorter pre-wetting procedure provides higher bond strengths and smaller scattering of the test results.

The polymer suspension was added to the cement :sand mortar in order to improve the bond strength between the mortar and the brick. However, the flexural tensile strength of mortar-brick bond obtained at the polymer-cement :sand mortar is considerably lower than at the ordinary cement :sand mortar. One of the reason for such behaviour could be the texture of the brick surface with the pore size below the diameter of the polymer particles.

Table 2: The average flexural tensile strength of bond between mortar and brick

MORTAR MIX STRENGTH [MPa]	C:S 1:3	C:S 1:3 + polymer	C:L:S 1:1:6	C:L:S 1:1:6 + PP fibres	C:L:S 1:1:6 + mesh	L:S 1:3
Age 120 days, coarse sand, 1st pre-wetting procedure	0,182	0,184	0,045	-	0,063	0,035
	st.dev. 35%	*st.dev. 73%*	-	-	*st.dev. 33%*	*st.dev. 13%*
Age 30 days, fine sand, 1st pre-wetting procedure	0,645	0,395	0,510	0,458	0,267	0,076
	st.dev. 32%	*st.dev. 26%*	*st.dev. 16%*	*st.dev 31%*	*st.dev. 24%*	*st.dev 18%*
Age 30 days, fine sand, 2nd pre-wetting procedure	0,731	0,561	0,450	0,457	0,427	0,081
	st.dev. 17%	*st.dev. 17%*	*st.dev. 13%*	*st. dev. 9%*	*st.dev. 16%*	*st.dev. 12%*

The flexural tensile strengths of the ordinary cement :lime :sand mortar, cement :lime : sand mortar with polypropylene fibres and cement :lime :sand mortar reinforced by glass-fibres mesh are approximately equal. The reason for about two times lower bond strength obtained at the masonry prisms made from cement :lime :sand mortar reinforced by the mesh and from bricks pre-wetted according to the first pre-wetting procedure could be in pre-damaged test specimens.

Due to the presence of the polypropylene fibres in the interface between brick and mortar, and therefore smaller interfacial zone between brick and binder, much lower flexural strength was expected for the specimens made with cement :lime :sand mortar with polypropylene fibres. The obtained results did not confirm this assumption. Obviously, in this kind of mortar the polymeric fibres had taken over the early tension stresses which arose during the first days of curing of the specimens, caused mainly by shrinkage of the mortar. Also the failure of bond was not so sudden, as it was the case for other mortars. The failure propagated a in more gradually manner.

3.2.   Diagonal tensile tests of wallettes

Table 3 presents the average values of the peak load and of global vertical and horizontal deformation at the peak load, obtained at the diagonal tensile tests of the wallettes made of five different mortars. The standard deviations give an information on the scatter of the test results. Since the bond strength between mortar and brick was not improved with the polymer addition, the tests of the wallettes made from polymer-cement mortar were not performed.

Table 3: The average peak loads and appertaining global vertical and horizontal deformations obtained at diagonal tensile tests of the wallettes

Mortar mix	Average peak load [kN]	Average global vertical deformation at peak load [mm]	Average global horizontal deformation at peak load [mm]
cement:sand mortar 1:3	11,48	1,074	-0,706
	st. dev. 35%	*st. dev. 85%*	*st. dev. 116%*
cement:lime:sand mortar 1:1:6	5,68	0,431	-0,201
	st. dev. 26%	*st. dev. 29%*	*st. dev. 83%*
cement:lime:sand 1:1:6 + PP fibres	6,25	0,412	-0,180
	st. dev. 58%	*st. dev. 20%*	*st. dev. 61%*
cement:lime:sand 1:1:6 + mesh	23,64	0,582	-0,371
	st. dev. 27%	*st. dev. 20%*	*st. dev. 57%*
lime:sand mortar 1:3	4,80	0,203	-0,218
	st. dev. 42%	*st. dev. 107%*	*st. dev. 137%*

The highest peak load was obtained in the case of the cement :lime :sand 1:1:6 mortar with the mesh embedded in bed joint. The mesh did not affect the bond since it was placed in the middle of the joint, but it had a significant effect on the tensile strength of the bed joint, preventing the masonry from early failure in tension. In the case of other mortars the magnitude of strengths corresponds to their flexural strengths.

Table 4: Experimentally obtained failure modes of wallettes

FAILURE MODES			
	SHEAR	COMBINED	SPLITTING
C:S = 1:3	-	4	1
C.L:S = 1:1:6	1	4	-
C.L:S=1:1:6+fibres	-	5	-
C.L:S=1:1:6+mesh	2	4	-
L:S = 1:3	3	3	-

The weak bond between mortar joints and bricks was a major reason for relatively low load-bearing capacity of wallettes achieved by diagonal tensile tests. The coarse sand particles reduced binder to brick contact surface, which results in lowering of the bond strength.

The size of specimens had a significant influence on the failure mechanism. Elastic FEM analysis of wallettes showed that in the case of two brick wide specimens the concentration of shear and tensile stresses within brick-mortar junction can anticipate the combined failure of the specimen before the splitting mechanism can occur. Therefore the most frequent failure modes observed were either shear or combined shear-splitting failure. Only one of the specimens failed by splitting, as it can be seen from Table 4.

The global horizontal and vertical deformations for specimens made with cement : sand and lime :sand mortars are presented in Figure 3. Specimens made with cement : sand mortar were more brittle (Figure 3-a) than the specimens made with lime :sand mortar (Figure 3-b). That was caused mainly by lower bond strength in the interface region. The failure of specimens made with lime :sand mortar propagated more gradually due to combined influence of low bond strength and low tensile strength of mortar.

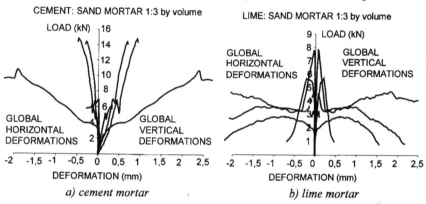

Figure 3: Results of diagonal tensile tests of wallettes made with different mortars

The influence of mesh embedded in mortar bed joints (macro-reinforcement) on the peak load of the wallettes is presented in Figure 4.

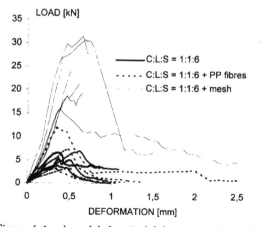

Figure 4: Load vs. global vertical deformations relationships

Due to the influence of macro-reinforcement the failure of the specimens occurred only along the interface zones of bed or head joints. Therefore, the influence of mortar tensile strength had negligible influence on the tensile and shear strength of specimens. An important parameter that influenced propagation of failure cracks was the even distribution of mortar in the head joints.

4. CONCLUSIONS

Several conclusions can be drawn from the analysis of experimental results. They all prove previous expectations about the effect of modifying mortars and about the influence of sand grading. The main influence of variation of mortar composition has been observed as an influence on brick to mortar interface zone. The variation of mortar to brick bond strength influenced the tensile and shear strength of masonry. The polypropylene fibres as well as coarser sand reduced the area of surface over which mortar binder acts on the brick surface. However, the open porosity of brick surface has also a significant influence on bond strength. In the case of experiments reported herein, the sizes of open pores on brick surface were relatively small in comparison to the size of some mortar particles. Therefore interlocking between mortar particles and brick surface was not very effective.

The use of coarser sand for the preparation of mortar decreases the shear strength of masonry. Therefore using it for the construction of masonry in earthquake prone areas can lead to overestimation of the expected earthquake resistance of masonry. This is the case in Slovenia, where suppliers do not pay sufficient attention on proper granulation of mortar sand. On the other hand, mortar made of coarser sand can be easily removed from bricks, which gives a chance for brick reuse after the decomposition of the existing masonry assemblage.

The used polymer improves the tensile strength of mortar in bed joints and at the same time reduces the bond strength between brick and mortar. The latter could be caused by larger diameter of the polymer particles compared to the open pore diameter of the brick surface. The filmed polymer inside the mortar structure limits also the evaporation of water from mortar and keeps it wetter for longer period. It can affect masonry if it is early loaded by extreme loading.

Polypropylene fibres influence positively the bond strength of brick to mortar interface because of the reduction of cracks developed during the mortar hardening. However, fibres may also reduce the bond area. Therefore further research is needed to examine the possibilities of proper use of polymer fibres.

So far, the most promising modification of mortars is by using the glass-fibre meshes placed in the middle of mortar bed joints. In this case the bond between brick and mortar is almost of the same quality as in the case of unmodified mortars, but the mortar tensile strength is increased by the action of mesh reinforcement.

The obtained results on the influence of different pre-wetting procedures have showed that the procedure with water immersion of bricks for 30 minutes and drying them for 30 minutes provides higher bond strength and smaller scattering of the results than the procedure with water immersion for 60 minutes and storage under PVC foil for 24 hours for all types of mortar.

5. ACKNOWLEDGEMENT

The paper presents the results of the research work at the University of Ljubljana. It is granted by the European Community through the project ATEM COPERNICUS and by the Ministry of Science and Technology of the Republic of Slovenia through the project Advanced Masonry. Their support is gratefully acknowledged.

6. REFERENCES

1. BOSILJKOV, V., "On modeling mechanical properties of the masonry ", *M.Sc. Thesis*, University of Ljubljana, Slovenia, October, 1996

2. BOSILJKOV, V., ŽARNIĆ, R., KRALJ, B. and PANDE, G., " Experimentally based computational modelling of masonry ", *Computer Methods in Structural Masonry-4*, Florence, Italy, September, 1997

MECHANICAL CHARACTERISTICS OF MODIFIED LIME MORTAR

S. Briccoli Bati Professor and **L. Rovero** PhD
Dipartimento di Costruzioni, Università di Firenze, Firenze, Italy

1.ABSTRACT

Experimental analyses of lime mortar were conducted with the aim of enhancing its mechanical properties through the addition of components commonly used in historical building practices prior to the introduction of Portland cement.

Specimens of different lime mortars were prepared using additives described in ancient architectural treatises and nineteenth-century technical handbooks. In particular, the study focused on recipes based on the addition of *cocciopesto*, tuff powder, iron slag, and fossil flour. The mechanical properties achieved through addition of the different components was evaluated experimentally and then compared against simple lime mortar. Such comparison reveals that addition of the materials under study significantly enhances the mechanical behavior of lime mortar.

2.INTRODUCTION

In the consolidation and restoration of historical buildings, the binding material chosen plays a critical role in the final results achieved, both aesthetically and mechanically. On the one hand, the use of cement mortar has often yielded poor results in both respects, as its properties are not well-matched to the materials commonly used in historical buildings. On the other, traditional lime mortar does not provide satisfactory results in terms of durability and strength [1],[2],[3] and [4]. Thus, in the search to improve the mechanical characteristics of lime mortars, there has been continued development and experimental assessment of such binding materials enhanced with additional constituents.

The present paper reports the results of an historical and experimental research project

Keywords: Lime mortar, Additive materials, Historical buildings

Computer Methods in Structural Masonry – 4, edited by G.N. Pande, J. Middleton and B. Kralj.
Published in 1998 by E & FN Spon, 11 New Fetter Lane, London EC4P 4EE, UK. ISBN: 0 419 23540 X

aimed at studying the mechanical characteristics of modified lime mortars. In particular, its goal was to assess how the mechanical properties of lime mortar might be improved by adopting archaic construction practices, common before the introduction of Portland cement, that called for the use of specific additives.

In order to draw from the wealth of past experience on using this type of mortar, a wide-ranging bibliographic research was performed on historical architectural treatises and technical handbooks on building practices.

Four distinct components were identified that used to be mixed with lime mortar in order to strengthen it: *cocciopesto* (brick powder), tuff powder, iron slag and fossil flour. Then, in experimental trials these additives were used in mixtures to manufacture laboratory samples for three point bending and uniaxial compression tests. Stiffness, strength and ductility of each mixture was experimentally tested and compared against plain lime mortars.

The results reveal that, while maintaining congruity with respect to the properties of historical buildings, such modified mortars generally enhance overall mechanical behavior.

3. MANUFACTURING RECIPES

Five different lime mortar recipes were followed in making the samples: four modified and one with no additives, as a reference.

The lime mortar without additive was manufactured with the following portions of components by weight:

    1/4 part hydrated lime mortar
    3/4 part aggregate
    1/1 water/binder ratio

The lime used was hydrated lime (Italcementi, Milano); the aggregate was siliceous sand from the Ticino River, with twelve grading classes ranging from 0.1 and 4.75 mm (fig.1).

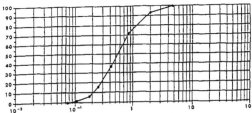

Fig. 1 Grading curve of siliceous sand from the Ticino River.

The modified lime mortars were manufactured following guidelines drawn from Vitruvius "De Architettura libri decem" [5]; the recipe for lime mortar reported in this treatise calls for substituting one part aggregate with one part additive.

The lime mortar with additive was thus manufactured with the following proportions by weight:

    - 1/4 part hydrated lime
    - 2/4 part aggregate
    - 1/4 part additive
    - 1/1 water/binder ratio

In these recipes, the types of lime and aggregate used were the same as those in the reference recipe. The additives employed in the different mixtures were: *cocciopesto* (i.e. brick powder), volcanic tuff powder, iron slag and fossil flour.

Cocciopesto, recommended as an additive by Vitruvius in his treatises, belongs to the class of artificial pozzolan and confers hydraulic properties on lime mortar. *Cocciopesto* was often employed by the Romans in areas far from pozzolan ore deposits. In fact, lime mortar with *cocciopesto* adheres to masonry even in humid and wet environments and was used for the subfloors of plaster coats bases of plaster-coated floors, in waterproof tank walling, scuttlebutts and pavements, as well as in outdoor masonry floors. In order to reproduce *cocciopesto* in the laboratory, brick fragments (San Marco Laterizi, Venice) were crushed and sifted with a N 60 sieve (0.25 mm).

The tuff powder additive was also suggested by Vitruvius in his treatises. In order to obtain tuff powder in the laboratory, blocks of yellow volcanic tuff quarried in caves near Sorano (Grosseto) were crushed and sifted across a N 60 sieve (0.25 mm). The Sorano volcanic tuff, a pyroclastic rock with highly diffuse porosity, is an aggregate of cemented ashes, lapilli and volcanic sand.

The use of iron slag as additive was advocated first by Belidor [6] in his 1729 treatise and later by Fonda (1764) [7], Sanvitali (1765) [8] and Milizia (1781) [9]. Iron slag, which belongs to the class of artificial pozzolan, contains silica, alumina and ferrous oxide, and imparts hydraulic properties to lime mortar. Laboratory samples were prepared employing the iron slag (found at the Etruscan Necropolis of Populonia - the remains of the extensive iron manufacturing carried out in ancient times).

The addition of fossil flour was recommended by Milizia in his 1781 work "Principi di architettura civile" [9]. Fossil flour is the siliceous fossil residue of diatoms - unicellular vegetable organisms that live in both fresh and sea water. Milizia maintained that mortar prepared with fossil material makes poor plaster, but very good binding material. In preparing the laboratory samples, commercial fossil powder (Celite Italiana, Milano) was employed.

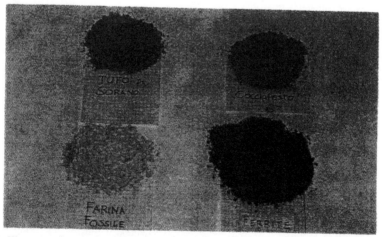

Fig. 2 The experimented additives: *cocciopesto* (i.e. brick powder), volcanic tuff powder, fossil powder and iron slag. Each additive was employed in multiple batches.

4. SPECIMENS

For each of the mixtures described in foregoing, 108 specimens were prepared and, after curing periods of 30, 90 and 180 days, subjected to a three-point bending test.

Specimens were fabricated using a mechanical mixer according to the methods prescribed by Italian regulations (D.M. 3/6/1968). The mixture was checked for its consistency, then cast in multiple steel formworks and compacted using a shaking table into two successive layers, each obtained with 60 blows. Specimens were kept 24 hours in a climate-controlled environment at 20+2°C and relative humidity of over 90%. Relative humidity was thereafter maintained at 50% for the entire curing period.

Fig.3 The *cocciopesto* lime mortar samples in the formwork.

5. MECHANICAL TESTS

For each of the different recipes and curing periods, 36 samples of dimensions 4x4x16 were subjected to bending tests (Fig. 4). The 72 pieces originating from the bending tests were then utilized for uniaxial compression tests with a 4x4 cm load divider.

Italian regulation UNI 6133 was followed in executing the bending tests: the load was applied at a constant rate of 0.8 daN/sec until the sample cracked, and the value of bending tensile strength derived from the fracture load.

Uniaxial compression tests were executed using a displacement control device. The applied load was recorded using a loading cell with a capacity of 1000 daN. Four transducers were arranged symmetrically on the loading plate to measure the overall longitudinal deformation of the specimen. Both the loading cell and the four transducers were connected via a data acquisition interface to a Personal Computer. This provided real-time recording and graphic display of the specimen's entire equilibrium pattern.

A thin layer of talcum powder was applied on the surfaces of the load applicator to reduce interface friction. The load was applied at a rate of 0.3 daN/sec, and load values and global longitudinal strain was recorded at each incremental step of 5 daN. In order to observe the ascending segment of the equilibrium curve, the test was prolonged until a load equal to 2/3 of the peak load was reached.

Fig.4 Three-point bending test.

Fig. 5 Compression test of a piece generated by the bending test.

6. RESULTS

Load-displacement diagrams recorded during tests were characterized in terms of a set of salient points taken to be those corresponding to the beginning and end of the linear segment, the point with abscissa corresponding to the peak load displacement, and the final point at 2/3 peak load, by convention (fig. 6(b)).

Compression strength was calculated using the peak load value, while the elastic modulus was derived from the global strain in the linear section. This latter value will be referred to as "apparent elastic modulus", as it was taken as a global measurement on a heterogeneous material.

Tables 1, 2 and 3 report mean values, standard deviations and variation coefficients of the mechanical parameters for specimens obtained with each different recipe with 30, 90 and 180 days' curing.

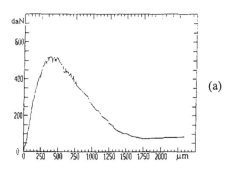
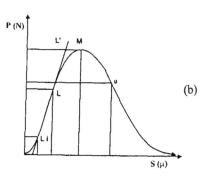

(a)

(b)

Fig. 6 (a) Load-displacement diagrams of a specimen of hydrated mortar. (b) Schematic load-displacement diagram.

CONCLUSION

Analysis of the experimental results obtained allows us to draw some significant conclusions with quite a high degree of confidence, given the large number of specimens and tests employed in the study.
All the additives tested significantly enhanced the mechanical properties of hydrated mortar. In particular, the *cocciopesto* yielded mechanical parameters values of /1.5 times greater than those of simple lime mortar.
The tuff additive, though less effective in the long run, provides significant shortening of the curing process: after 30 days' curing, specimens reach 89% of the mechanical properties that they attain after a curing period of 180 days. This suggests that the effect of the addition of tuff consists basically of accelerating the strengthening process.

		traction strength 4x4x16 (Mpa)	compression strength 4x4x4 (Mpa)	Young modulus 4x4x4 (Mpa)	cinematic ductility 4x4x4
reference	mean	0,31	0,93	133	1,63
lime	std.dev	0,055	0,17	26,6	0,41
mortar	var.coef.	0,18	0,18	0,20	0,25
mortar	mean	0,58	2,45	307	1,26
with	std.dev	0,099	0,39	67,5	0,24
tuff powder	var.coef.	0,17	0,16	0,22	0,19
mortar	mean	0,65	3,62	367	1,35
with	std.dev	0,096	0,69	99,1	0,34
cocciopesto	var.coef.	0,15	0,19	0,27	0,25
mortar	mean	0,51	3,15	228	1,14
with	std.dev	0.08	0.63	52.4	0,27
fossil flour	var.coef.	0.16	0.20	0.23	0.24
mortar	mean	0,42	1,7	159	1,72
with	std.dev	0.07	0.35	31.8	0.37
iron slags	var.coef.	0.17	0.21	0.20	0.22

Tab. I Mechanical parameter values after 30 days' curing.

		traction strength 4x4x16 (Mpa)	compression strength 4x4x4 (Mpa)	Young modulus 4x4x4 (Mpa)	cinematic ductility 4x4x4
reference lime mortar	med	0.47	1.07	133	2.56
	dev.std	0.056	0.14	21.3	0.59
	coef.var	0.12	0.13	0.16	0.23
mortar with tuff powder	med.	0.68	2.66	342	1.33
	dev.std	0.61	0.40	54.7	0.24
	coef.var	0.09	0.15	0.16	0.18
mortar with cocciopesto	med.	0.83	3.64	375	1.56
	dev.std.	0.091	0.51	78.7	0.39
	coef.var	0.11	0.14	0.21	0.25
mortar with fossil powder	mean	0.51	3.36	340	1.13
	std.dev	0.05	0.43	68	0.29
	var.coef.	0.10	0.13	0.20	0.26
mortar with iron slags	med.	0.51	1.80	170	1.94
	dev.std.	0.06	0.23	25.5	0.44
	coef.var	0.12	0.13	0.15	0.23

Tab. II Mechanical parameters after 90-days' curing.

		traction strength 4x4x16 (Mpa)	compression strength 4x4x4 (Mpa)	Young modulus 4x4x4 (Mpa)	cinematic ductility 4x4x4
reference lime mortar	mean	0.614	1.65	151	1.39
	std.dev	0.049	0.18	25.7	0.29
	var.coef.	0.08	0.11	0.17	0.21
mortar with tuff powder	mean	0.74	2.74	352	1.59
	std.dev	0.044	0.27	66.9	0.25
	var.coef.	0.06	0.10	0.19	0.16
mortar with cocciopesto	mean	1.11	3.96	402	1.39
	std.dev	0.013	0.51	84.4	0.29
	var.coef.	0.12	0.13	0.21	0.21
mortar with fossil powder	mean	0.95	3.50	380	1.10
	std.dev	0.10	0.42	76.1	0.22
	var.coef.	0.11	0.12	0.20	0.20
mortar with iron slags	mean	0.78	2.09	191	1.24
	std.dev	0.07	0.37	40.1	0.27
	var.coef.	0.10	0.13	0.21	0.22

Tab. III Mechanical parameters after 180-days' curing.

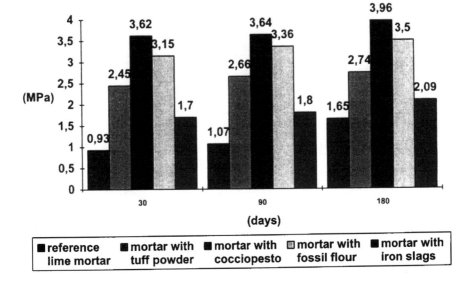

Fig. 7 Histogram of compressive strength.

REFERENCES

1. Menicali, U., I materiali dell'edilizia storica, Roma 1992.
2. Tufani, A., Le malte nel restauro; Perugia 1987.
3. Toniolo L., Bugini R., Alessandrini G., Rapporti composizionali di una malta antica, Arkos, 15, 1991.
4. Briccoli Bati S., Experimental knowledge of the traditional building materials, Protection of the Architectural Heritage Against Earthquakes, International Center for Mechanical Sciences Courses and Lecture, CISM,Udine,1995.
5. Vitruvio, De Architettura libri decem, (I sec a.C.), Edizione curata da L. Mingotto, Pordenone, 1990
6. Belidor, B. Forest de, La science des ingenieurs, (1729), Fratelli Negretti, 1832.
7. Fonda G., Elementi di architettura civile e militare ad uso del collegio Nazareno, Roma, 1764.
8. Sanvitali F., Elementi di architettura civile, 1765.
9. Milizia, F., Principi di architettura civile, (1781), Remondini, Bassano, 1825.

ESTIMATION OF THE IN SITU STRENGTH OF MORTARS

S. Kolias Associate Professor[1], **C. Douma** Civil Engineer[2] and
K. Papadopoulos Civil Engineer[2]
[1]*National Technical University of Athens (NTUA), Iroon Politechniou 5,
157 72 Zografou, Greece and* [2]*Ex Graduates of NTUA*

1. ABSTRACT

The possibilities of estimation of the in-situ mortar strength by measuring a) the total water porosity and b) the time needed for a masonry drill to penetrate a certain depth into the mortar, are investigated in the laboratory and the results are shown to be promising.

2. DESCRIPTION OF THE METHODS

The first method is based on the measurement of the total water porosity of small fragments of mortars (1-2 cm^3) taken from the structure. The measurement is very simple requiring [1]saturation under vacuum for 2 hrs and soaking for 24hrs, of the small mortar pieces and subsequently determining their volume (pecnometer method) and their water content (drying at 105 $^\circ$C for 24 hrs). From the total porosity the paste porosity is calculated according to the formula :

$Pt = pPp + sPs$ where:

Pt = total water porosity as measured

Pp = calculated paste porosity

Ps = porosity of the sand in case of mortars

p and s the volumetric proportions of the paste(volume of binder + total water) and the sand in the mix respectively

In a previous work by Kolias [1], this method has given satisfactory results for testing cement mortar and concrete.

Keywords : Mortars, in - situ strength, non - destructive testing, porosity

Computer Methods in Structural Masonry – 4, edited by G.N. Pande, J. Middleton and B. Kralj.
Published in 1998 by E & FN Spon, 11 New Fetter Lane, London EC4P 4EE, UK. ISBN: 0 419 23540 X

The second method is based on the time needed for an ordinary masonry drill, (D=4mm) pressed at right angles onto the mortar surface with a constant force (180N) and constant speed of rotation (75 or 175 rev/min), to penetrate a certain depth (4mm) into the mortar mass. Reference in this case is made to the work reported by Cucci and Barsotti [2] involving the energy consumed by an ordinary hand-held electric-driven masonry drill, to penetrate a certain depth into the mortar mass.

In this work a special simple equipment, Fig 1, was developed by which the masonry drill bit, held firmly and pressed with a constant force onto the specimen , can be rotated with a predetermined number of revolutions per minute irrespectively of the resistance to the rotation encountered. This is achieved by a variable speed electric driven motor and a combination of cogwheels. The specimen, placed onto steel rollers to eliminate friction, is pushed on the drill with the help of a mechanical wrench and the force applied is measured by a proving ring. When the required load is reached, the electric motor and a stop watch are switched on simultaneously and the drilling process starts. During the drilling process, as the drill bit is penetrating the specimen, the force should be slightly adjusted by small rotation of the handle of the mechanical wrench so that the force remains within ± 3% of the required load. The drilling process is terminated and the drilling time is noted, when the dial gauge indicates that the required depth of penetration has been reached. Steel blocks of known weight can be placed on top of the specimen so that the required compressive state of stress can be achieved.

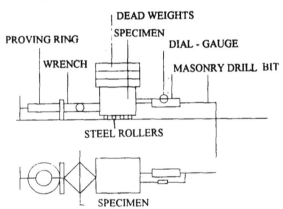

Fig.1 : Equipment for testing samples with the drilling method

For the time being, this method can be used for the estimation of the in situ strengths of mortars by taking small fragments of mortar from the structure and embedding them into the mass of a quick setting binder in such a way that a small prism is formed with the fragment occupying part of one of its faces. The prism can be placed on the testing device and the drilling test can be carried out with or without compressive state of stress.

3. METHODOLOGY OF THE INVESTIGATION

The potential of the two methods is investigated in the laboratory on 70 mm cubes prepared with various types of mortars. On each cube the drilling method is first applied on two opposite faces and subsequently the cube is crushed for strength determination. From the fragments of the crushed specimen several pieces are collected and the total water porosity is measured. It is thus possible to correlate the strength from a cube specimen, with the porosity and the drilling time. Since, however, the strength is related to a much greater mass of mortar than the mass involved in the drilling and the porosity tests, the measurements of drilling time and porosity are based on at least 4 drillings and 4 porosity pieces per specimen. The mean value of these measurements is correlated to the specimen strength.

The cube is placed onto the compressive testing machine so that the specimen faces which had the drilling cavities come into contact with the machine platens. By doing so the influence of the small drilling cavities on the compressive strength is eliminated (due to the triaxial state of stress developed in this area.). This was confirmed by a secondary investigation, Fig. 2, involving tests on specimens with two or three drilling cavities on each face of the specimen (which came into contact with the machine platens) and twin specimens without any drilling cavity.

The work included the following types of mortars prepared with crushed limestone sand:
• Cement mortars (only for drilling tests)
• Lime mortars
• Cement-lime mortars
• Lime-pozzolana mortars

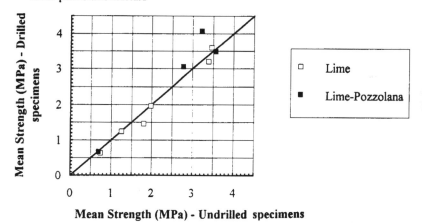

Fig. 2 : Comparison between drilled and undrilled specimen strength.

288, 70mm cube specimens were prepared (Table 1) with different mix proportions so that a large range of strengths could be investigated covering a range of types of mortars which can be found in the present types of structures. It is believed that this work could also give some evidence as far as mortars of historical structures are concerned.

Additional research work involving sands of various types and of various hardness is being currently carried out at the NTUA.

For the drilling method the parameters which were investigated included the effect of
1   the existence or not of compressive state of stress

2   the rotation speed
3/   the force by which the drilling bit was pressed onto the specimen
4/   the effect of water saturation before testing
5/   the depth of penetration
6/   the diameter of the drill
The results of 1,2 and 3 are only shown in this presentation.

Table 1. Mix Proportions of the Mortars

Lime Mortars		Lime-Pozzolana Mortars		Cement-Lime Mortars		Cement Mortars (drilling only)	
Nbr of specs	Lime/Sand (by vol)	Nbr of specs	Lime/ Pozzolana/ Sand(by vol)	Nbr of specs	Cement/ Lime/Sand (by vol)	Nbr of specs	Water/ Cement (by mass)
12	1/0.5	12	1/0.35/0.85	12	1/1/6	24	0.6
12	1/0.75	12	1/0.8/3	10	1/2/8	4	0.7
24	1/1	12	1/1/2.5	10	1/4/13	4	0.8
12	1/1.5	12	1/1.5/2.5			8	0.9
12	1/2.5	12	1/2/4			4	1
12	1/3	12	1/3/3			8	1.1
12	1/4	12	1/4/3			4	1.2
12	1/5					8	1.3

4. PRESENTATION AND DISCUSSION OF THE RESULTS

4.1 Porosity method

The relationship between strength and calculated paste porosity is shown in Fig 3 for the mortar mixes examined. It can be seen that the lime mortars and the lime pozzolana mortars seem that are grouped together forming a curvilinear relationship between strength and porosity. The cement -lime mortars follow, for the range of strengths examined, a straight line relationship similar to that reported by Kolias [1] on cement mortars . The correlation coefficient is 0.84 for the cement mortars (linear relationship) and 0.64 for lime and lime-pozzolana mortars (curvilinear relationship).

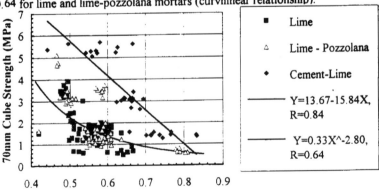

Fig. 3 : Relationship between calculated paste porosity and strength

The disadvantage of the method is that the calculated paste porosity should be determined and this calculation is based on the volumetric proportions of the aggregates/sand in the mix. The volumetric proportions are practically not constant in the mass of the mix and in most of the times are not known, especially in cases of old or historical masonry mortars. These proportions are also dependent on the amount of the binder that has been in fact hardened which is not always known. Particularly this becomes more important in case of lime-pozzolana mixes for which it is difficult to estimate the percentage of pozzolana that has reacted with lime and the percentage of pozzolana which acts as inert material in the mix. To this effect may also be attributed the discrepancy of the results of these mixes and why some indicate that they group together with the lime mortars while there is also an indication that they follow a separate linear relationship.

There is also a possibility that some of the calcium hydroxide could be dissolved and leached out from the porosity sample during the 24hr saturation period and thus altering the results of porosity measurements.   This possibility was examined with a limited number of tests , in which twin porosity samples were subjected to saturation with either distilled or with lime-saturated water. No significant difference was detected.

Obviously there is enough ground for further improvement but the results are believed to be encouraging.

4.2 drilling method

The effect of compressive state of stress on the drilling time-strength relationship is shown in Fig 4. It can be seen that the application of compressive stress equal to $\sigma y=0.1$ MPa moves the relationship to the right to longer drilling times, while the inclination becomes slightly less steep. The correlation coefficient is satisfactory, 0.97,   in both cases inspite of the fact that different mortar types were used. It should be noted that it would be more representative of the real stress conditions if a confining pressure was applied on the specimen instead of the vertical stress, but this condition is more difficult and increases the cost of the equipment considerably.

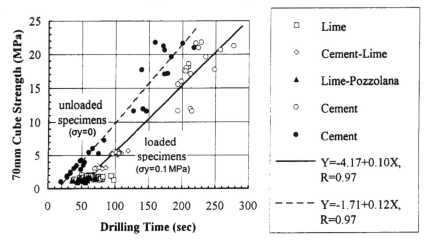

Fig. 4 : Influence of the compressive state of stress (175 rev/min)

Fig 5 shows the influence of the rotation speed of the drill for stress free specimens ($\sigma y=0$) on the relationship between drilling time and strength for various types of mortars. It can be seen that increased rotation speed makes the relationship steeper. In both cases the linear relationship has a satisfactory correlation coefficient higher than 0.9 inspite of the fact that different mortar types were used.

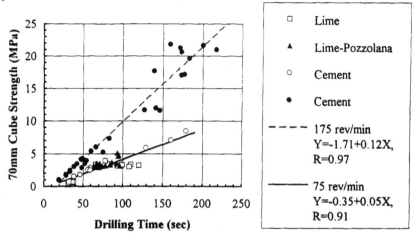

Fig. 5 : Influence of the rotation speed ($\sigma y=0$)

Fig 6 shows the influence of the compressive force with which the drilling bit is pushed onto the cement-mortar specimens. This force is an important parameter since in cases of mortars with low strength, such as lime mortars or mortars of historical structures, it should be reduced in order to avoid mortar fractures due to the concentrated stress applied by the drilling bit. On the contrary in cases of mortars with higher strengths the force should be increased in order to avoid excessive drilling times. It is also interesting to note that, as the work [2] indicates, the influence of the compressive force and speed of rotation may be practically minimised if the energy consumed during the drilling process is measured instead of the time required to reach a certain depth.

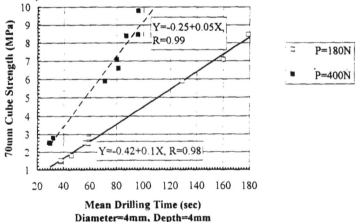

Fig. 6 : Influence of the force applied to the drilling bit on the compressive strength

5. CONCLUSIONS

The following broad conclusions can be drawn:
- Both methods are promising.
- The porosity method is influenced by the type of the binder of the mix.
- An improvement is needed for the calculation of the paste porosity in order to enhance the predictive accuracy of the porosity method.
- The drilling method seems to be independent from the type of the binder but is influenced by a number of parameters such as, rotation speed, state of stress, diameter of the drilling bit , force applied on the drilling bit, moisture conditions of the mortar at the time of testing. However, this work shows that by keeping these parameters constant, a very satisfactory correlation can be achieved and it is therefore believed that the method could find practical use in the future. More research could reveal the best combinations of the above parameters to suite different levels of mortar strength.

The type and hardness of the sand used for the preparation of the mortars is an important parameter which needs investigation

It is thought that the two methods could be used in combination in order to improve the accuracy of the strength estimation and this is a field of further research.

It is also believed that these methods could be extended for the estimation of the mortar strength of historical-ancient structures and a preliminary research work on this field in the near future is scheduled.

6. REFERENCES

1. S.Kolias, "Investigation of the possibility of estimating concrete strength by porosity measurements", Materials and Structures,1994, 27 , pp 265-272.
2. N.Gucci, R. Barsotti, "A non-destructive technique for the determination of mortar load capacity in situ", Materials and Structures, 1995, 28, pp 276-283.

A FINITE ELEMENT MODEL OF ARCH RING BEHAVIOUR

T.G. Hughes Senior Lecturer[1] and **M.J. Baker** Design Engineer[2]
[1]*Cardiff School of Engineering, Cardiff University of Wales, PO Box 917, Cardiff, CF2 1XH, Wales, UK and* [2]*Quadrant Consulting, 38 Cathedral Road, Cardiff CF1 9SU, Wales, UK*

1. ABSTRACT

This paper contains details of the development of a macro finite element model designed specifically to deal with masonry arch ring behaviour. The detailed steps in the development of the model and its application to a previously tested arch bridge are presented.

A series of tests were initially undertaken on stack bonded masonry. These tests were performed on masonry composed of brick units and mortar that had previously been tested for a range of mechanical properties. The individual component properties and the stack wall tests were used to develop a micro model of this masonry behaviour. This micro model was then run in a series of numerical experiments to develop a suitable macro failure model.

Following validation of the macro model in a number of biaxial numerical experiments the model was tested against the full scale results from the physical testing to destruction of the Bridgemill arch.

2. INTRODUCTION

Considerable effort has been expended in the UK in recent years in gaining a better understanding of the behaviour of masonry arch bridges. This work has been driven by the economic need to maintain the significant infrastructure which these bridges, many of them over 100 years old, constitute.

Keywords: Arches, Finite Elements, Masonry

Computer Methods in Structural Masonry – 4, edited by G.N. Pande, J. Middleton and B. Kralj. Published in 1998 by E & FN Spon, 11 New Fetter Lane, London EC4P 4EE, UK. ISBN: 0 419 23540 X

The UK masonry arch research programme has three strands, testing of real structures, laboratory modelling at a range of sizes and numerical modelling[1]. The large cost of testing existing real structures coupled with the difficulty of intrusive monitoring and sometimes lack of definition has limited this programme of tests. Considerable effort has been expended in laboratory modelling and these tests have provided qualitative information on strength generating parameters. The numerical modelling has generally been undertaken using 1-D models. This paper describes the development of a 2-D model necessary if progress is to be made in numerically modelling at full scale the features identified in experimental scale models.

3. FINITE ELEMENT MODELLING

The finite element analysis of masonry is especially difficult as the analyses of masonry structures require the complex nature inherent in brittle fracturing materials to be considered in conjunction with the effects of jointing, and how the joints , as planes of weakness change the way the composites respond to load.

The concept behind the development of the masonry arch ring model was to establish a detailed finite element analysis method for application solely to masonry arch bridges. The software incorporates a material model that describes, in detail, the behaviour of masonry in arch rings and also allows the use of other standard plasticity soil models to model the behaviour of the backfill material.

3.1 Development of the arch ring material model.

As a result of the curvature and jointing inherent in masonry arches the material model required that any analysis that would identify the true behaviour would need to be developed based on a local co-ordinate system and anisotropic behaviour. Consequently, the development of the masonry arch macro model required the orientation between local and global directions to be considered for each element; along with anisotropic behaviour, the effect of jointing, the brittle nature of the component materials and the non uniform elasticity constants. The model was developed using two main assumptions, based on consideration of overall arch behaviour.

i)   The shear force component of the stress in the arch ring would be minimal when the stresses were rotated to the local directions (i.e. parallel to the arch intrados). When considering an arch ring, the structure is relatively slender and the angle between the line of thrust and intrados at any point will be small. This is clear from Fig. 1 which shows, at failure, the effective thinning of the arch and the line of thrust produced by a typical cracking elastic analysis. The line of thrust is parallel with the centreline geometry at the points of high stress where both the tensile and compressive failure criteria are most used. At the locations between hinges where the angle between the geometry and the line of trust is largest the behaviour is likely to remain elastic.

ii)   Multi-ring behaviour would need to be considered using a separate material layer for each arch ring and also for the joint between the rings. Alternatively, joint elements may be incorporated to allow for sliding and ring separation.

The co-ordinate transformations required during the development of the model were, local to global D-matrix, global to local strains, global to local stress and local to global stress. These were required to allow the plasticity relationships to be determined in terms of local co-ordinates and then rotated to the global directions. The development of the transformation matrices is relatively standard and is described in detail elsewhere[2].

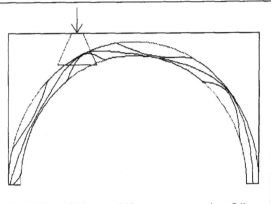

Fig. 1 Line of Thrust within a masonry arch at failure

The first stage in the homogenisation of the material properties was to determine the values of the local direction elastic material parameters from the individual brick and mortar material properties. The combined material values were generated using a procedure of homogenisation and is based on the process developed by Pande et al[3]. With only the first stage of the two stage homogenisation process used. The composite strength components used within the model are developed based on consideration of the geometry of the masonry and the orientation of jointing, to give local directional material strengths. The local direction D-matrices used this paper include formulations that are dependent on the local direction elastic moduli and Poisson's ratios, for both the orthogonal directions.

The properties required to generate the local D-matrices are the homogenised elastic moduli, poissons ratio and shear modulus. The anisotropic D-Matrices are developed from the standard stress-strain relationships in three dimensions. The elastic constitutive relationships based on both plane stress and plane strain were formulated with the elastic material properties developed from the individual material properties of the units and joints and from the fundamental principles of elasticity.

3.1.2 Development of arch ring failure surface.

In order to determine the onset of failure within a material and its nature after the elastic limit is reached a failure surface is required The failure surface was generated in terms of the local co-ordinate system and is based on a stack bonded arrangement of brick units and mortar joints. Analyses were undertaken using a state of the art concrete model and the material properties taken from the laboratory experiments. Using a procedure of micro analysis of the stack bonded ring section with units and joints as separate elements with different material properties assigned for each section. Biaxial loading states were applied to the model to generate a true two dimensional failure surface for the composite material. The material properties used for the units and joints in the stack bonded model are shown in Table 1.

The results from the analyses are given in Figure 2, in which the compression-compression, tension-tension and the tension-compression zones are shown. It may be seen that the failure of the arch ring arrangement is

Stack Material Property	Mortar	Brick	Units
Elastic Modulus	9500	13750	N/mm²
Poisson's Ratio	0.145	0.175	
Compressive Strength	10.2	21.5	N/mm²
Tensile Strength	0.4	1.2	N/mm²
Strain at Peak Stress	0.0017	0.0022	
Fracture Energy	0.03	0.019	N/mm

Table 1 Micro model component mechanical properties

not isotropic, but clearly dependent on the angle of the principal stress with the intrados of the arch ring. Subsequently, a numerical description was required, to allow the

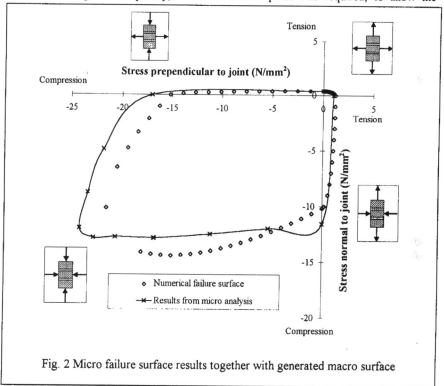

Fig. 2 Micro failure surface results together with generated macro surface

surface to be incorporated into the material model. The failure surface is inclined in the compression-compression zones and has differing uniaxial tensile and compressive strengths in both the local x and y-directions. The failure surface fitted is shown in Figure 2 and was generated from unit and joint strength properties. It was described by four different equations, a separate one for each segment of the surface and it is shown graphically, overlaying the results from the micro model analysis of the stack bonded units. Generally, the surface was similar to the results from the numerical analysis programme, although there is a slight overprediction of the compressive strength in the local x-direction.

The failure surface is described for the four zones:

Compression-Compression Zone, for both local x and y-directions;

$$F_{cc}\left(\sigma'_x,\sigma'_y\right) = \frac{\left(\sigma'_x - \dfrac{\sigma'_y.f_{cx}}{f_{cy}}\right)^2}{f_{cx}^{\;2}} + \frac{\sigma'^2_y}{f_{cy}^{\;2}} - 1 \tag{1}$$

Tension-Tension Zone, for both local x and y-directions;

$$F_{tt}\left(\sigma'_x,\sigma'_y\right) = \frac{\sigma'^2_x}{f_{tx}^{\;2}} + \frac{\sigma'^2_y}{f_{ty}^{\;2}} - 1 \tag{2}$$

Tension in local x-direction, Compression in local y-direction

$$\text{Diff} = C_x - f_{tx}$$
$$F_{tc}\left(\sigma'_x,\sigma'_y\right) = \frac{\left(\sigma'_x - \text{Diff}\right)^2}{C_x^{\;2}} + \frac{\sigma'^2_y}{D_y^{\;2}} - 1 \tag{3}$$

Compression in local x-direction, Tension in local y-direction

$$\text{Diff} = C_y - f_{ty}$$
$$F_{ct}\left(\sigma'_x,\sigma'_y\right) = \frac{\sigma'^2_x}{D_x^{\;2}} + \frac{\left(\sigma'_y + \text{Diff}\right)^2}{C_y^{\;2}} - 1 \tag{4}$$

where:

σ'_x, σ'_y	local direction stress components
$f_{cx}', f_{cy}', f_{tx}', f_{ty}'$	local direction compressive and tensile strengths
C_y, C_x	minor elliptical radius.
D_x, D_y	major elliptical radius.

The values of the two constants for the tension compression zones are calculated from consideration of the position of the intercept on the tension axis and the position and gradient of the intercept on the compression axis.

3.1.3 Stress-strain relationships.

Two separate types of stress-strain relationship were incorporated into the masonry formulation, one based on the compressive behaviour and one on the tensile behaviour.

Due to the nature of the loading in a masonry arch bridge, it was decided that a simple compressive material stress-strain relationship would adequately describe the behaviour of the masonry arch ring. In this case an elastic-perfectly plastic relationship was used. This meant that the stress on reaching the failure surface did not cause the surface to translate or expand, but remained the same size with the compressive stress always being returned to the same surface, consequently the material does not lose strength on crushing.

Experimental work on testing of masonry material properties showed that both mortar and brick have considerable post peak tensile strength. The behaviour shows that after the materials tensile strength has been reached, and a crack begins to form, the tensile strength gradually reduces to zero. To incorporate this into the model it required that the tensile strength in the two orthogonal directions i.e. parallel to the arch intrados and normal to the arch intrados be considered separately. The tensile failure normal to the intrados is dominated by the mortar joints, all cracks forming in them. In the direction parallel to the intrados the failure must pass through the brick or masonry units, which generally have much higher tensile and compressive strengths. This has resulted in two different tensile strengths being required in the analysis of the masonry ring structure, with both being based on the mortar and brick properties.

The softening sections of the stress-strain relationships require that a curve be fitted which allows the tensile strength to reduce as the plastic strain increases, with the quantity of softening dependent on the initial tensile strength and the fracture energy. A number of different tensile fracture material models have been developed for use in the analysis of concrete structures with a major contribution in the work by Rots[4]. More recently exponential softening curves have been developed by William et al[5] which more closely represents the real softening behaviour of brittle fracturing materials. The exponential softening response has been used exclusively in this work as is shown in Fig. 3.

The exponential softening curve formulation requires the maximum tensile strength, the crack length (i.e. the integration point length) and the Griffiths Fracture energy, G_F, be known for the material. Firstly the strain level, ε_0, at which the tensile

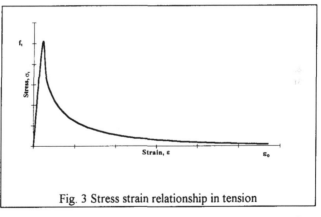

Fig. 3 Stress strain relationship in tension

stress returns to zero is required. The peak strain value is determined at the gauss point, where the crack length is the square root of the integration area, ensuring that the fracture energy per unit area of the model is constant across the structure.

$$\varepsilon_0 = \frac{C_1 \, G_F}{l_e \, f_t} \tag{5}$$

Once the value of the peak strain, ε_0, has been determined the exponential softening function may be used to evaluate the new reduced tensile strength value. The softening function is defined in such a way so that the reduced tensile strength is calculated from the quantity of plastic strain that has occurred. As a result of the different properties in the two directions, two tensile stress-strain relationships were required to describe the masonry ring failures. Resulting in the requirement for two individual independent softening directions to be developed for the tensile fracturing of the masonry ring. The model required independent values of the peak strain, plastic strain, tensile strength and fracture energy to be used, one set for each of the two local directions. The plastic strain values were calculated individually by separating the flow equation into two component parts, one for the local x direction and one for the local y direction:

The tensile fracture formulation allows the tensile strength to reduce in one direction leaving the strength in the normal direction unaffected. Resulting in the yield surface reducing in size in one direction but retaining full tensile strength in the opposite direction. If the tensile strength in the second direction is subsequently exceeded the surface may contract and the value of the tensile strength reduce independently of the first direction. This formulation for the biaxial tensile softening of the masonry arch ring is termed directionally independent plastic material softening.

3.1.4 Main formulation.

The main elements of the model therefore are;
i)   Rotation matrices for stress and strain components and the D matrix.
ii)   Homogenisation processes to determine the combined elastic moduli, Poisson's ratios, compressive and tensile strengths and fracture energy of the arch ring.
iii)   Pre-failure constitutive relationship, an elastic D matrix relationship, with the matrices being developed based on the anisotropic material behaviour applicable to the arch ring.
iv)   Strain decomposition into elastic and plastic strain components.
v)   A failure criterion, in the case of the masonry model, separate criteria were used for the four zones of the failure surface.
vi)   The flow rule developed for the masonry arch model is associated to the failure surface, as with the failure criteria four different formulations were required for the four different zones of the failure surface.
vii)   The softening and hardening relationships for the masonry arch model is separated into compression and the two local tension directions. The compressive model being an elastic-perfectly plastic relationship and the tension failure occurring using directionally independent tensile softening.
viii)   The consistency condition states that the stress must remain in or on the yield surface at all times and the elastic strain is recoverable and plastic is irrecoverable. This relationship is true when the differential of the yield function equals zero.

4 APPLICATION

The macro arch model was applied to the analysis of the full scale test results from the loading to destruction of Bridgemill Arch. This arch was tested as part of the UK arch research programme and with a 16 metre span and a rise of only 2.8m it was both the longest and shallowest arch tested. This arch is suitable for analysis using the developed model as it was of single ring masonry construction. The arch failed by snap-through intervening in advance of the full formation of the hinges that form the mechanism. Fig. 4 shows a comparison between the experimental and arch ring model predictions.

5. CONCLUSIONS

There are a number of innovative parts to the model that have been developed specifically for this application to the masonry arch rings. These include the tensile directional softening, the anisotropic D-matrix formulation and the non-symmetric failure surface described in four separate zones. Also included is compression failure

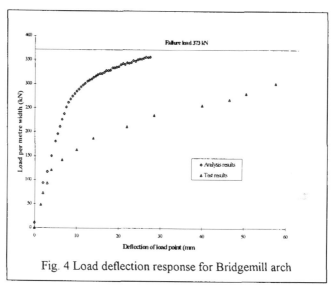

Fig. 4 Load deflection response for Bridgemill arch

with an elastic-perfectly plastic stress-strain relationship and tensile failure with exponential softening.

6. ACKNOWLEDGEMENTS

The authors acknowledge the financial support of both the UK Engineering and Physical Sciences Research Council and the European Union through the Copernicus contract the *Advanced Testing of Masonry*

7. REFERENCES

[1]   Page, J.(Ed.) 1993. State of the art review, masonry arch bridges. *Transport Research Laboratory, Department of Transport,* HMSO.

[2]   Baker, M.J. 1996. Finite element analysis of masonry with application to arch bridges. PhD Thesis, University of Wales, College of Cardiff.

[3]   Pande, G.N., Kralj, B. & Middleton, J. 1994. Analysis of the compressive strength of masonry given by $f_k = K(f_b)^\alpha (f_m)^\beta$. *The Structural Engineer.* Volume 71. No. 1. pp. 7-12.

[4]   Rots, J.G. 1988. Computational modelling of concrete fracture. Civil engineering department, Delft University of Technology.

[5]   William, K., Hurlburt, B. & Sture, S. 1985. Experimental and constitutive aspects of concrete failure. *Proceedings of the American Society of Civil Engineers, Seminar on Finite Element Analysis of Reinforced Concrete Structures.* Tokyo, Japan.

THREE DIMENSIONAL FINITE ELEMENT ANALYSIS OF MASONRY STRUCTURES USING HOMOGENISATION: APPLICATION TO ARCH STRUCTURES

J. Middleton, G.N. Pande, B. Kralj and **F. Gazzola**
Deaprtment of Civil Engineering, University of Wales, Singleton Park, Swansea SA2 8PP, Wales, UK

1. ABSTRACT

This paper presents an accurate and efficient technique for the structural analysis of masonry arch structures subject to dead and imposed loads. Masonry is treated as a composite material and a two-stage homogenisation technique is used to model its structural response. The first stage homogenisation considers modelling of masonry units including the bed and perpend joints into a homogeneous, orthotropic material. Second stage homogenisation is invoked when cracking of any of the masonry constituents, i.e. units or mortar joints, is detected. Here, the position and orientation of the crack is calculated and the crack is subsequently smeared into the surrounding homogenised material. By adopting this approach the computational effort necessary for standard non-linear finite element (FE) analysis of large three-dimensional masonry structures is considerable reduced and an accurate analysis becomes tractable using a standard workstation or PC. The method is validated using a physical model of an arch structure. The technique is used to model an historic bridge and the results indicate that the approach adopted is highly efficient and permits the accurate analysis of general 3D masonry structures.

2. INTRODUCTION

In recent years, the finite element method has gained immense popularity amongst engineers and a large number of packages for undertaking such analyses are commercially available. If, however these packages are applied to masonry, each unit and joint has to be meshed separately in order to allow for only one material to appear within each element. It can be shown that the number of finite elements, N_{elem}, required to analyse a simple masonry panel in this way is equal to:

$$N_{elem} \approx 8 n_1 n_2 n_3 \tag{1}$$

Computer Methods in Structural Masonry – 4, edited by G.N. Pande, J. Middleton and B. Kralj.
Published in 1998 by E & FN Spon, 11 New Fetter Lane, London EC4P 4EE, UK. ISBN: 0 419 23540 X

where n_1 and n_2 and n_3 are the number of units in the horizontal, vertical and through the thickness directions of the panel, respectively. It is therefore obvious that analysing large, complicated three-dimensional structures using this approach is not feasible. A number of researchers have adopted finite element analyses in which elements span over a number of units and joints and assume average properties of masonry. However such analyses are not valid and give erroneous results.

Using the homogenisation technique, masonry having two constituent materials (units and mortar) can be substituted by a single material. Equivalent (homogenised) material properties of this substitute material, satisfying an equal strain energy principle, can be derived. Also, unique transformations between stresses in the homogenised material and the constituents can be derived. Therefore, instead of analysing the F.E. model for masonry, a substitute, or homogenised equivalent material is analysed.

The computational advantage of using the homogenisation technique, is based on the fact that discretisation does not have to follow the internal structure of masonry (i.e. spatial distribution of units and joints). Here the rules that apply for any homogeneous structure apply so that one finite element can stretch over several units and joints. This advantage has been recognised by several authors leading to a variety of different approaches to homogenisation[1,2].

Initially, the homogenisation technique was introduced in the analysis of jointed rock masses, [3-5] and was applied in a non-rigorous way leading to several steps for the homogenisation procedure, each step corresponding to one set of parallel joints. More rigorous approaches have recently been proposed by several authors such as Anthoine[2] and Urbanski[1]. These approaches, though theoretically sound, have the disadvantage of being rather cumbersome to implement efficiently. This is due to the non-existence of explicit relationships between the constituents' properties and the homogenised properties for the general case of masonry.

Failure of masonry is usually associated with the development of a number of cracks in a brittle manner. Detecting and tracking the growth of cracks can pose an insurmountable task even for high performance computers. In the approach presented here, a second stage of homogenisation is introduced whereby cracks are homogenised with the surrounding material thus removing the need for remeshing required by the standard approaches. In this way, the required computational effort is again reduced whilst the pattern of failure is still preserved.

3. FIRST STAGE OF HOMOGENISATION - EQUIVALENT PROPERTIES OF MASONRY

The homogenisation technique adopted here is based on references[3,5,6,7]. The procedure is based on the condition of equal strain energies of the parent and the homogenised materials and only a brief review is presented here.

The basic assumptions made to derive the equivalent material properties of masonry are:- (1) masonry units and mortar joints are perfectly bonded.
(2) that the perpend mortar joints are continuous.

The first assumption may appear to be restrictive since there is considerable debate on the bond strength of mortars and the mortar unit interface. The bond strength is generally likely to be less than the tensile strength of mortar, especially taking into account the effect of poor workmanship, build quality and missing joints etc,. It will be shown in the next section how this restriction is taken into account in this model. The second assumption is necessary for the homogenisation procedure adopted here and it has been shown[2] that the assumption of continuous instead of staggered perpend joints does not significantly effect the stress distributions in the constituent materials.

Let the compliance matrix of the orthotropic equivalent homogenised material be denoted by $[\overline{C}]$. The Stress-strain relationship of the equivalent homogenised masonry material is respresented in incremental form by

$$\dot{\overline{\varepsilon}} = [\overline{C}]\dot{\overline{\sigma}}$$

(2)

where

$$\dot{\overline{\sigma}} = \left\{\dot{\overline{\sigma}}_{xx}, \quad \dot{\overline{\sigma}}_{yy}, \quad \dot{\overline{\sigma}}_{zz}, \quad \dot{\overline{\tau}}_{xy}, \quad \dot{\overline{\tau}}_{yz}, \quad \dot{\overline{\tau}}_{zx}, \right\}^T$$

$$\dot{\overline{\varepsilon}} = \left\{\dot{\overline{\varepsilon}}_{xx}, \quad \dot{\overline{\varepsilon}}_{yy}, \quad \dot{\overline{\varepsilon}}_{zz}, \quad \dot{\overline{\gamma}}_{xy}, \quad \dot{\overline{\gamma}}_{yz}, \quad \dot{\overline{\gamma}}_{zx}, \right\}^T$$

(3)

and

$$[\overline{C}] = \begin{bmatrix} \dfrac{1}{\overline{E}_x} & -\dfrac{\overline{v}_{xy}}{\overline{E}_x} & -\dfrac{\overline{v}_{xz}}{\overline{E}_x} & 0 & 0 & 0 \\[2mm] -\dfrac{\overline{v}_{yx}}{\overline{E}_y} & \dfrac{1}{\overline{E}_y} & -\dfrac{\overline{v}_{yz}}{\overline{E}_y} & 0 & 0 & 0 \\[2mm] -\dfrac{\overline{v}_{zx}}{\overline{E}_z} & -\dfrac{\overline{v}_{zy}}{\overline{E}_z} & \dfrac{1}{\overline{E}_z} & 0 & 0 & 0 \\[2mm] 0 & 0 & 0 & \dfrac{1}{\overline{G}_{xy}} & 0 & 0 \\[2mm] 0 & 0 & 0 & 0 & \dfrac{1}{\overline{G}_{yz}} & 0 \\[2mm] 0 & 0 & 0 & 0 & 0 & \dfrac{1}{\overline{G}_{zx}} \end{bmatrix}$$

(4)

Using the equivalent strain energy requirement together with equilibrium and kinematic compatibility conditions for the constituents, the exact expressions for the 9 elements of the compliance matrix $[\overline{C}]$, i.e. $\overline{E}_x, \overline{E}_y, \overline{E}_z, \overline{\mu}_{xy}, \overline{\mu}_{yz}, \overline{\mu}_{zxy}, \overline{G}_{xy}, \overline{G}_{yz}, \overline{G}_{zx}$ in a closed form can be found, see [6,7]. These parameters depend on the following:-

• unit size and thickness of mortar joints

- elastic properties of units, and
- elastic properties of joints

The homogenisation procedure outlined above gives the *orthotropic* properties of masonry in a local coordinate system where the x axis is aligned along the length of the unit, y axis along its height and the z axis through the thickness of the panel. In the case of curved structures like arches, the orientation of the principal material directions changes from point to point and it has to be taken into account using a transformation of the type:

$$\left[\overline{C}_g\right] = [T]^T \left[\overline{C}\right][T] \tag{5}$$

Here $\left[\overline{C}_g\right]$ stands for transformed (global) compliance matrix and $[T]$ is a transformation matrix depending on the normal vector of the plane of the bed joints.

The homogenisation procedure used to calculate the expressions for the equivalent material properties also gives unique relationships between stresses in the equivalent material and the stresses in masonry constituents, i.e. units, bed joints and perpend joints in the form

$$\dot{\sigma}_u = [S_u]\dot{\overline{\sigma}}$$
$$\dot{\sigma}_{bj} = [S_{bj}]\dot{\overline{\sigma}}$$
$$\dot{\sigma}_{pj} = [S_{pj}]\dot{\overline{\sigma}} \tag{6}$$

where [S] is a structural matrix and subscripts u, bj and pj stand for bricks, bed joints and perpend joints, respectively. Explicit expressions for those structural matrices are given in [6,7].

The procedure described so far is applicable for the linear analysis of any masonry structure and can be described in the following steps:

(i) Global compliance matrix is calculated for each element (Equation 5)

(ii) Equivalent material properties are calculated from mechanical and geometrical properties of the constituents

(iii) Standard FE solution procedure is undertaken giving displacement and stress fields of the homogenised material. It should be noted that displacements calculated are equal to the displacement of a non-homogeneous structure, but the stresses are equivalent or notional average stresses and are different from the stresses in constituents.

(iv) Using the relationship given by Equation 6 stresses in each constituent material can be calculated.

4. SECOND STAGE OF HOMOGENISATION - MODELLING OF CRACKED MASONRY

Here masonry is modelled as an elastic brittle material with tensile cracking being the only non-linearity considered. In order to achieve further computational efficiency another stage of homogenisation, this time involving cracks and the intact masonry is introduced [6,8].

It is assumed that cracks occur in any constituent if the major principal stress, σ_1 in that constituent equals its tensile strength f_t, i.e. the failure criterion F is:

$$F = \sigma_1 - f_t = 0 \tag{7}$$

The orientation of the plane of crack in three dimensions is normal to the direction of the major principal stress. Once cracking occurs in the material, its effect is smeared onto the neighbouring material through a homogenisation technique and equivalent properties of the cracked masonry are developed. Here, an averaging procedure based on the work of Pietruszczak and Niu[9] is adopted for the three-dimensional case and is described below:

The cracks in masonry are treated as a constituent of masonry. Thus damaged, or cracked masonry consists of two constituents - intact masonry and cracks which are assigned properties of a weak material.

These stress/strain rates in cracked material can be taken as volume averages of the stress/strain rates in the two constituents of composite material,

$$\dot{\sigma} = \mu_i \dot{\sigma}^i + \mu_j \dot{\sigma}^j$$
$$\dot{\varepsilon} = \mu_i \dot{\varepsilon}^i + \mu_j \dot{\varepsilon}^j \tag{8}$$

where, μ_i and μ_j represent the volume fraction of the constituent materials where subscript i is used to denote quantities relating to intact material and j to denote quantities relating to cracks.

Assuming perfect bonding at the *interface* of the crack and surrounding material the equilibrium and kinematic conditions along the interface can be established. It is assumed that the volume occupied by the crack is negligible compared to the volume of the element (which follows from the relatively small width of the cracks). The response of cracks can be conveniently described by introducing a velocity discontinuity $\{\dot{g}\}$ (measure of crack width and tangential movements) which is a function of the strain field and the crack width:

$$\{\dot{g}\} = \left\{ \dot{g}_y, \dot{g}_x, \dot{g}_z \right\}^T \tag{9}$$

Based on the assumption of the negligible crack width and by incorporating the kinematic conditions. strain rate can be rewritten as:

$$[\delta]\dot{\varepsilon} \approx [\delta]\dot{\varepsilon} + \mu\{\dot{g}\} \qquad (10)$$

where

$$[\delta] = \begin{bmatrix} 0 & 1 & 0 & 0 & 0 & 0 \\ 0 & 0 & 0 & 1 & 0 & 0 \\ 0 & 0 & 0 & 0 & 1 & 0 \end{bmatrix} \qquad (11)$$

and μ is a volume fraction of the crack.

After some algebraic manipulations[6,8] the constitutive relationship for the cracked masonry can be obtained as:

$$\dot{\sigma} = [D^{eq}]\dot{\varepsilon} \qquad (12)$$

where,

$$[D^{eq}] = [D][S_i^i] \qquad (13)$$

with $[S_i^i]$ being structural matrices relating strains between the homogenised cracked material and either of its constituents. To take into account the orientation of the crack a rotation of $[D^{eq}]$ of the form

$$[D_r^{eq}] = [T]^T[D][T] \qquad (14)$$

is necessary, where component of the transformation matrix $[T]$ depend on the normal vector of the plane of the crack.

In a FE realisation, loads are applied incrementally and within each load increment, this procedure is implemented as follows:

(i)   After stresses in the constituents are determined, occurrence of cracks in each constituent is checked following Equation 7. For each point under consideration three checks have to be completed, one for each constituent i.e. units, bed joints and head joints. It is important to note that tensile strengths of the constituent materials are generally different.

(ii) If cracking is detected the orientation of the crack is calculated together with the velocity discontinuity vector.

(iii) New, homogenised constitutive relationship for the cracked masonry is evaluated using Equations 13 and 14.

(iv) Out of balance residual stresses are calculated and iteration is performed until all the forces are in equilibrium.

In the above, no checks for failure of bond between mortar joints and units are made. Consequently a pragmatic approach is adopted. Here it is assumed that there is a chain type of link between the joint and the units which will fail when subjected to tensile loading. In computer implementation it means that a tensile strength value which is lesser of the strength of mortar and the bond strength is entered into Equation 7. The two-stage homogenisation procedure described above has been implemented in an in-house 3D FE program STRUMAS (Structural Masonry Analysis System) and this program was used for the solution of the following problems.

5.   EXAMPLES

5.1   Model arch bridge

A model masonry arch bridge which was tested by Melbourne and Gilbert[10] has been simulated using the above described approach.

	Units	Joints
E [N/mm^2]	9000	4000
v	0.2	0.15
f_t [N/mm^2]	11.5	0.9
f_c [N/mm^2]	22.7	22.7

Table 1: Material properties for model arch bridge.

During the construction of the model a deliberate defect of spandrel detachment was made. The effect of this is that there is no coupling between the spandrel and the arch barrel and hence the arch barrel was the only load bearing element. The loads applied to the model were as follows:

- Dead load consisting of the self weight of the arch and the backfill pressure. This was applied as the first load step in the numerical analysis.

- Live load - an incrementally increasing concentrated load was applied at ¼ span until collapse occurred.

Figure 1 shows the geometry of the arch, and Figure 2 the finite element mesh which consists of 396 elements and 2295 nodes. It is emphasised here that the number of elements used is small since an homogenisation approach was adopted. The stresses obtained from the analysis are average stresses and are converted to stresses in constituents according to Equation 6 to assess cracking and damage.

Material properties used in the analysis are given in Table 1. It should be noted here that due to the lack of information, tensile strengths of the materials were assumed according to Grimm[11]. Due to this only a qualitative comparison of the results is possible. Figure 3 shows contour plots of crack widths when the load applied was close to the ultimate load. Formation of a typical four-hinge mechanism is obvious which corresponds to the experimentally observed failure mode.

In Figure 4, the applied load - displacement curve for a single monitoring point (quarter span point on the intrados) is given. This shows a very close agreement to that obtained experimentally. It should be noted that relatively few elements (396) have been used in this analysis. If a finite element model with one element for each unit and one element for each joint were used, the number of elements required would be about 15000 for modelling of the barrel only.

5.2 Pontypridd Bridge

This historic masonry bridge has been the subject of various technical papers and an interesting overview and analysis is given by Coventry[12]. Pontypridd bridge, which spans the river Taff, in Wales, was built in 1755 under the supervision of a local stone mason Thomas Edwards. With a span of 42.7m it was, for three quarters of a century, the largest arch bridge in the UK. The existing bridge was actually his third attempt to span the river. The first bridge with three arches stood for about two years before being destroyed by the floods in 1748-49. The second bridge was a single 42.7m span arch. Even before being finished, due to high self weight of the spandrels walls, the slender middle portion of the arch sprung up destroying the bridge. In his third attempt, in order to avoid the same problem, Edwards introduced cylindrical openings in the spandrel, thus reducing the weight on the haunches. Also, there are some reports stating that charcoal was used as infill in order to reduce dead load. Having such an unusual history, being one of the largest arches in the world for a long time and still being in use today this bridge has attracted much attention. This structure has been chosen by the authors to demonstrate the capabilities of the 'STRUMAS' package in solving complex masonry structures such as historic bridges and buildings.

The 3D finite element mesh of the bridge structure is shown in Figure 5. This consists of 3608, 20 noded isoparametric elements and 19260 nodes. The masonry arch, spandrel walls and cylindrical cavities have been modelled quite accurately. The two-stage homogenisation technique was used for the analysis of the bridge subject to dead and live loads. The live load was taken as a uniformly distributed load acting over one half of the span of the bridge. The foundations of the arch were not modelled due to the lack of reliable information. Soil structure interaction aspects of the fill were considered and two extreme conditions: (a) extrados of the arch perfectly smooth (b) extrados of the arch perfectly rough being considered.

The results of the finite element studies are shown in Figures 5 and 6. As is well known, such studies produce information on deformed shape, strains and stresses. In this particular case, when the homogenisation technique is used, stresses in the stone units, bed joints and perpend joints are also obtained along with crack widths, if indeed cracking takes place. As mentioned earlier, cracking in masonry is one of the most important aspects to be studied. Hence, contours of crack widths have been plotted for all zones where cracking takes place. Insignificant cracking at the springing levels is noticed under the conditions of dead loading only. However, cracking takes place at the intrados and extrados of the arch ring when dead load as well as live loads are considered.

Load Case	Maximum stress Mpa	Minimum stress Mpa
Dead load Back fill action considered	0.09	-12.8
Back fill not considered	0.15	-30.6
Dead load + live load Back fill considered	0.15	-32.1
Dead load + live load Back fill not considered	0.09	-18.8

Table 2: Maximum and minimum stress in masonry subject to various loading conditions

The maximum tensile and compressive stresses computed are 0.15 and 32.1 MPa respectively for the dead load case. Table 2 shows maximum and minimum stresses for various load conditions and also for the cases when the structural action of the infill is considered. The analysis shows that only small regions of the structure are subject to cracking under loading conditions considered.

To perform the computations for the Pontypridd Bridge a Sun Ultra 170 workstation with a UNIX operating system was used. The computer times were of the order of 20 hours for 100 load increments.

6.   CONCLUSIONS

It is shown in this paper that a rational three-dimensional finite element analysis of a large-scale masonry structure is feasible using a moderate sized computer. Since masonry is a composite material, it is essential that the method of homogenisation is used to provide accuracy of analysis. A two-stage homogenisation technique was used to obtain the constitutive properties of intact and cracked masonry. The analysis of the model arch shows that this technique is accurate, effective and viable for the analysis of complex masonry structures. The results of the numerical model matched closely those of the physical experiments. The analysis of William Edwards' bridge in Pontypridd, Wales, has been undertaken for heuristic purposes.

In view of the ease of setting up computer models compared to physical models, it would appear that computer models based on homogenisation technique have an important role to play in the analysis, design and assessment of masonry structures.

It is emphasised that no formal assessment of the integrity of the Edwards Bridge as been undertaken up by the authors. The analyses reported here are purely to demonstrate the effectiveness and validity of the homogenisation technique proposed by the authors.

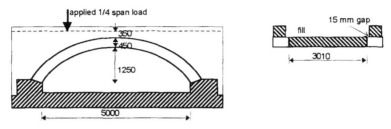

Figure 1. Dimensions and loading conditions of model arch bridge

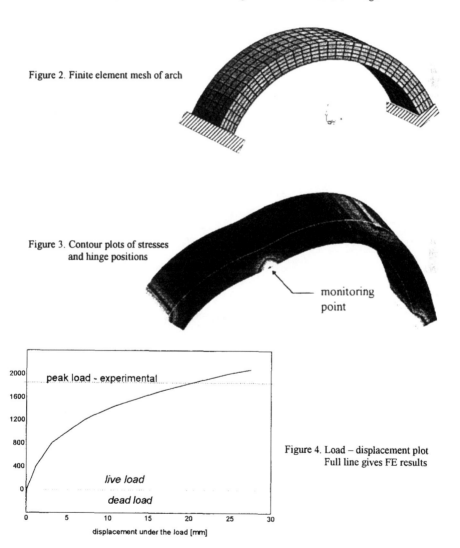

Figure 2. Finite element mesh of arch

Figure 3. Contour plots of stresses and hinge positions

monitoring point

Figure 4. Load – displacement plot
Full line gives FE results

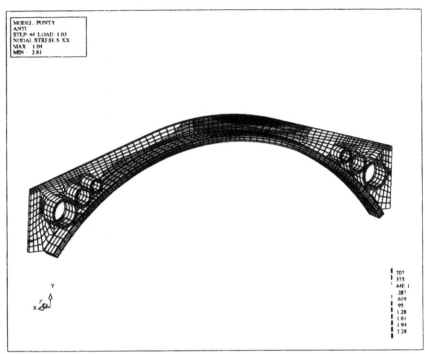

Figure 5. Contour plots of stresses in masonry in the x direction

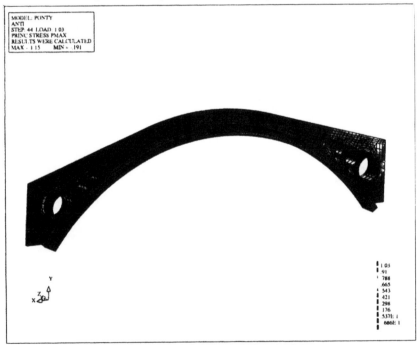

Figure 6. Contour plots of major principal stresses in masonry

7. REFERENCES

1. Szarlinski, J., Kordecki, Z. & Urbanski, A. Finite element modelling of the behaviour of masonry walls and columns by the homogenisation approach. In Middleton, J. & Pande, G.N., editor, *Computer Methods in Structural Masonry - 3*: 32-41, Swansea, Books & Journal International. 1995.

2. Anthoine, A.. Derivation of the in-plane elastic characteristics of masonry through homogenisation theory. *International Journal of Solids and Structures*, 32: 137-163, 1995.

3. Salamon, M.D.G.. Elastic moduli of stratified rock mass. *International Journal of Rock Mech. Min. Sci.*, 5: 519-527, 1968.

4. Gerrard, C.M. Equivalent elastic moduli of a rock mass consisting of orthorhombic layers. *International Journal of Rock Mech. Min. Sci. & Geomech. Abstr.*, 19: 9-14, 1982.

5. Gerrard, C.M. Elastic models of rock masses having one, two or three sets of joints. *International Journal of Rock Mech. Min. Sci. & Geomech. Abstr.*, 19: 15-23, 1982.

6. Pande, G.N., Middleton, J., Kralj, B. & Lee, J.S. Numerical modelling of brick masonry panels subject to lateral loadings. *Computers & Structures*, 61: 734-745, 1996.

7. Pande, GN., Kralj, B. & Middleton, J. On the compressive strength of masonry. *The Structural Engineer*, 72: 15-24, 1995.

8. Lee, J.S. *Finite element analysis of structured media*. Ph.D thesis, University of Wales Swansea, 1993.

9. Pietruszczak, S. & Niu, X. On the description of localised deformation. *International Journal of Numerical Anal. Meth. Geomech.*, 17: 791-805, 1993.

10. Melbourne, C & Gilbert, M. The behaviour of multiring brickwork arch bridges. The Structural Engineer. Vol. 73, No.3/7. 39-47, 1995.

11. Grimm, C.T. Strength and related properties of brick masonry. *ASCE Journal of the Structural Division*, 101: 217-231, 1975.

12. Coventry, W,M. A history of Pont-y-ty-Pridd with an investigation of the stability of the arch. London: Sands & Co., 1902.

DISCRETE ELEMENT MODELLING OF THE SEISMIC BEHAVIOUR OF STONE MASONRY ARCHES

J.V. Lemos Research Officer
Laboratório Nacional de Engenharia Civil (LNEC), Lisboa, Portugal

1. ABSTRACT

A numerical study of the seismic behaviour of stone masonry arches was conducted with a discrete element model, using a three-dimensional rigid block representation. Time domain analyses were performed to obtain the structural response and simulate the failure process. Applications to circular and pointed arches, subject to various levels of in-plane and out-of-plane dynamic loading, are presented.

2. INTRODUCTION

Modelling tools capable of an effective representation of the seismic behaviour of stone masonry are increasingly demanded by projects aimed at the preservation of historical buildings and monuments. Safety assessment and evaluation of restoration actions rely on a correct understanding of the mechanics of these structures, governed to a large extent by their discontinuous nature. Discrete element methods, based on an explicit representation of blocks and joints, supply a powerful tool for the detailed modelling of architectural components [1]. In the present study, these techniques were applied to the analysis of unconfined stone masonry arches under earthquake loading. Free-standing arches are often found in partially ruined structures, and their safety is always a cause for concern. In the analyses, in-plane and of out-of-plane actions and failure modes were considered, and the effect of partial confinement was investigated.

The comprehensive analysis of historical masonry structures still poses a significant challenge to engineers, owing to their complexity and particular patterns of behaviour linked to traditional construction techniques [2]. While the increase of computational

Keywords: Stone masonry, Earthquake engineering, Discrete elements

Computer Methods in Structural Masonry – 4, edited by G.N. Pande, J. Middleton and B. Kralj.
Published in 1998 by E & FN Spon, 11 New Fetter Lane, London EC4P 4EE, UK. ISBN: 0 419 23540 X

power available makes possible the development of models composed of many discrete blocks, for large structures it may be preferable to follow a strategy based on the articulation of global and detailed models. In this approach, a simplified global representation is achieved by standard structural element techniques; discrete element models, such as those presented in this paper, are used to study the behaviour of sub-structures, taking into account failure modes due to separation and sliding of the blocks. Boundary conditions for the detailed models may be supplied by the global analysis, e.g., in terms of acceleration histories at support points. On the other hand, the discrete element models may provide the mechanical characteristics of the individual components (e.g. columns, arches) to be introduced in the global model.

3. BLOCK AND CONTACT REPRESENTATION IN 3DEC

The study was conducted with the discrete element programme 3DEC [3,4] which incorporates both rigid and deformable blocks. For dynamic analysis, the latter, involving an internal element mesh for stress analysis, lead to longer run-times. As the deformation and failure of masonry structures built with competent stone are mostly determined by the behaviour of the joints, a rigid block representation was selected.

The solution is obtained by the numerical integration in time of the equations of motion of the system by means of an explicit algorithm. This technique requires small time steps for numerical stability, but enables a general analysis, accounting for joint separation and sliding, and extending into the large displacement regime. Detection and update of contact between the blocks are performed automatically when the structure undergoes significant changes in geometry.

To model large, complex discontinuous systems with the desired generality, namely involving large displacements, it is convenient to use a local representation of contact. In this approach, no joint elements are defined, the interaction between blocks being represented by sets of point contacts, of either vertex-to-face or edge-to-edge type (Fig. 1). The force at each contact is determined solely by the relative displacement between blocks at that location. In order to simulate correctly the stiffness of a face-to-face contact, each point contact is assigned an area, allowing standard joint constitutive relations, formulated in terms of stresses and displacements, to be applied. Changes of contact type and position may take place during the analysis, requiring a careful treatment of contact force updates.

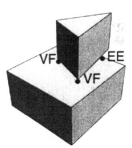

Fig. 1 - Types of points contacts: vertex-to-face and edge-to-edge.

4. DESCRIPTION OF THE MODELS

4.1 Geometry

The first model represents a circular arch of 7.5 m span, with a rectangular cross-section, 0.5 m thick and 1.0 m wide (Fig. 2). The arch was decomposed into 17 blocks

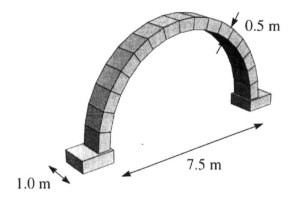

Fig 2 - Circular arch model.

and 2 additional support blocks. The pointed arch model (Fig. 3) has a similar span and a height of 5.3 m. The cross-section was approximated by a T-shape with a height of 0.55 m, a width of 1.0 m at the extrados and 0.4 m at the intrados. This shape was formed by 2 joined blocks (with cross-sections of 1.0x0.2 m and 0.4x0.35 m). In 3DEC, blocks must be convex, so other shapes are created by joining blocks, i.e., constraining them to move as a single rigid body. The keystone is thus composed of 4 joined blocks, leading to a model of 19 independent blocks, plus 2 support blocks.

4.2 Material properties

The blocks were idealised as rigid bodies, with a unit weight of 27 kN/m^3. A Mohr-Coulomb constitutive model was adopted for the joints, with the following properties:

Joint normal stiffness	10^3 - 10^4 MPa/m
Joint shear stiffness	10^3 - 10^4 MPa/m
Friction angle	35°
Cohesion	0
Tensile strength	0

Both stiffness values used are low, representing very deformable joints. Experimental data, however, indicate that irregularities of the block contact surfaces may lead to such ranges of stiffness [5]. Low stiffnesses reduce the system eigenfrequencies, making structures of this size vulnerable to seismic action. The dynamic runs were performed with a low value of mass-proportional damping: 0.3% of critical at 1 Hz. The stiffness-proportional component of Rayleigh damping, which would be desirable to reduce high frequency noise, was not used, as it would require smaller time steps.

4.3 Loading procedure

The arches were first subject to self-weight, and then the dynamic loading was applied in the form of a prescribed motion at the 2 support blocks. The same input history was

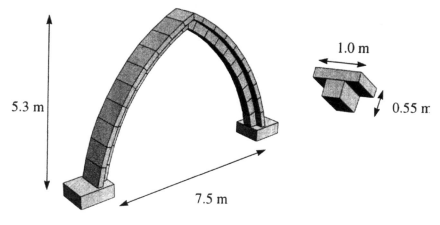

Fig 3 - Pointed arch model.

applied to both supports. In order to analyse the various modes of behaviour, in-plane and out-of-plane actions were applied independently, only in the horizontal direction. The input was provided by the main horizontal component of the 1986 Kalamata earthquake, with a peak ground acceleration of 0.27 g, a peak velocity of 0.24 m/s, and a peak displacement of 0.04 m. The displacement history is displayed in Fig. 4. This input record was scaled by increasing factors until failure ensued. Static analysis with horizontal mass forces, acting either in-plane or out-of-plane, were also performed, in order to compare the equivalent static methods with the dynamic results.

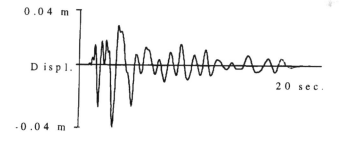

Fig. 4 - Horizontal displacement record, Kalamata earthquake.

5. ANALYSIS OF RESULTS

5.1 Elastic behaviour

Assuming the joints to be elastic, i.e., disallowing separation and sliding, the natural vibration frequencies of the system may be determined. Table 1 lists the values for the lowest symmetric in-plane and out-of-plane modes, for each joint stiffness case. Higher stiffnesses would imply eigenfrequencies above the range of practical interest.

However, it should be kept in mind that the non-linear behaviour of the joints, with block rocking and uplift, leads to a softer system, which may become vulnerable to the frequency content of earthquakes.

Table 1 - Elastic natural frequencies

	k_n (MPa/m)	In-plane (Hz)	Out-of-plane (Hz)
Circular arch	10^3	3.9	3.4
	10^4	12.4	10.8
Pointed arch	10^3	3.1	2.2
	10^4	9.9	6.8

5.3 Circular arch

The earthquake record applied at the base blocks was scaled to produce the arch collapse. The peak accelerations required are listed in Table 2, for the circular arch. The first line indicates the equivalent static mass force that produces failure. The in-plane failure mode, a typical 4-hinge mechanism, is illustrated in Fig. 5. A very low value is obtained, 0.06 g, demonstrating the well-known weakness of unconfined circular arches. For this cross-section, the out-of-plane strength is actually higher. The dynamic analyses lead to higher failure accelerations, but still quite low for the more deformable joints. It is interesting to note that the out-of-plane toppling failure mode (Fig. 5, right) was only attained for static loading; in all dynamic runs, the in-plane 4-hinge mechanism occurred even for out-of-plane loading, because it corresponds to a much lower excitation level. The development of such physical instability is triggered, in the numerical model, by the round-off approximations in the time stepping scheme.

Table 2 - Circular arch model - Failure peak accelerations

		k_n (MPa/m)	in plane loading		out of plane loading	
	static		0.06 g	(I)	0.20 g	(O)
Unconfined	dynamic	10^3	0.09 g	(I)	0.23 g	(I)
		10^4	0.16 g	(I)	0.27 g	(I)
Partially	static		0.20 g	(I)	0.25 g	(O)
confined	dynamic	10^3	0.47 g	(I)	0.61 g	(I)

(I) In-plane failure mode   (O) Out-of-plane failure mode

5.4 Circular arch with partial confinement

Free-standing arches existing in partially ruined monuments often display some lateral constraint at the abutments. This effect was simulated by placing 3 blocks on either side of the arch, to a height of 4.1 m. These blocks were not laterally restrained, resisting the collapse mechanism only by their weight. However, the results indicate a

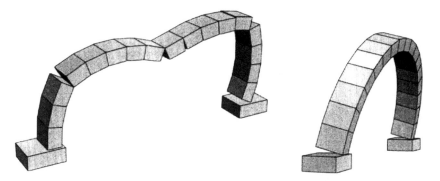

Fig. 5 - Circular arch. In-plane and out-of-plane failure modes.

substantial increase in strength, particularly in the dynamic runs (Table 2). The failure mode is shown in Fig. 6.

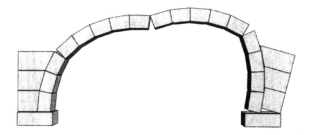

Fig. 6 - Failure mode for partially confined circular arch.

5.4 Pointed arch

The results of the pointed arch analyses are summarised in Table 3. This structural shape and cross-section lead to a higher strength than the circular arch. One of the in-plane failure modes obtained is shown in Fig. 7. Several variations on this mode were obtained, some with significant slip near the key block, leading to disengagement of the blocks. The out-of-plane loading for this arch produced always out-of-plane collapses, such as the one in Fig. 8, which involves significant block rotations.

Table 3 - Pointed arch model - Failure peak accelerations

	k_n (MPa/m)	in plane loading		out of plane loading	
static		0.19 g	(I)	0.20 g	(O)
dynamic	10^3	0.47 g	(I)	0.47 g	(O)
	10^4	0.61 g	(I)	0.67 g	(O)

(I) In-plane failure mode   (O) Out-of-plane failure mode

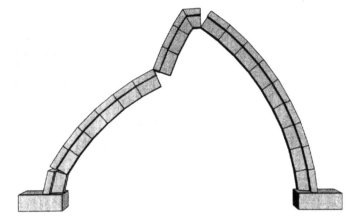

Fig. 7 - Pointed arch model. Failure mode for in-plane loading.

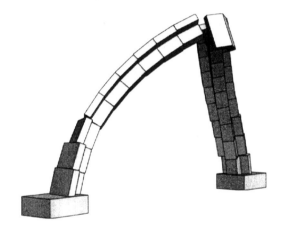

Fig. 8 - Pointed arch model. Failure mode for out-of-plane loading.

5.5 Cross circular arches

Cross vaults are often supported on circular ribs. A simple, conservative estimation of the coupling effect between arches, neglecting the web filling, was performed with a model of two orthogonal arches, each with the dimensions given in Fig. 2. Results for horizontal loading along the direction of either the arch or the bisector planes are given in Table 4, and the three-dimensional failure mechanisms in Fig. 9.

6. FINAL REMARKS

Discrete elements enable the detailed modelling of the dynamics of stone masonry. For complex structures, the assignment of realistic boundary conditions, which reflect the overall behaviour, requires further development. For example, in the analyses reported herein, the same dynamic input was applied at both arch supports, not accounting for the differential movements of support pillars or walls. Joint stiffness is an important parameter for dynamic analysis; further experimental data for various conditions of the contact surface, namely from insitu vibration tests, is required.

Table 4 - Cross circular arches - Failure peak accelerations

		k_n (MPa/m)	arch plane loading	bisector plane load.
Cross arches	static		0.19 g	0.17 g
	dynamic	10^3	0.27 g	0.34 g

Fig. 9 - Failure modes for cross circular arches. Loading on arch or bisector planes.

7. REFERENCES

1. Lemos J.V., 'Discrete element modelling of historical structures', Proc. Int. Conf. New Technologies in Structural Engineering', (Eds. S.P. Santos and A.M. Baptista), LNEC, Lisbon, vol. 2, 1996, pp. 1099-1106.
2. Kordecki Z. and Szarlinski J., 'Structural problems of the gothic art of building', Computer Methods in Structural Masonry - 3 (Eds. Pande and Middleton), Books, & Journals Int., 1995, pp. 262-271.
3. Cundall P.A., 'Formulation of a three-dimensional distinct element model - Part I: A scheme to detect and represent contacts in a system composed of many polyhedral blocks', Int. J. Rock Mech. Min. Sci., vol. 25, 1988, pp. 107-116.
4. Hart R.D., Cundall P.A. and Lemos J.V., 'Formulation of a three-dimensional distinct element model - Part II: Mechanical calculations', Int. J. Rock Mech. Min. Sci., vol. 25, 1988, pp. 117-125.
5. Carydis P.G., Mouzakis H.P., Papantonopoulos C., Papastamatiou D., Psycharis I.N., Vougiokas E.A. and Zambas C., 'Experimental and numerical investigations of the earthquake response of classical monuments', Proc. 11th WCEE, Acapulco, Mexico. 1996.

SHAPE OPTIMIZATION PROBLEMS OF MASONRY DOMES

I. Feletti Post-doctoral Fellow and **M. Rapallini** Post-doctoral Fellow
Dipartimento di Costruzioni, Università di Firenze, Piazza Brunelleschi 6, 50121 Firenze, Italy

1. ABSTRACT

This paper deals with some problems about the static solutions of masonry domes, where the shape profile is spherical or unknown. In particular it's pointed out the effect of the lantern in the stress state of masonry domes. The ogival shells optimum profile of minimum weight under the self-weight and the weight of the lantern in the top of dome are also obtained

2. INTRODUCTION

The great vaulted structures erected before XIX century were built with masonry, a material which, as well known, is not able to resist against tensile stresses.
The classical analysis of these structures by means of membrane theory - that is assuming the hypothesis that the thrust surface does coincide with the neutral surface of the dome - often shows that tensile stresses are present in the lower parallels of domes.
Introducing the hypothesis that masonry is a No-Tension Material, in this research an inverse problem is examined for revolution masonry domes, studied as optimization problems with membrane theory [1,2]. In a previous paper an optimization problem was solved to obtaine the shape of revolution domes with constant thickness and minimum weight; in that case the solution was given only for the particular case of a dome subjected to its own weight and whose shape might be represented by means of a polynomial [3]. The constraint introduced in this problem, deriving from the no-tension characteristic of the masonry, was that the tensor of membrane stresses had to be positive semidefinite.

Keywords: Optimization, Ogival Dome, Masonry

Computer Methods in Structural Masonry – 4, edited by G.N. Pande, J. Middleton and B. Kralj.
Published in 1998 by E & FN Spon, 11 New Fetter Lane, London EC4P 4EE, UK. ISBN: 0 419 23540 X

In this paper, the problem of pointed domes is examined in different weight conditions, also comprehensive of the case of the lantern weight on their top. The ogival shape of masonry domes was often adopted for combining the ease in construction techniques of the spherical shape to higher verticality of springers, that allows a lower thrust on the supporting structures.

A solution has been investigated, always in the context of constrained optimization, that might give the "optimum shape" for the meridian of a masonry pointed dome, in case of minimum thrust and compressive stresses everywhere.

3. BASIC RELATIONSHIPS

3.1 Notation

The following notations has been used: the bar over the symbols indicates the not yet adimensional parameters.

$\overline{N}_\phi$ = Membrane meridional forces per unit of lenght, $N_\phi = \dfrac{\overline{N}_\phi}{\overline{\gamma}\, \overline{s}\, \overline{K}}$;

$\overline{N}_\theta$ = Membrane circumferential forces per unit of lenght, $N_\theta = \dfrac{\overline{N}_\theta}{\overline{\gamma}\, \overline{s}\, \overline{K}}$;

$\overline{W}$ = Weight of the dome ; $\overline{Q}$ = Weight of the lantern;

λ = ratio coefficient between the weight of the dome and the lantern one;

$\overline{q}$ = Weight per unit of lenght of the lantern on the parallel of the top hole;

$\overline{R}$ = radius of curvature of the meridian of the dome;

$\overline{s}$ = thickness of the dome; $\overline{\gamma}$ = specific weight of structural material;

$\overline{K}$ = diameter at the springer of the dome; $\overline{\phi}$ = angle at the springer of the dome;

ϕ_i = angle at the springer of the lantern;

ϕ_0 = angle which expresses the ogivality of the dome.

3.2 Equilibrium equations

The equilibrium equations for a shell in the membrane theory are written using polar coordinates where $\phi,\ \theta \in [0,\pi/2]$.

The integral equilibrium equations for spherical domes with lantern are [4]:

$$\begin{cases} N_\phi(\phi) = -\dfrac{R\,(\cos\phi_i - \cos\phi)}{\text{sen}^2\,\phi} - q\,\dfrac{\text{sen}\,\phi_i}{\text{sen}^2\,\phi} \\[3mm] N_\theta(\phi) = -N_\phi(\phi) - R\cos\phi + q\,\dfrac{\text{sen}\,\phi_i}{\text{sen}^2\,\phi} \end{cases} \tag{1}$$

In these equations, starting from the relation $\overline{\gamma}\, \overline{s}\, \overline{R} = \lambda\, \overline{q}$ and introducing the nondimensional term $R = \dfrac{\overline{R}}{\overline{K}}$ of the radius, we obtain that the load q depends only on the quantities R and λ : $q = \dfrac{R}{\lambda}$. So the integral equilibrium equations for ogival domes with lantern may be written as:

$$
\begin{cases}
N_\phi = -\dfrac{R\left[(\cos\phi_i - \cos\phi) - \text{sen}\,\phi_0(\phi - \phi_i)\right]}{\text{sen}\,\phi(\text{sen}\,\phi - \text{sen}\,\phi_0)} - \dfrac{R}{\lambda}\,\dfrac{(\text{sen}\,\phi_i - \text{sen}\,\phi_0)}{\text{sen}\,\phi(\text{sen}\,\phi - \text{sen}\,\phi_0)} \\[2mm]
N_\theta = -\dfrac{R}{\text{sen}^2\,\phi}\left[\text{sen}\,\phi_0(\phi - \phi_i) - (\cos\phi_i - \cos\phi) + \cos\phi\,\text{sen}\,\phi(\text{sen}\,\phi - \text{sen}\,\phi_0)\right] + \quad (2)\\[2mm]
\qquad + \dfrac{R}{\lambda}\,\dfrac{(\text{sen}\,\phi_i - \text{sen}\,\phi_0)}{\text{sen}^2\,\phi}
\end{cases}
$$

In this case the radius of curvature of the meridian is a constant value, $r_1 = R$, while the the second radius of curvature of the shell is variable, i.e.:

$$
r_2 = \frac{r}{\text{sen}\,\phi} = R\left(1 - \frac{\text{sen}\,\phi_0}{\text{sen}\,\phi}\right) \tag{3}
$$

3.3 No-Tension Material constitutive hypothesis

In the context of membrane theory of shells of revolution, no-tension assumption implies:

$$
\begin{cases}
\overline{N}_\phi(\phi) \le 0 \\[2mm]
\overline{N}_\theta(\phi) \le 0
\end{cases} \tag{4}
$$

4. THE EFFECT OF A LANTERN ON A SPHERICAL DOME

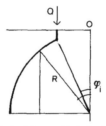

Fig.1 The spherical dome with the load of the lantern

Let's examine the relation existing between the weight $\overline{W}$ of the dome and the weight $\overline{Q}$ of the lantern under our hypothesis:

$$
\begin{cases}
\overline{W} = 2\pi\,\overline{\gamma}\,\overline{s}\,\overline{R}^2(\cos\phi_i - \cos\phi) \\[2mm]
\overline{Q} = 2\pi\,\overline{q}\,\overline{R}\,\text{sen}\,\phi_i \\[2mm]
\overline{\gamma}\,\overline{s}\,\overline{R} = \lambda\overline{q}
\end{cases} \tag{5}
$$

from these three relations we obtain:

$$\overline{W} = \frac{\lambda \overline{Q} \left(\cos\phi_i - \cos\phi\right)}{\operatorname{sen}\phi_i} \qquad (6)$$

which allowed us to express the weight of the lantern in relation with the weight of the dome. Now we want to verify the effect of a lantern weight on the stress state of a spherical masonry dome. In particular it's pointed out the relation between the inversion angle of circumferential stress and the weight and the radius of the lantern [5]. Infact, if we impose that $N_\theta(\alpha)=0$, we can obtain the relation among the admissible width α and the parameters λ and ϕ_i :

$$N_\theta(\alpha) = \lambda \left(\cos\phi_i - \cos\alpha\right) - \lambda\left(\cos\alpha \operatorname{sen}^2 \alpha\right) + \operatorname{sen}\phi_i = 0 \qquad (7)$$

By means of this expression, it's possible to reckon which is the maximum value among the radiuses at springer of the dome, once the multiplicative ratio λ and the value of ϕ_i have been imposed.

Otherwise we can allott the dimension of the dome, through the angle α, and the size of the lantern -through ϕ_i - and, in this way, we can find its limit weight, that may give a measure of its maximum height.

In the next table the corresponding values are given, as well as the curve which expresses the variation of $\alpha(\lambda)$ for a lantern with radius at springer equal to 0,1736R ($\phi_i =10°$).

Tab. 1: $\phi_i =10°$			
λ	α	λ	α
2	42,97°	15	52,13°
3	47,51°	20	52,37°
4	49,16°	25	52,51°
5	50,04°	30	52,60°
6	50,60°	35	52,67°
7	50,98°	40	52,72°
8	51,26°	45	52,76°
9	51,47°	50	52,79°
10	51,64°	100	52,93°

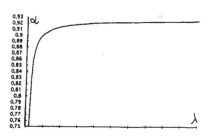

Fig.2 : Curve expressing the variation of $\alpha(\lambda)$

We can note that if $\lambda=10$, or if the lantern weight is $Q=1/57 \ P$, for $\phi_i<\phi<90°$, the inversion angle of the circunferential stress is $\alpha = 51,64°$.

If we increase the lantern weight, the range of admissibility for parallel stresses decreases until, at limit, is $\alpha =42,97°$, which happens when is almost $Q =1/6 \ P$.

Analogously, if we consider an angle $\phi_i=20°$, we can obtain admissible values for $3<\lambda<100$, with corresponding α: $46,05°<\alpha<56,02°$. In this case the curve flattens at $\lambda>6$.

We can point out a common characteristic of these curves, calculated for the several values of the angle ϕ_i : infact, we note a quick increase of the admissibility angle when λ

grows -which corresponds to a decrease of the weight of the lantern-, but for a limited range of values. Infact the curve flattens (decreases its slope) at $\lambda > 5$, and this behaviour signs out the decrease of the influence of the weight Q with respect to the structure below: the stress solution of this case is that of the only self-weight.

If we compare these two cases, we can observe that the inversion point of the circumferential stress decreases, when the dimension of the lantern radius at springer rises: infact for $\lambda = 50$ and $\phi_i = 10°$ we find $\alpha = 52,79°$; and for $\lambda = 50$ and $\phi_i = 20°$, we obtain $\alpha = 59,94°$. It's evident that the weigth of the lantern makes lower the value of the angle α, giving a positive contribution to the stress state. It's to be pointed out that, in these cases of spherical shape of the dome and presence of a lantern, we can't choose the springer angle of the dome.

5. SHAPE OPTIMIZATION FOR REVOLUTION OGIVAL MASONRY DOMES.

5.1 General problem

To improve the results and to obtain the shape of a revolution masonry dome subjected to compressive stress only, in the presence of a hole and a lantern, we propose to solve an optimization problem: the solution will be searched within the ogival domes, i.e. those domes generated by the rotation of a circumference with its center of curvature out of the symmetry axis. In this way the meridian shape of an ogival dome will be determined, with minimum weight and compressed everywhere under different conditions of weight.

We must solve a problem of constrained optimization [6], with presence of equality and inequality contraints: the form of the choosen solution method adopts the construction of the Lagrangian function.

As design variable we generally assume the parameter ϕ_0 which expresses the dome "ogivality", and the parameter $\overline{\phi}$ which expresses the spring of the dome, so that to minimize the weight of a dome with maximum diameter K, assigned at the springer.

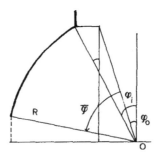

Fig.3 The shape of ogival domes with lantern

In the general problem we impose that the circumferential stresses must be zero at the parallel where we want $\left(\phi = \overline{\phi} \right)$ -that is at the springer of the dome- and we also impose some inequalities describing the constraints for the parameters which rules on the geometry of the dome. We can express the general problem as:

$$\left[\begin{array}{l} \min W\left(\bar{\phi},\phi_0\right) \\[2mm] \text{sub } N_\theta\left(\bar{\phi},\phi_0\right) = 0 \\[2mm] f\left(\phi_0,\bar{\phi}\right) \le 0 \end{array}\right. \qquad (8)$$

where $f\left(\phi_0,\bar{\phi}\right)$ represents a set of inequality constraints, depending on the specific case by the geometry and the weights.

In the next specific problem we'll consider the presence of the hole and of the lantern at the top of the dome.

5.2 The problem for an ogival dome, opened at the top, subjected to its self-weight and to the weight of a superior lantern

The choosen design variables are ϕ_0 and $\bar{\phi}$; the optimization problem becomes:

$$\left[\begin{array}{l} \min W = R^2 \int\limits_{\phi_0}^{\bar{\phi}}\left(\operatorname{sen}\phi - \operatorname{sen}\phi_0\right)d\phi \\[4mm] \text{sub } N_\theta\left(\bar{\phi}\right) = 0 \\[3mm] N_\theta\left(\phi_i\right) \le 0 \\ \phi_0 \ge 0 \\[3mm] \phi_0 - \bar{\phi} + I = 0 \\ \left(\operatorname{sen}\phi_i - \operatorname{sen}\phi_0\right) \ge 0 \end{array}\right. \qquad (9)$$

The Lagrangian function of the problem is:

$$L = W + l_1 N_{\theta\left(\bar{\phi}\right)} + l_2\left(-\phi_0 + s_2^2\right) + l_3\left(\phi_0 - \bar{\phi} + I\right) + l_4\left(R\left(-\operatorname{sen}\phi_i + \operatorname{sen}\phi_0\right)\right) + \\ l_5\left(N_{\theta\left(\phi_i\right)} + s_5^2\right) \qquad (10)$$

The respect of the sufficient conditions of the second order shows that the stationary points are also minimum points for the objective function.

Once solved, the conditions of Kuhn-Tucker provided the several values of ϕ_0 and $\bar{\phi}$ when ϕ_i,λ,I vary.

It may be observed that for every ϕ_i and I exists a range of admissible values of λ, which expresses the fulfill of the limit on $\bar{\phi}$ and ϕ_0 throught the inequality constraints.

When λ grows, that is when the weight of the lantern decreases, the parameters ϕ_0 and $\bar{\phi}$ increase till a limit value, which depends on the size of the lantern radius, and beyond which a further decrease of the weight of the lantern doesn't change the optimum shape of the dome.

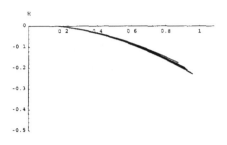

Fig.4: Profiles of the ogival vaults in the cases of $\phi_i=10°$ and $\lambda=7,9,11,20,30,40,50,60,70,80,90,100$

It is possible to analyze the results obtained for example in the case of $I = 50°$, $\phi_i = 10°$ and $\lambda = 6$. The stress state in the shell is explained in the next table: it may be pointed out that the meridional stress decreases from -0,4942 at the apex to -0,3157 at the springs; the circumferential stress decreases from -0,0107 to -0,0139. In the case of $\lambda=100$ may be observed that the stress state is similar to the state of a dome with an hole and without lantern. In the case of $I = 50°$, $\phi_i = 20°$ and $\lambda = 5$ the stress state in similar to the previous one.

Tab. 2: $\phi_i=10°$ $I=50°$ $\lambda=6$		
ϕ	$N\phi$	$N\theta$
10°	-0,4942	-0,0107
15°	-0,3656	-0,1179
20°	-0,3084	-0,1604
25°	-0,2958	-0,1606
30°	-0,2903	-0,1464
35°	-0,2913	-0,1282
40°	-0,2973	-0,0923
45°	-0,3034	-0,0636
51°	-0,3157	-0,0139

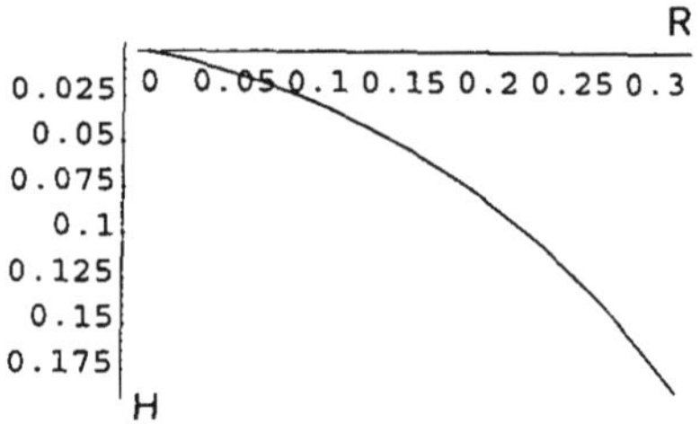

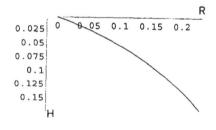

Fig. 5: Profile of the dome for $\phi_i=10°$, $\lambda=6$.

Tab. 3: $\phi_i=20°$ $I=50°$ $\lambda=5$		
ϕ	$N\phi$	$N\theta$
20°	-0,3024	-0,1657
25°	-0,2913	-0,1662
30°	-0,2872	-0,1498
35°	-0,2889	-0,1313
40°	-0,2955	-0,0652
45°	-0,3020	-0,0633
51°	-0,3145	-0,0165

Fig.6: Profile of the dome for $\phi_i=20°$, $\lambda=5$.

In the next figure the curve expressing the relation among P/Q and ϕ_0, is plotted. A range of admissible values for λ exists for every ϕ_i (lantern dimension). It may be observed that as λ grows, i.e. for the decrease of the lantern weight, the parameter ϕ_0 grows too, till a value where the optimum shape doesn't modify.

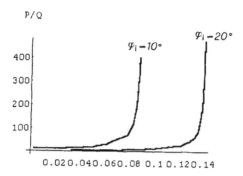

Fig. 7: Relation among P/Q and ϕ_0, ϕ_i

In this case of shape optimization it's possible to state a range I for the difference between the parameters ϕ_0 and $\bar{\phi}$ and the dimension of the hole at the top: a solution always exists, while the lantern weight has a limit.

To generalize this problem, the next step of the research will be the solution in the case where the shape of the dome is approximate through cubic splines.

6. REFERENCES

1. Prager W.and Rozvany G.I.N., 'Optimal spherical cupola of uniform strenght', Ing. Arch., 49, 1980, pp.287-293

2. Pesciullesi C., Rapallini M., Tralli A. and Cianchi A., 'Optimal Spherical Masonry Domes of Uniform Strenght', ASCE Jou. of Str. Engng., 123 n.2, 1997, pp.203-209

3. Rapallini M., Tempesta G. and Tralli A., 'Cupole in muratura: problemi inversi', Atti del Conv. Naz. 'La meccanica delle murature tra teoria e progetto', Messina, Italy, 1996, pp.255-264

4. Flugge W., 'Stresses in shells', Springer Verlag, 1962

5. Rapallini M., 'La statica delle cupole in muratura da Giovanni Poleni a William Prager', Tesi di dottorato, Firenze, Italy, 1994

6. Banichuk N.V., 'Introduction to optimization of structures', Springer Verlag, 1990

THE BEHAVIOUR OF ANCIENT MASONRY TOWERS UNDER LONG-TERM AND CYCLIC ACTIONS

G. Mirabella Roberti Assistant Professor, **A. Anzani** Research Fellow and **L. Binda** Full Professor
Department of Structural Engineering, Politecnico di Milano, Italy

1. ABSTRACT

Masonry specimens sampled from ancient buildings have been subjected to long term compressive tests, compressive tests at constant load steps and cyclic load tests. The development of creep behavior at relatively low stress level and a creep-fatigue interaction in the case of cyclic load tests has been observed. A constitutive rheological model is proposed for the interpretation of the experimental results.

2. INTRODUCTION

The safety of historic masonry buildings has become a concern in Italy and in Europe after the collapse of some monumental constructions (the Civic Tower of Pavia, 1989; the bell-tower of St. Magdalena in Goch, 1992; the Noto Cathedral, 1996). An appropriate calculation of safety factors can hardly be carried out in the case of historic buildings due to the lack of information on the mechanical properties of the material and of suitable models which can take into account the specific characteristics of masonry (e.g. non-homogeneity, anisotropy, inelastic behavior, etc.).

Nevertheless, some of the above mentioned collapses have pointed out that ancient structures under the action of heavy dead loads show a time-dependent behavior capable to worsen the strength and deformation properties of the material. The effects of this behavior on the safety of historical buildings should be enabled to be accounted for. To this purpose, experimental studies have to be carried out and appropriate constitutive models have to be implemented in order to predict possible danger of collapse, and hopefully prevent them.

3. EXPERIMENTAL PROGRAM

Masonry prisms belonging to the Pavia Tower (XI cent.) and to the crypt of the Cathedral of Monza (XVI cent.), a building of the same age as the bell-tower annexed to the Cathedral, have been investigated by compressive tests, trying to simulate on service

Computer Methods in Structural Masonry – 4, edited by G.N. Pande, J. Middleton and B. Kralj.
Published in 1998 by E & FN Spon, 11 New Fetter Lane, London EC4P 4EE, UK. ISBN: 0 419 23540 X

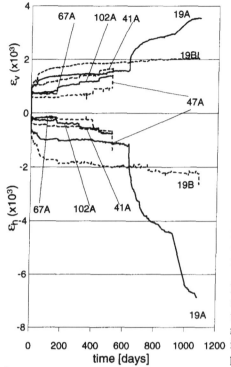

a)

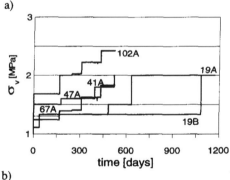

b)

Fig. 1 - Long term tests on the masonry from the Tower of Pavia: average strain (a) and nominal stress (b) vs. time.

conditions. Given the impossibility of collecting samples directly from the bell-tower of the Cathedral of Monza, the chance was exploited of taking material from the masonry of the crypt, which was built with the same material and in the same period of the tower. Following previous experimental research (Anzani et al. 1995), long term tests were carried out on prisms cut from the ruins of the Pavia tower and a series of constant load step and cyclic load compression tests were carried out on the prisms cut from the crypt of Monza. Although such accelerated tests are not equivalent to true creep tests, nevertheless they appear to be useful in detecting the creep behavior and fatigue-creep interaction of the material.

Representative parameters of the material behavior have been obtained from the results according to a simple rheological model, which may be implemented in structural analysis to predict the lifetime of the structure.

3.1 Long term tests

6 prisms of $(300{\times}300{\times}510)$ mm^3, cut from the ruins of the Tower of Pavia, have been tested in controlled conditions of $20°C$ temperature and 50% RH using hydraulic machines able to keep constant a maximum load of 1000 KN. Individual readings of vertical and horizontal displacements were taken directly on the sample with LVDTs on a base of 30 cm; overall readings were also taken in the vertical direction. The load was applied in subsequent steps, kept constant until either the creep strain reached a constant value or a steady state was attained. The initial stress was chosen between 40% and 50% of the static strength which was estimated according to the results of sonic tests previously carried out. Of course, due to the different response of the material, each prism had a different load history.

4 prisms have already been tested to failure whereas two of them (41A and 102A) are still under loading. In fig. 1 the results obtained on all prisms are shown.

3.2. Constant load steps tests

Compression tests in load control were carried out on 4 prisms of dimensions (200×200 ×300) mm³, loading the prisms monotonically up to a stress value of 2.25 MPa and then applying the load in subsequent steps kept constant for defined periods of time of about 180 min., during which creep strain took place.

In fig. 2 the vertical stress vs. vertical strain and the vertical strain vs. time are shown, obtained on one of the prisms tested. Creep strain can be clearly observed when the load is kept constant, with the appearance of tertiary creep during the application of the last load step.

3.3. Cyclic tests

Cyclic tests were carried out in load control out on 3 prisms of dimensions (200×200× 300) mm³. The samples were loaded monotonically up to a stress value of 2.25 MPa, equal to the 65% of the average static compressive strength previously obtained with monotonic tests; cyclic actions of ± 0.15 MPa at 1 Hz were then applied for a period of 1.5 hours. The vertical stress was subsequently increased of 0.25 MPa, and a new cyclic phase was applied; this sequence was then repeated until failure.

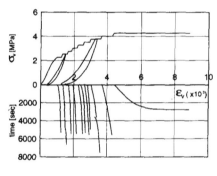
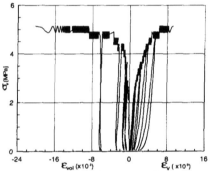

Fig. 2 - Results of a test with constant load steps carried out on prism Ip4.

Fig 3- Results of a cyclic test carried out on prism Ip1.

In fig. 3 vertical stress vs. vertical and volumetric strain diagrams obtained on one of the prisms tested are shown as an example. It can be observed that, during the application of the cyclic load, large deformations take place. Moreover, considering the volumetric strain, it appears that dilation occurs to the material after very low stress values are reached.

4. ANALYSIS OF THE RESULTS

In order to gain more information on the mechanical behaviour of the masonry under study, some parameters were calculated from the experimental data as described in the followings.

The individual deformations measured on the four sides of the same specimen in the vertical and the horizontal directions reported in fig. 4 may look very scattered. Nevertheless, all readings are showing the same trend, therefore taking an average value in both directions and assuming homogeneous and isotropic behaviour of the material, looks convenient and allows to refer to volumetric and deviatoric strains.

In a standard compression test only vertical stress σ_v is different from zero, and if isotropic behaviour is assumed the two horizontal deformations ε_h are equal: so in this case volumetric and deviatoric deformations can be expressed respectively by:

$$e_{vol} = \varepsilon_v + 2\varepsilon_h \qquad (1)$$

$$\varepsilon_h' = \varepsilon_h - \tfrac{1}{3}e_{vol} = \varepsilon_h - \tfrac{1}{3}\varepsilon_v - \tfrac{2}{3}\varepsilon_h = \tfrac{1}{3}(\varepsilon_h - \varepsilon_v)$$
$$\varepsilon_v' = \varepsilon_v - \tfrac{1}{3}e_{vol} = \varepsilon_v - \tfrac{1}{3}\varepsilon_v - \tfrac{2}{3}\varepsilon_h = \tfrac{2}{3}(\varepsilon_v - \varepsilon_h) \qquad (2)$$

and the stress components are expressed by: $p = \tfrac{1}{3}\sigma_v$, $\sigma_v' = \tfrac{2}{3}\sigma_v$, $\sigma_h' = -\tfrac{1}{3}\sigma_v$.

4.1 Long term tests

Looking at fig. 4b some interesting aspects can be outlined: (i) despite the apparent scatter due to the sensitivity of the calculated value to random reading errors, the volumetric strain seems nearly constant during the first part of the test; (ii) the creep behaviour is evident from the deviatoric strain plots.

Generally speaking, after a considerable amount of time, and particularly when failure is approached, volumetric deformation decreases and becomes negative (negative deformation corresponds to dilation due to fracturing and crack opening) while the slope of the plot, i.e. the strain rate, increases.

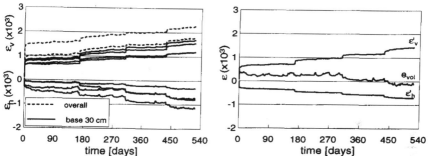

Fig. 4 - Prism 102A from the tower of Pavia: (a) vertical and horizontal strain components vs. time; (b) volumetric and deviatoric strain components vs. time.

4.2 Constant load steps tests

Due to the testing procedure explained above, steady state cannot be completely separated from primary creep, because the effect of the first load step is not completely exhausted when the second step is applied.

Nevertheless, some interesting remarks can be made from the values of the strain rate calculated at the end of each load step, as shown in fig. 5 for specimen Ip4 and in fig. 6 for all tests. An increasing trend is evident after a certain stress level, indicating that the material undergoes mechanical damage; the relationship between strain rate and stress departs from linearity after a certain stress level is reached and this is more evident for volumetric strain rate, as shown in fig. 6 where it is negative from the beginning of the test. This probably indicates that the material from the crypt of Monza was already in bad conditions when started to be tested, having probably undergone a damaging process during its previous load history.

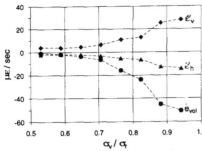
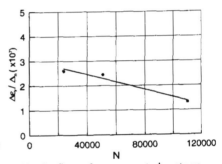

Fig. 5 - Prism Ip4 from the crypt of Monza: secondary creep strain rate vs. normalised stress level.

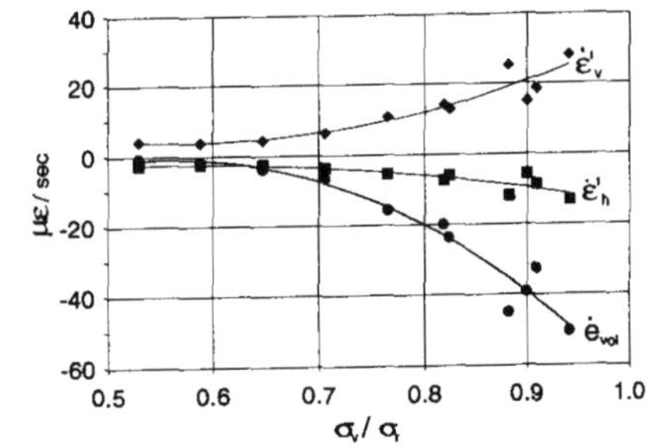

Fig. 6 - All prisms from the crypt of Monza: secondary creep strain rate vs. normalised stress level.

4.3 Cyclic tests

The strain rate per cycle was calculated for the prisms tested cyclically, as presented by Taliercio & Gobbi (1996) relatively to cyclic tests on concrete specimens. It was interesting to notice that plotting the strain rate vs. the number of cycles allows to distinguish primary, secondary and tertiary creep phases quite clearly.

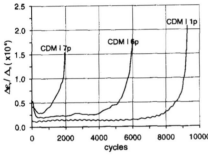

Fig. 7 - Vertical strain rate vs. no. of cycles at a stress level equal to $\sigma_m \pm 0.15$ MPa.

Fig. 8 - Secondary creep strain rate vs. fatigue life N (total number of cycles)

Fig. 7 shows the results relative to the last series of cycles obtained on each prism: though the number of test results is not particularly significant, anyway it is interesting to notice that a relationship between the strain rate of the secondary creep phase, corresponding to the diagram branch having horizontal tangent, and the fatigue life of the material, corresponding to the total number of cycles at failure, can be found. In particular, the higher the strain rate of secondary creep phase, the shorter the fatigue life, as appears also from fig. 8.

5. RHEOLOGICAL CONSTITUTIVE MODEL

Once the time-dependent and fatigue behaviour of the material has been outlined, appropriate constitutive laws can be assumed. Given the difficulty of interpreting the non-homogeneity, anisotropy and non-linearity of the material response, the Authors attempted to apply a rheological model based on classical linear viscoelasticity.

Referring to uniaxial loading (see Flügge 1967), the strain at time t is defined by:

$$\varepsilon(t) = \int_{-\infty}^{t} J(t-\tau)\frac{d\sigma(\tau)}{d\tau}d\tau \qquad (3)$$

where $\sigma(-\infty) = 0$, and $J(t)$ is the creep compliance function.

The creep compliance can be obtained from a one-step compression test; in this case $\sigma(t) = \sigma_0 H(t)$, where $H(t)$ is the Heaviside step function, defined as $H(t) = 0$ for $t < 0$ and $H(t) = 1$ for $t > 0$; the strain is obtained as:

$$\varepsilon(t) = \int_{-\infty}^{t} J(t-\tau)\sigma_0\delta(\tau)d\tau = J(t)\sigma_0, \qquad (4)$$

being $\delta(t) = \dfrac{dH(t)}{dt}$ the Dirac delta, so that $J(t)$ can be obtained as $J(t) = \dfrac{\varepsilon(t)}{\sigma_0}$.

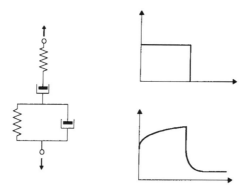

Fig 9 - Bürger model, stress history and strain response

5.1 Rheological model

A simple way of accounting for primary and secondary creep is to use a rheological model, as already proposed in previous research where special attention was paid to the visco-elastic behavior of masonry prisms (Anzani et al., 1993). A Bürger model is here adopted, composed by a Maxwell element connected in series with a Kelvin element (Fig. 4). The total deformation $\varepsilon(t)$ of the Bürger model is:

$$\varepsilon(t) = \varepsilon_M(t) + \varepsilon_K(t). \qquad (5)$$

where $\varepsilon_M(t)$ is the deformation accounted for by the Maxwell element and $\varepsilon_K(t)$ is the deformation accounted for by the Kelvin element. The governing equations of Maxwell and Kelvin elements have respectively the form

$$\dot{\varepsilon}_M = \dot{\sigma}/E_M + \sigma/\mu_M, \qquad \sigma = E_K\varepsilon_K + \mu_K\dot{\varepsilon}_K, \qquad (6,7)$$

where E corresponds to the stiffness of the spring, μ is the coefficient of viscosity of the dash-pot and the superposed dot means time derivative (see Hult, 1966).

The viscoelastic equations can be conveniently generalised to multiaxial stress states, referring to bulk (or *volumetric*) and to shear (or *deviatoric*) behaviour instead of uniaxial, and the following quantities can be defined for $0 < t < t_1$

$$e_{vol}(t) = J_V(t)p_0, \qquad \underline{\varepsilon}'(t) = J_S(t)\underline{\sigma}'_0 \qquad (8,9)$$

where:

$e_{vol} = \text{tr}(\underline{\varepsilon})$ is the volumetric strain;

$p_0 = -\frac{1}{3}\text{tr}(\underline{\sigma}_0)$ is the volumetric stress;

$\underline{\varepsilon}' = \underline{\varepsilon} - \frac{1}{3}e_{vol}\mathbf{I}$ is the deviatoric strain tensor;

$\underline{\sigma}'_0 = \underline{\sigma}_0 + p_0\mathbf{I}$ is the deviatoric stress tensor.

If a Bürger model is chosen to represent the creep behaviour of the material, the volumetric J_V and deviatoric J_S creep compliance can be written as:

$$J_V(t) = \frac{1}{K_0}\left(1+\frac{t}{\tau_0^v}\right)+\frac{1}{K_1}\left(1-e^{-t/\tau_1^v}\right) \qquad (10)$$

$$J_s(t) = \frac{1}{G_0}\left(1+\frac{t}{\tau_0^s}\right)+\frac{1}{G_1}\left(1-e^{-t/\tau_1^s}\right) \qquad (11)$$

It can be noticed that to calibrate the model eight material parameters have to be experimentally determined from the same test. From a one-step test, the volumetric and the deviatoric creep compliance, J_V and J_S can be respectively obtained by:

$$J_V = \frac{e_{vol}}{p} = 3\frac{\varepsilon_v + 2\varepsilon_h}{\sigma_v}, \quad J_s = 2\frac{\varepsilon_h'}{\sigma_h'} \equiv 2\frac{\varepsilon_v'}{\sigma_v'} = 2\frac{\varepsilon_v - \varepsilon_h}{\sigma_v} \qquad (12, 13)$$

5.2 Time-step integration

The rheological model has been implemented into a FE code and a time-step integration method has been adopted. In the uniaxial case, during the time increment Δt_k, the stress increment $\Delta\sigma_k$ is assumed constant, so it can be calculated as: $\Delta\sigma_k = E_k(\Delta\varepsilon_k - \Delta\eta_k)$ where, for the Bürger model:

$$\frac{1}{E_k} = \frac{1}{E_0}\left(1+\frac{\Delta t_k}{2\tau_0}\right)+\frac{1}{E_1}\left[1-\left(1-\exp\left(-\frac{\Delta t_k}{\tau_1}\right)\frac{\tau_1}{\Delta t_k}\right)\right]$$

$$\Delta\eta_k = \left[1-\exp\left(-\frac{\Delta t_k}{\tau_1}\right)\right]\varepsilon_{k-1}^* + \frac{\sigma_{k-1}}{E_0}\frac{\Delta t_k}{\tau_0} \qquad (14, 15)$$

which can be calculated at the previous step *k-1* via the recursive formula:

$$\varepsilon_{k-1}^* = \Delta\sigma_{k-1}\frac{1}{E_1}\left[1-\exp\left(-\frac{\Delta t_{k-1}}{\tau_1}\right)\right]\frac{\tau_1}{\Delta t_{k-1}}+\exp\left(-\frac{\Delta t_{k-1}}{\tau_1}\right)\varepsilon_{k-2}^* \qquad (16)$$

5.3 Calibration of the model

To calibrate the model, the volumetric and deviatoric mechanical parameters expressed in eqs. 10, 11 were calculated from the experimental results for all the prisms subjected to long term tests. In fig. 10 the relationship between the deviatoric parameters and the stress values are shown as an example in the case of one of the prisms tested.

Preliminary results, obtained by the application of the rheological model to numerically simulate the long term tests experimentally carried out, are shown in fig. 11. It is encouraging to notice that from a qualitative point of view the trend of the strain vs. time diagram of the initial tests phase is quite similar to that

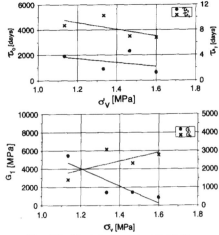

Fig. 10 - Deviatoric parameters vs. stress level from prism 102A.

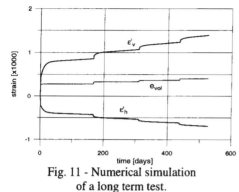

Fig. 11 - Numerical simulation
of a long term test.

obtained in the laboratory. The research is still in progress in order to widen the model applicability, so to interpret also the material behaviour when failure is approached.

6. CONCLUSIONS

The time-dependent behaviour of old masonry has been experimentally shown, together with the fatigue-creep interaction under cyclic loading. Due to the low numerosity of the sample, a statistical data elaboration cannot be made. Nevertheless, since two different types of masonry have shown a similar behaviour, the proposed interpretation gains larger support.

The action of cyclic and persistent loads have proved to cause a severe damage on the mechanical properties of ancient masonry. The fatigue life of masonry under uniaxial cyclic compression is related to the secondary creep strain rate, which is the strain rate during the phase of stable cyclic damage growth.

The creep behaviour of masonry can be suitably studied through simple rheological models: it was shown that the mechanical parameters governing the viscous behaviour of masonry depend on the stress level, a decrease of their value and a damage development being associated to an increase of the stress level.

7. ACKNOWLEDGEMENTS

ENEL-CRIS is gratefully acknowledged for carrying out the long term tests; archh. P. Garau and L. Giannecchini for elaborating the experimental data.

8. REFERENCES

1. Anzani A., Binda L., Melchiorri G., 'Time dependent damage of rubble masonry walls', Proceedings of the British Masonry Society, n. 7, vol. 2, pp. 341 - 351, 1995
2. Taliercio A. L. F, Gobbi E., 'Experimental investigation on the triaxial fatigue behaviour of plain concrete, Magazine of Concrete Research', 48, No. 176, Sept., 1996, pp. 157-172.
3. Flügge W., 'Viscoelasticity', Blaisdell Pub. Co., Waltham, Ma, 1967.
4. Anzani A., Binda L., Mirabella Roberti G., 'Time Dependent Behaviour of Masonry: Experimental Results and Numerical Analysis', 3rd STREMA, Bath, Computational Mechanics Publications, 1993, pp. 415-422.
5. Hult J.A., Creep in Engineering Structures, Blaisdell Pub. Co., Waltham, Ma, 1966.
6. Abrams D.P., Noland J.L., Atkinson R.H., 'Response of clay-unit masonry to repeated compressive forces', Proceedings of the 7th IBMaC, Melbourne, Australia, 1985, pp. 565-576.
7. Binda L., Mirabella Roberti G., Poggi C., 'Il campanile del Duomo di Monza: valutazione delle condizioni statiche', Editoriale L'Edilizia, n. 7/8, 1996, pp. 44-53.

LIMIT ANALYSIS OF MASONRY DOMES

M. Rapallini Post-doctoral Fellow and **G. Tempesta** Associate Professor
*Dipartimento di Costruzioni, Università degli Studi di Firenze, Firenze,
Italy*

1. ABSTRACT

The peculiarity of using methods of graphical statics lies in its outstanding flexibility and in the related possibility of easily modelling a wide variety of structural shapes. These methods are furthermore valid whenever applied to the analysis of existing masonry structures. This paper discusses an analytical interpretation concerning the graphical static analysis of axial-symmetrical masonry domes described by M. Lévy .

1 INTRODUCTION

By 1740, that is, 150 years after the magnificent dome of S. Pietro in Rome was completed, some of the cracks, which are always present in masonry structures, had widened so much that concern was felt for its safety (Fig. 1). A number of experts were consulted in 1742 and 1743. One of them was Giovanni Poleni, professor of experimental philosophy at the University of Padua. He submitted his report in 1743 and published it in a book form [1] in 1748. Poleni's theory was based on *'Linae Tertii Ordinis Neutoneanae '*,a book published in Oxford in 1717 by J.Stirling. In this book was included Gregory's theory with the suggestion that the catenary might be formed by balancing smooth and frictionless spheres. In his analysis Poleni reproduced this saggestion and observed that an arch subject to uniform loading should be proportioned to have the shape of an inverted catenary. He used the same idea for the analysis of the dome (Fig. 2). Considering thin *orange slices* segments of the dome, each forming an arch, he defined the line of thrust and showed that it lay within the masonry thickness of the dome. Poleni made up a string of heavy balls, the weight of each ball being proportional to the weight of a unit portion of the orange-segment, determining its catenary shape by hanging the string from to supports (Fig. 3). Turning this curve and superimposing it on the cross section of the dome, he founded that the catenary lay entirely within the thickness and that the *orange slices arches* were therefore quite safe.

Computer Methods in Structural Masonry – 4, edited by G.N. Pande, J. Middleton and B. Kralj.
Published in 1998 by E & FN Spon, 11 New Fetter Lane, London EC4P 4EE, UK. ISBN: 0 419 23540 X

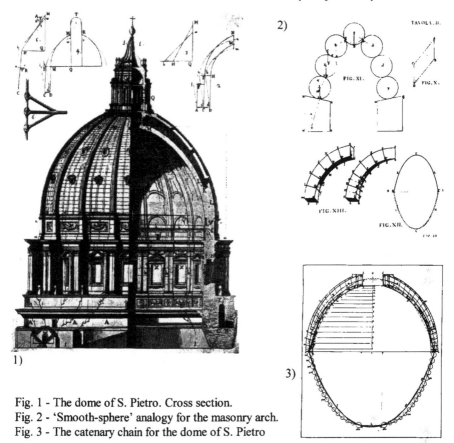

Fig. 1 - The dome of S. Pietro. Cross section.
Fig. 2 - 'Smooth-sphere' analogy for the masonry arch.
Fig. 3 - The catenary chain for the dome of S. Pietro

In another investigation [2] T. Le Seur, F. Jacquier and R.G. Boscovich arrived at a similar conclusion. They also noted that the crack pattern involved not only the vault, but the whole structural supporting structures.

The study of masonry domes has been the object of several research projects starting from some eighteenth century memoirs of the *Academiae Royale des Sciences* in Paris. One of the first aspects of the problem to be examined, was the search for the optimal shape to be given to a masonry dome. Pierre Bouguer [3], analysing a masonry dome subject to self-weight, stated that in order to have equilibrium, the meridian must have the same shape of the curve that represents the funicular of the loads which are relative to the dome lune. Subsequently, a similar solution to the same problem was found by many others scholars. All the suggested solutions presuppose the hypothesis that the dome have a one-dimensional behaviour in order to be observed as a series of "orange slices" or lunes of variable width, tapering to zero at the crown, placed in simple contact with each other. Under these assumptions, the analysis of masonry domes presents clear similarities with the analysis of masonry arches. The problem of the search for the optimal shape to be assigned to a masonry arch or dome, is replaced in time, by the issue of the definition of the theoretical question basically reversed, that is how to determine the stress state of an arch or dome with assigned dimensions and shape. Among the multitude of methods proposed for the analysis of domes, the ones suggested by Maurice Lévy [4], Schwedler [5] and, in more recent times, by Jacques Heyman [6] and Irving Oppenheim et al. [7], are very significant.

The principal aspect in the analysis of masonry structures, is the consideration that the reacting structure is the main unknown element of the problem. Only by discarding the hypothesis that the thrust curve has to coincide with the axis of the structure, can we effectively describe the behaviour of a masonry structure. Through the resolution of the membrane equilibrium equations we obtain, in a hemispherical dome acted upon by its own weight, a compressive stress state all along the meridians, and along the parallels up to $\varphi = 51\ 82°$, while tensile stresses develop along the parallels from that value to the support at $\varphi = 90°$. Generally, in-fact, a dome whose shape has a positive index of gaussian curvature, displays tensile hoop stress in the lower portion of the structure. This phenomenon is not admissible as long as we are dealing with masonry in the hypothesis of no-tension material. For this reason, the methods that were proposed in order to find a solution, had to account for two, so to speak, opposite phenomena we can observe in the existing masonry domes. We are referring to the fact that the hemispherical masonry domes get damaged along the meridians without any problem for the stability of the structure; and, on the contrary, there are masonry domes which do not present any damages in those areas in which the parallels are stretched.

Poleni's approach anticipates, in a way, what will later be the application of the limit analysis of masonry structures, recently set up by J. Heyman. The solutions that Heyman and Oppenheim proposed, which were both obtained from the context of the limit analysis, are based on the hypothesis that the hoop stress resultant between the contiguous lunes of the dome is zero. What is hypothesised, is therefore a one-dimensional behaviour. Under these assumptions, the analysis of the dome stability is obtained through a comparison between the funicular curve related to the loads which operate on a lune and the shape of the cross section of the dome. This method presupposes that the reacting structure is unknown. On one hand Heyman an analytical solution to the problem related to the search for the shape that Bouguer had proposed. On the other hand, the limit analysis allows us to formulate the problem of the analysis of a masonry arch or dome by removing the condition that the thrust line must be contained in the third middle of the section, as we will later see, Lévy still maintained. Oppenheim, in the wake of Heyman, suggests as a solution to the problem of ogival and hemispherical masonry domes, the comparison between the shape of the thrust line, deriving from the thrust surface, and the cross section of the dome lune we want to analyse. Schwedler's method, instead, involves the identification of two parts of the dome that have a different behaviour, starting from the point in which circumferential stresses become zero, which is obtained through the conventional membrane theory. In this case, we accept the coincidence between the thrust line and the middle line in the portion of the dome in which we have a two-dimensional state of stress. This method accounts for the fact that, being a masonry dome a double curvature surface, it displays a bi-axial state of stress: the thrust line changes only in the second portion of the dome where the parallels would be stretched. It is worth noticing that, for the hypothesis that were chosen, in the second portion of the dome we can usually observe the thrust line coming out of the thickness; in this method no further condition is imposed on the thrust surface of the second portion.

After all, we can see how the analysis of masonry domes has dealt with two substantially different approaches: either considering the two-dimensional stress state, without introducing the thrust line among the unknowns of the problem, or considering the unknown reacting structure hypothesising a one-dimensional state of stress. Lévy, in his book on graphical statics [4], proposes a very interesting approach to the analysis of axisymmetric masonry domes. The solution provided by Lévy belongs to the sphere of graphical statics. The original idea is to accept the fact that the circumferential stress would result equal to zero in those areas where tensile stress would develop, and at the same time, to consider a two-dimensional state of stress, typical of double curvature structures such as domes, where this turns out to be compatible.

These assumptions are explained by the fact that usually a dome displays two kinds of behaviour: one corresponding to the portions of the cup where there are compressive stresses both along the meridians and the parallels; the other relative to the portions where the parallels are stretched: in this case we must assume that the stress is taking place only along the meridians. From this approach comes the fact that the reacting structure is one of the unknown elements of the problem, which is typical in the analysis of masonry structures. The double behaviour was already hypothesised by Schwedler's solution: in which case we accepted the result that the membrane equilibrium equations provided up to the inversion parallel; Lévy, on the other hand, introduces some hypothesis that modify the overall solution. In a masonry dome, to the indeterminate position of the funicular curve relative to the part of the cup in which the parallels are subjected to tensile stress, is added the unknown location of the point where we have the crossing between the two different behaviours, that is what Lévy calls the *"point neutre"*.

Lévy obtains the solution of the problem through means that are typical of graphical statics.

2. LÉVY'S GRAPHICAL SOLUTION

In the fourth section of his book, Lévy deals with the problem related to the stability of masonry domes, and in particular he resolves the static problem of a hemispherical masonry dome of variable thickness. There are two essential hypotheses that the author assumes: the vault is formed by a first series of fictitious joints which are continuos and endlessly close to one another, according to the meridian plans of the vault, and by a second series of joints with similar characteristics obtained from sectioning the vault at the intrados according to the conical cross sections; we disregard the adherence of the mortars, since the joints can only tolerate compressive stress. As previously stated, the fundamental idea behind this procedure is based on the awareness of the existence of two unknown elements within the problem: the actual reacting structure and the location of the *"point neutre"* to which a *"parallele neutre"* corresponds. Since masonry, for instance, is not capable of carrying tensile stresses, Lévy claims that the portion of the dome which is located above the *"point neutre"* tolerates the $q'd\theta$ horizontal compressive stress that are transmitted by the adjacent sections of the vault, while the one located below such a point, cannot tolerate any hoop stress. Lévy takes into account the contribution that the circumferential stress gives to stability, for the entire section of dome where such stress is admissible. The *"point neutre"* is the principal unknown factor regarding the analysis of a masonry dome. Lévy also points out that the thrust line of that portion of the dome which is located below the cracking joint, can be nothing but the funicular curve of the acting loads; the one relative to the upper portion can, on the other hand, be any funicular curve as long as it is contained inside the thickness of the dome. The funicular curves are determined by using graphical statics methods.

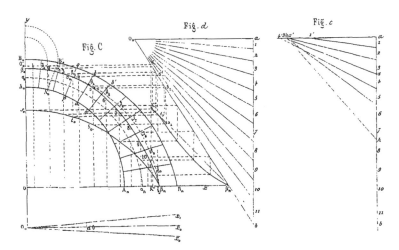

Fig. 4 - Lévy's method. Graphical construction.

In order to draw the thrust line relative to a dome lune $A_0B_0A_nB_n$ (Fig. 4), Lévy considers separately the two parts that are above and below the pq joint, where the *"point neutre"* is located. Starting from the upper part pq A_0B_0, after having arbitrarily assigned the pq position, from an a point, we draw the polygon of forces; the compressive force exerted on each joint, consists of the load of the dome which is positioned above such a joint, and of the $q'd\theta$ a horizontal forces. We then continue by arbitrarily drawing the $B_0\sigma_0$ curve, which represents the thrust line of the upper portion. In the lower portion of the dome pqA_0B_0 we disregard the $q'd\theta$ horizontal forces, since they become tensile forces. The compressive force exerted on each joint of the dome consists of the one exerted on pq and that of the load of the portion of the lune which is comprised between pq and the joint we are considering.

The first of these forces is represented, on the polygon of forces, by $O\sigma$, which is parallel to the tangent in σ_0, the second is represented by σ_h. Consequently, the stress on this joint is shown by the vector Oh (Fig. 4) drawn from the fixed point O. Let's assume that the $\beta_n\sigma_0$ curve, drawn with a full line in the dome section, represents the funicular curve: this is necessarily tangent in σ_0 to the $\beta_n\sigma_0$ that has been adopted as the thrust line of the upper portion. The thrust line relative to the lower portion of the dome is, therefore, entirely known, if we are aware of pq, the $\beta_0\sigma_0$ curve, and the O pole on the aO straight line. In a dome, on the other hand, an entire portion of curve, precisely $\beta_0\sigma_0$, is arbitrary and therefore: *"an infinite number of joints (all above pq) can be on the verge of opening at the same time"*. In this section of the dome the $\beta_0\sigma_0$ curve, which corresponds to the condition of limit equilibrium, is therefore nothing but the one that passes through the extrados of the third middle. Consequently the $\beta_n\sigma_0$ funicular curve must:

- be tangent to $\beta_0\sigma_0$ in an undetermined σ_0 point;
- pass, generally through the β_n extrados of the third middle, over to the spring or, be tangent, in an undetermined point, to the a_0a_n curve, which is the place that defines the lower edge of the third middle;

- have its own pole on the horizontal aO drawn through the a origin of the polygon of the forces.

After drawing a tentative polygon starting from the external point of the dome base, we look for the location of the *"point neutre"*. Such a request has its explanation in the search, on the funicular curve, for a t_0 point that can allow the tangents of the two curves in t_0 and σ_0 to meet on the $O\beta n$ horizontal line, after having drawn the $t_0\sigma_0$ vertical line far enough to meet the $\beta_0\beta_n$ curve in σ_0.

In order to solve this problem, Lévy suggests the use of a method, that he describes in detail, which is attributed to Eddy [7]. Eddy writes that the point in which the compressive stress along the meridian becomes zero is that point to which corresponds *"the equilibrium curve, in departing from the crown, first becomes more nearly vertical than the tangent of the meridian section"* Up to this point, we are dealing with the same geometrical considerations which led to Bouguer's inequality. Eddy, however, adds a statement concerning the thrust: *"for above that point the greatest thrust that the dome can exert, cannot be so great as at this point where the thrust of the arch-lune is equal to that of the dome"*.

The graphical construction that Lévy proposed, going back to Eddy's procedure, is as follows: determination of the stone weight of the dome and of the straight lines action of the loads; outline of a tentative funicular polygon which is satisfactory only as long as it passes through the extrados of the third middle over to the dome impost; drawing the horizontal lines for the ordinates of the extrados of the third middle to meet the O_0b. Then draw the vertical lines passing through these points to meet the respective vertexes of the tentative funicular polygon; these ordinates create a curve which passes through the β_n' point. Draw, passing through β_n', a tangent to the curve: the point of tangency is t_0'; from that point, draw the horizontal line to meet the tentative funicular polygon in t_0.

The vertical line drawn from t_0, intersects the extrados curve of the third middle in σ_0: this point represents the parallel in which the circumferential stresses are zero and in which we identify the passage between the two parts of the dome that have different behaviour. In order to draw the funicular curve that is relative to the lower portion of the dome, it is convenient to find the new polar distance. To identify this distance, we should draw from O a segment with an OO_0' length equal to the O_0a polar distance. We then draw a horizontal line that passes through this point until it reaches the O_0b in K''. Let's then draw a vertical line that cuts the $\beta n't_0'$ tangent in K' and the $O\beta_n'$ horizontal line in K''. The new polar distance is $K'K''$, which is represented in the polygon of forces by the Oa segment. The funicular curve of the entire dome then consists of two distinctive parts, one located above and the other one below the section in which the circumferential stresses are zero, that is the PQ segment. The pressures exerted below σ_0 are expressed by the polar radii drawn from O, which is, as previously stated, the new pole of the polygon of the forces; those exerted on the joints above σ_0 are expressed by those portions of straight lines which are parallels to the tangents to the $\beta_0\sigma_0$, comprised within the Oaf right angle, passing through the points of separation among the forces, and terminating in the 1',2',3',.. points.

The graphical procedure described has been used for the set up of a software program which allows to apply Lévy's method easily. The following figure (5) shows the graphical constructions and the outlining of the funicular curve, inside the third middle, in a section of spherical dome with constant thickness, that has an external radius equal to 10 meters. We have considered as operating element only the self-weight.

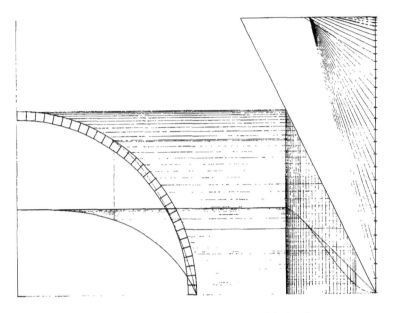

Fig. 5 - Lévy's method. Plotting from original software.

3. LÉVY'S METHOD. ANALYTICAL MEANING

The conditions that are at the basis of Lévy's graphical procedure, can also be satisfied by using an easy analytical approach. In order to simplify the procedure, let's assume that the section of the dome has a spherical shape and a constant thickness, and that R is the medium radius. Observing the cross section of the dome, the funicular curve related to the upper part is represented by the equation of the medium surface. In such a region the behaviour is assumed two-dimensional. In the lower part, where the behaviour has to be assumed one-dimensional, the funicular curve depends on the line of thrust related to self-weight loads. The required limit condition is that such a funicular curve passes through the point of extrados of the dome at the base. In addition it is necessary to assume that the tangent to the two curves is equal at the point in which they connect. Such a point, which is the unknown of the problem, is precisely the *"point neutre"* that Lévy refers to.

The equation of the funicular curve related to the upper part is :

$$yc = \sqrt{R^2 - x^2} \qquad (1)$$

The general equation of the funicular curve related to the lower part is:

$$yp = \frac{Rx}{2H}\sqrt{R^2 - x^2} + \frac{R^3}{2H} arcsin\left(\frac{x}{R}\right) + c_1 x + c_2 \qquad (2)$$

Assuming the condition of symmetry we have:

$$y' p_{(x=0)} = 0 \qquad \rightarrow c_2 = -\frac{R^2}{H} \tag{3}$$

The condition which expresses the passage of the funicular curve through the point at the base of the dome ($r_p, 0$) allow us to determine :

$$c_1 = \frac{R^2}{H}\left(r_p - \frac{\pi R}{4}\right) \tag{4}$$

The actual funicular curve related to the lower part is:

$$yp = \frac{Rx}{2H}\sqrt{R^2 - x^2} + \frac{R^3}{2H}arcsin\left(\frac{x}{R}\right) + \frac{R^2}{H}\left(r_p - \frac{\pi R}{4}\right)x - \frac{R^2}{H} \tag{5}$$

Assuming that the tangent to the two curves is equal at the point in which they connect, we obtain the value the horizontal thrust H as a function of x, for $0 < x \le r_p$:

$$y'_p = y'_c \quad \rightarrow \quad H = \frac{R^2\sqrt{R^2 - x^2}}{x} - \frac{R(R^2 - x^2)}{x} \tag{6}$$

The x-axis position of the *"point neutre "* x_n is determined by imposing $y_p = y_c$. Substituting such a value in (6) we obtain the constant H relative to the funicular curve y_p. As an example, in figure 6 the numerical solution for two domes of different thickness is plotted. We can notice that in the second case a limit solution is achieved.

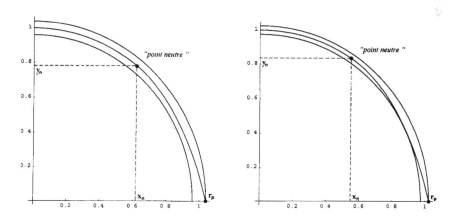

$R=1 \quad r_p =1.04 \quad x_n =0.613 \quad H=0.27 \quad \dfrac{t}{R}=0.08$ $\qquad$ $R=1 \quad r_p =1.025 \quad x_n =0.555 \quad H=0.252 \quad \dfrac{t}{R}=0.05$

Figure 6: Dome sections with thickness ratio as 0.08 and 0.05.

4. REFERENCES

1. Poleni G, 'Memorie Istoriche delle Gran Cupola del Tempio Vaticano', Padua, 1748.
2. Le Seur T., Jacquier F.,Boscovich R.G, 'Parere di Tre Matematici Sopra i danni, che si sono trovati nella cupola di San Pietro sul fine dell'anno MDCCXLII. Dato per ordine di Nostro Signore Papa Benedetto XIV', Roma, 1742.
3. Bouguer P., 'Sur les lignes courbes qui sont propres à former les voûtes en dôme, Memoires Academiae Royale des Sciences', Parigi, 1734, 149-166.
4. Lévy M. 'Cupoles en maçonnerie, Chapter II, La statique graphique et ses applications aux constructions', IV Partie, Gauthier-Villars Imprimeur - Librarie, Paris, 1888, pp.42-51.
5. Santarella L. 'Lastre curve sottili di rivoluzione, cap. X, Il cemento armato', vol II, Hoepli, Milano, 1972, pp.672-679.
6. Heyman J. 'On shell solution for masonry domes, International Journal of Solids and Structures', 1967, 3, 227-241.
7. Oppenheim I.J. & Gunaratnam D.J., Allen R.H., 'Limit state Analysis of Masonry Domes, ASCE Journal of Structural Engineering', 1989, 115, 868-882.
8. Eddy H.T. 'Spherical Dome of Masonry', Chapter XVI, Researches in Graphical Statics, Van Noostrand, New York, 1878, pp. 56-59.
9. Rapallini M.,Tempesta G.,Tralli A., 'Cupole in Muratura : Problemi Inversi', Atti del Convegno Nazionale La Meccanica delle Murature tra Teoria e Progetto, Messina, 1996, pp. 255-294.
10.Rapallini M., Tempesta G., 'Lévy's Method for the Analysis of Masonry Domes', Statica Grafica : Un Linguaggio Matematico per la Scienza delle Costruzioni, to be printed.

EARTHQUAKE PERFORMANCE OF ST CONSTANTINOS CHURCH IN KOZANI, GREECE: CORRELATION OF OBSERVED DAMAGE WITH NUMERICAL PREDICTIONS

G.C. Manos Professor[1] and **M. Triamataki** Research Assistant
Laboratory Strength of Materials, Department of Civil Engineering,
Aristotle University, Thessaloniki, Greece
[1]*E-mail: gcmanos@civil.auth.gr*

1. ABSTRACT

The earthquake performance of the masonry walls of a church, damaged during a recent earthquake, is the subject of this paper. These damaged masonry walls included a considerable number of door and window openings. Initially, a numerical study is performed on the stiffness properties of representative simple cases of masonry walls with a combination of openings. Based on the findings of this study a 3-D numerical simulation of the church's seismic response is next performed representing the masonry walls with linear "beam type" elements and utilizing the recording of the ground acceleration obtained during this earthquake at a close distance from the church. The deformation levels obtained from this 3-D seismic analysis are then used again, in order to simulate the masonry walls' earthquake performance through a more refined 2-D FEM representation. The numerical results obtained from these numerical studies agree well with the observed masonry wall damage. However, certain discrepancies are also observed.

2. INTRODUCTION

On May 13th, 1995, a Ms=6.6 earthquake occurred in the prefecture of Kozani in the Northwest of Greece. The epicenter of this earthquake was quite close (20km) to Kozani, the capital of the prefecture. Of the many churches to sustain structural damage in the area one of the largest is located in the city of Kozani. This relatively new church is formed by a three-aisled structure with a central spherical dome; the overall dimensions are 39m in length, 15m in width and the central dome height reaches 20m (figures 1 and 2). The mid-aisle, apart from the central part covered by the spherical dome, is covered by a semi-circular cylindrical dome that peaks at a height of 12.25m from ground level, supporting a two-sided inclined wooden roof. The side aisles are formed in two levels, with the upper level serving as the women's area (gynekonitis). This upper part is formed by a flat floor slab, supported on an internal colonnade and the external side wall, and an inclined slab as its ceiling, which peaks at 8.2m, and serves to support a one-sided wooden roof cover (figure 2). Initially, a brief description of the structural system is presented. Next, the damage observed in this church during this strong earthquake sequence is also presented and discussed. Then, the most significant aspects of the numerical simulation, which were used to study the observed earthquake performance are presented

Computer Methods in Structural Masonry – 4, edited by G.N. Pande, J. Middleton and B. Kralj.
Published in 1998 by E & FN Spon, 11 New Fetter Lane, London EC4P 4EE, UK. ISBN: 0 419 23540 X

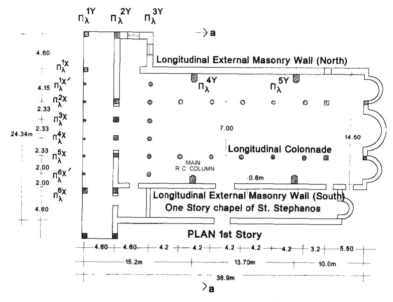

Figure 1. Plan of St. Constantinos (1ˢᵗ Story).

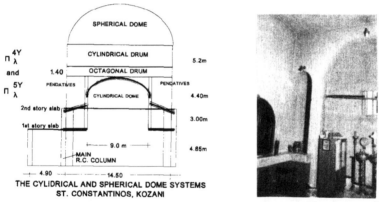

Figure 2. St. Constantinos (Section a-a)

Figure 3. Damage of the masonry walls

3. THE STRUCTURAL SYSTEM. AND OBSERVED STRUCTURAL DAMAGE.

The structural system which is employed in this church is a mixed one; the central domed part is composed of reinforced concrete whereas the remainder of the external shell is formed by thick un-reinforced masonry (figure 1). More specifically, the central semi-spherical dome is supported through a system of a cylindrical and octagonal drums that transfer the gravity loads through thick spherical-triangular pendatives on four main reinforced concrete columns at the corners. These columns are of rectangular cross-section (0.8m by 1.0m) with a height of 7.85m from the ground level to the base of the central dome-system (figures 1 and 2). The central dome is made of reinforced concrete as is the remaining part of the central nave, which is covered by a circular cylindrical dome and supported on the internal side by two longitudinal series of two-story colonnades also made of reinforced concrete together with the slabs (top and bottom) of the side aisles. The height of the bottom story is 4.85m whereas the height of the top story is 3.0m. The side aisles, apart from the internal support in the form of colonnades, are also formed by external longitudinal two-story high thick masonry walls with door and window openings. These side masonry walls, being 0.6m thick, together with the system of the three-fold east apses, are the parts of the structure that sustained the major earthquake damage, which is presented in the next section. Figure 3 shows a typical damage pattern and figure 4 depicts these walls with their overall dimensions together with the observed damage. Moreover, indicated within circles and rectangles are performance ratio values (shear and flexure) which are discussed in more detail in section 4.3. The observed damage is well documented and can be classified into two main types. The first type of damage is of minor structural significance and signifies the initiation of separation of the central reinforced concrete part from the surrounding structure. The main reinforced concrete elements did not sustain in themselves any noticeable structural damage. Thus no sign of distress could be seen either in the four large central columns supporting the heavy central spherical dome system or at the columns forming the colonnades that support the central-aisle semicircular cylindrical dome or the side aisle slabs. The second type of damage is of structural significance and is related to the masonry walls. There was widespread damage at the longitudinal masonry walls as well as at the apses. This damage could be clearly seen on the surface plaster at the internal wall surface concentrated at the piers of these walls between the opening in both the first and second story (figure 3). The external surface of these walls was covered by special decorative masonry units and mortar joints, without any surface plaster, and thus it was more difficult to identify the damage there, unless the level of distress was sufficient to also dislocate the decorative facade, as can be seen in figures. The damage was more noticeable at the masonry piers of the 1st story level; however, the masonry piers of the 2nd story level also exhibited signs of distress.

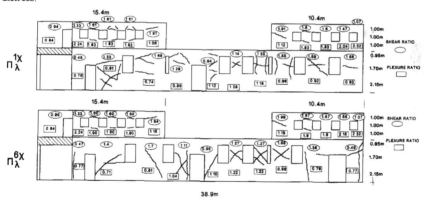

Fig. 4. Observed damage and correlation with results from 3-D numerical simulation (Section 4.2)

4. NUMERICAL SIMULATION OF THE OBSERVED PERFORMANCE

The threefold attempt to numerically simulate the observed earthquake performance of this church is briefly presented below as well as in sections 4.1., 4.2., 4.3. No attempt was made at present to measure, in-situ or at the laboratory, elastic properties and strengths of the prototype masonry.

- Initially, three simple masonry wall formations with door and window openings, that represent small parts of the complex masonry walls of the church, are studied numerically (section 4.1).
- The conclusions drawn from the previous study are included in a 3-D numerical simulation of the response of the whole church under the combined action of the gravitational and earthquake forces. Axial and shear stress resultants are derived, that, when compared to empirical strength estimates, lead to conclusions that can be correlated with the observed damage (section 4.2.).
- Finally, the complex longitudinal masonry wall of this church is again studied numerically, when its overall deformations are similar to those derived in the previous 3-D study. This time the numerical approximation of the masonry wall is more refined than that which was employed in the 3-D study (section 4.3.). From the distribution of the axial and shear stress resultants conclusions can again be drawn related to the observed damage.

4.1. Numerical study of the stiffness of simple masonry wall formations.

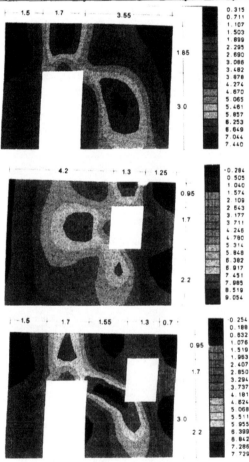

Contours of Sxy (dimensions in m, t/m)

Figure 5. Simple masonry wall configurations

Three simple masonry wall structural formations were studied prior to analyzing the relatively complex overall earthquake response of this church. These simple formations, depicted in figure 5, had the same overall dimensions; 6.75m length and 4.85m height. They were taken to represent segments of the outer masonry walls of St. Constantinos. The first formation (A) included a door, the second (B) a window and the third (C) a door and a window. The dimensions of these openings were again representative of the openings of St. Constantinos masonry walls; certain brief information is listed in the first two columns of Table 1. The following three distinct steps were utilized in the subsequent analysis:

a. Initially, each formation was discretized with quadrilateral elements; for these masonry walls the values of E=2500Mpa, G=1100Mpa and an equivalent thickness of 240mm, were assumed to be valid for the prototype masonry. The load was applied in the form of prescribed horizontal displacement 0.5mm at the top horizontal wall boundary. From the reaction forces and the shear stress distribution the total wall horizontal stiffness as well as the stiffness of the individual masonry piers in each formation were derived, listed in column 3 of Table 1.

b. Next, each structural formation was again numerically simulated with linear "beam type" finite elements having the same elastic and geometric properties as before. This simulation also employed rigid zones at the top and the bottom of each masonry pier according to an approximation of this type traditionally employed in analyzing masonry walls. These rigid zones were represented with linear elements of very large stiffness. Again the total horizontal stiffness and that of the individual piers was derived for each formation. The basic element dimensions and the results for step b are listed in the subsequent four columns of Table 1; the stiffness variation in a percentage basis, that can be observed when comparing the results from the b-step solution with that of the step a, is also included. As can be seen from this comparison the variation in the total stiffness is 42.4% for formation A, 242.6% for formation B and 66.3% for the formation C.

c. An effort was made in this third step to lower these discrepancies in the stiffness approximation as they resulted by the linear beam and the rigid zone-type elements approximation, which was described in step b. This was done by adjusting the height of the rigid zone at the top and the bottom of these piers per formation that exhibit the largest discrepancy. In this attempt the objective was to match the total stiffness of the linear pier with that of the 2D-FEM approximation; no extra attempt was made to also match exactly the partial stiffness of each individual pier per formation. This adjustment is listed in the last three columns of Table 1. As can be seen from the results obtained from the previously described numerical approximations it can be concluded that for all studied formation the linear pier is stiffer than the 2D-FEM approximation; moreover, it can be seen that :
- The linear pier stiffness approximation is quite reasonable for formation A with only one door (42.4% variation).
- This approximation results in large discrepancies for formation B that includes only one window (242.6% variation).
- For formation C, with both a window and a door, the stiffness discrepancy is mostly confined in the pier that lies between the two openings. The total discrepancy for this formation is 66.3%.

Table 1		2D-FEM	Initial	Linear	Pier	Solution	Adjusted	Linear	Pier
Name Code	Component	2-D FEM Stiffness (t / m)	Pier Height (m)	Rigid Zone Height (m)	Linear Pier Stiffness (t / m)	Variation 2DFEM (%)	Adjusted Stiffness (t / m)	Variation 2DFEM (%)	Pier Height Variation (m)
A with one door	Total	10866	4.85	–	15504	42.4	10866	0.0	–
	Left Pier (h/l=2)	2352	3.0	1.85 top	2760	17.3	3026	28.9	–
	Right Pier (h/l=0.85)	8514	3.0	1.85 top	12744	49.7	7840	-7.9	+0.9top
B with one window	Total	13660	4.85	–	46800	242.6	14200	3.95	–
	Left Pier (h/l=0.40)	9060	1.7	0.95 top 1.40 bot	38200	321.6	12240	35.1	0.80 top 1.50 bot.
	Right Pier (h/l=1.36)	2480	1.7	0.95 top 2.20 bot	7600	206.5	1960	-21.0	0.80 top 0.80 bot.
C one door and one window	Total	8600	4.85	–	14300	66.3	8500	-1.1	–
	Left Pier (h/l=2)	2440	3.0	1 85	3160	29.5	3660	50.0	–
	Mid Pier (h/l=1.9)	2720	1.9	1.85 top 1.10 bot	8700	219.9	2240	-17 6	+0.9 top +1.1 bot.
	Right Pier (h/l=1.36)	1780	1.7	0.95 top 2.20 bot	2420	36.0	2600	46.0	–

4.2. Numerical Simulation of the earthquake response of the whole church

The objective here was to obtain the overall seismic response of this church, employing a three-dimensional numerical model, as will be described in the following. As can be seen in figures 1 and 4, the outer masonry walls include mainly window openings with a limited number of doors. Thus it was decided to use in the 3-D numerical model the linear pier approximation, with elastic beam-elements and rigid zones with axial, shear and flexural stiffness for both the R.C. members and the masonry walls, introducing at the same time a stiffness reduction of the order of 100% for the masonry elements, in order to conform in this general way with the findings of the numerical study of section 4.1. This was done by introducing a value for the Young's modulus of the masonry walls equal to 1250 Mpa. The following are additional assumptions:

- The R.C. slabs were assumed to act as diaphragms. This was also adopted for the central spherical dome system starting at the base of the massive triangular pendatives and upwards.
- This behavior of a horizontal diaphragm was also assumed to be a valid approximation for the circular cylindrical dome of the mid-aisle in the longitudinal direction.
- The elastic properties of the masonry walls were obtained by assuming the cross-sectional properties of their mortar joints with a value for the modulus of elasticity determined empirically for low-mortar strength masonry construction. Due to the damaged condition of the structure no attempt was made to estimate these properties from in-situ measurements.
- All elements were assumed to be fully fixed at the ground level.
- The various elements resisting the seismic forces were grouped in linear frames along the longitudinal and transverse axes, designated in figure 1 with the symbol $\Pi\lambda,\chi$ and $\Pi\lambda\psi$, respectively. This numerical model employed all these frames together with full continuity in their interconnections to form the final three-dimensional simulation of the church. This 3-D structural system was subjected to the combined longitudinal and transverse seismic actions, which were based on the recorded earthquake ground motions, together with the gravitational forces. In this way, for each pier, belonging to the damaged masonry walls, its state of stress was obtained in terms of axial (N), shear (Q) and flexure (M) resultants, for the simultaneous action of the seismic and gravitational forces (see figures 6).

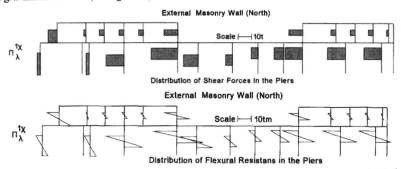

Fig. 6. Shear and moment stress resultants in longitudinal masonry walls from the 3-D simulation.

Next, by assuming the usual stress distribution patterns along each pier's cross-section, the amplitude of peak axial and shear stress values for each masonry pier were derived. These peak values were then divided by approximations of the corresponding peak strength value for each pier. The axial strength was derived by assuming a level of mortar strength and by accepting that the appearance of tensile stresses indicated the initiation of flexural damage; the shear strength also depended on the assumed quality of mortar as well as on the axial state of stress. Empirical formulas that are accepted by concrete block masonry construction were employed in deriving these axial and shear strength values. These ratio values of the obtained peak axial and shear stress level of stress over the corresponding strength values for each pier, are the values listed in figure 4 together with the observed damage patterns, designated as flexural and shear ratios. From the correlation of these ratio values it can be deduced that when, in a masonry pier, the flexural ratio is greater than one, this signifies that this pier should fail in flexure whereas when the shear ratio is greater than one, then shear failure is indicated for this pier. When both values are smaller than one then no damage, either shear or flexural, is expected to have occurred in this pier whereas when both values are greater than one then the dominant mode of failure should correlate to the greater value of the two ratios (either flexure or shear). If the damage patterns of figure 4 are examined under this light, good agreement can be seen between the observed performance and that which could be predicted utilizing the assumed seismic force levels, the peak numerical stress results with the corresponding strengths and the described approach, based on the flexural and shear ratio values.

4.3. 2-D FEM numerical simulation of the earthquake behavior of the longitudinal masonry wall.

Apart from the study of the damaged masonry walls, described in section 4.2., an additional 2D-FEM numerical simulation is performed here. This time each one of the almost identical longitudinal damaged masonry walls are simulated by 125 thin shell elements per wall (figure 7); the original (not reduced) value of the Young's modulus is used, equal to 2500 Mpa. The total wall stiffness, derived in this way, is found to be in good agreement with the stiffness of the linear pier approximation incorporated in the 3-D simulation, outlined in section 4.2. In this way, the overall displacement response, which resulted from this 3-D dynamic analysis, should also agree well with that of the current 2D-FEM simulation. Consequently, the maximum displacement levels, that resulted from the 3-D dynamic analysis for the 1st and 2nd story levels for the outer masonry walls, are used here as prescribed imposed displacements.

Figure 7. Finite Element Simulation of the Masonry Walls (North)

Full fixity at the ground level. was assumed again for the masonry walls. The following two actions were considered to occur simultaneously: first the gravitational forces were considered, which this time were approximated more accurately than the approach used in section 4.2. Here they were introduced by applying line loads at the levels of the 1st and 2nd story slabs as well as by assigning gravity loads in each masonry wall element. Next, the seismic action was introduced by imposing horizontal displacements equal to the maximum levels obtained from the numerical simulation of the 3-D earthquake response of the whole church (section 4.2). Figure 7 depicts the deformed masonry wall, under the combined action of gravity loads and imposed horizontal displacements. The shear (Nxy) and flexural (Ny) stress resultants per unit length (t/m) were obtained in this way for the combination of the described loading. These shear (Nxy) stress resultants are depicted in figure 8. As can he seen , distress is again indicated here at the masonry piers located at a distance from the two ends of these complex longitudinal walls . For these piers there is reasonably good correlation between distress patterns predicted according to the simulation of section 4.2 and the current simulation. However, there are noticeable discrepancies between the behavior of the piers located at the end of the masonry walls. The current 2-D FEM simulation results in lower stress levels than the ones obtained from the numerical simulation presented in section 4.2.

Figure 8. F.E Shear Stress Resultants of North Masonry Wall (Configuration 1)

5. CONCLUSIVE REMARKS
-The employed linear elastic beam element representation for both the R.C. and the masonry parts to form a 3-D structural configuration for this complex church, seems to be quite successful in predicting the observed damage.
-The previous conclusion, however, must be seen under the light of the employed stiffness approximation for the masonry piers that was based on a separate study as well as on realistic assumptions regarding the levels of seismic forces and the strength of the masonry piers. These issues are not generally resolved in having a definitely prescribed procedure.
-Subsequent more accurate finite element simulation of the masonry walls exhibits good agreement for the interior piers with the previous simulation; however, it also exhibits certain discrepancies for the piers at the ends of the masonry walls.
-These discrepancies may be due to influences coming from the transverse walls. These effects need to be further researched for such complex masonry wall formations.
-Efficient methods and procedures for establishing with confidence in-situ fundamental mechanical properties of masonry walls is of paramount importance.

REFERENCES
- Earthquake Design of Concrete Masonry Buildings, R.E. Englekirk and G. C. Hart, Prentice Hall, Inc., 1984, ISBN 0-13-223156-5.
- Engineered Masonry Design, J. I. Glanville and M. A. Hatzinikolas, Winston House Winnipeg, ISBN 0-929004-00-0.
-G.C. Manos, M. Triamataki, B. Yasin "Correlation of the observed earthquake performance of the church of St. Constantinos in Kozani-Greece with numerical predictions", STREMAH V, pp. 309-318, 1997.

MECHANICAL BEHAVIOUR OF THICK MORTAR JOINTS UNDER COMPRESSION: EXPERIMENTAL AND ANALYTICAL EVALUATION

C. Ignatakis Assistant Professor[1], **M. Zygomalas** Dr. Civil Engineer[1],
S. Liapis Civil Engineer, **C. Poulioulis** Civil Engineer and
J. Roussos Civil Engineer
[1]*Department of Civil Engineering, Aristotle University of Thessaloniki, Greece*

1. ABSTRACT

The first results of a research program on the mechanical behaviour of hand-made flat bricks, mortar and masonry with thick mortar joints, used for restorations of Roman and Byzantine monuments in Northern Greece, are presented. Compressive and tensile strength of bricks and mortars are measured and stress-strain curves of bricks using strain gages are plotted. The compressive strength curves of small masonry piers are measured for three values of joint thickness. Finally, the analytical results under compression of twelve masonry piers, covering a wide spectrum of joint thicknesses, modeled by a specific nonlinear F.E. computer code are also presented.

2. INTRODUCTION

The masonry of Roman and Byzantine monuments, in Greece in particular and in all the Mediterranean region in general, is characterized by the flat clay bricks and the thick mortar joints. For the restoration of these monuments there is a need of hand-made flat bricks and mortars compatible with the original ones. There is also a need of studying the mechanical behaviour of masonry with thick mortar joints masoned by using materials of this kind. A lot of work to these directions has been done at the Reinforced Concrete Laboratory of Aristotle University of Thessaloniki [1, 2, 3]. In the present paper the first results of a research program studying the mechanical properties of hand-made bricks, mortars and masonry with thick mortar joints are presented.

Keywords : Byzantine Masonry, Bricks, Mortars, Compressive strength, Deformations

Computer Methods in Structural Masonry – 4, edited by G.N. Pande, J. Middleton and B. Kralj.
Published in 1998 by E & FN Spon, 11 New Fetter Lane, London EC4P 4EE, UK. ISBN: 0 419 23540 X

3. EXPERIMENTAL RESEARCH

3.1 Bricks

The 9th Ephorate of Byzantine Antiquities is using, for the restoration works, handmade flat clay bricks manufactured by small local factories in the outskirts of Thessaloniki. <u>Physical characteristics of the bricks</u> : External cracks are visible in some of the bricks. The presence of internal hair cracks was detected in many bricks by ultrasonic tests.
- Dimensions : l_x x l_y x t = about 40 x 30 x 5cm
- Dry specific gravity : $16KN/m^3$
- Furnace temperature : 800 - 900 C^0

<u>Mechanical characteristics of the bricks</u> : Strong anisotropy was detected in strength and Young Modulus along the directions l_x and l_y, by both destructive and ultrasonic tests.
- Compressive strength - deformations : Small cubic specimens (18 pieces, 4x4x4cm) and two groups of prismatic specimens (2x9=18 pieces, 4x4x12cm) were cut along the l_x and l_y direction from macroscopically uncracked bricks. Four prismatic specimens of each group were instrumented using strain gages on two opposite surfaces of each specimen (fig.1). The results from the relative tests are summarized in the table below :

Specimens	Loading direct.	Compressive strength (MPa) min-max →mean / st. deviation	Young Modulus Initial val. (MPa)	Poisson ratio Initial values
Cubic	Random	$f_{bcc} = 10.64 \div 15.84 \rightarrow 13.20/s=1.70$	-	-
Prismatic	l_x	$f^x_{bc} = 6.88 \div 12.22 \rightarrow 10.50/s=1.83$	$E^x_{bo} = 3400 \div 8000$	$v^x_{bo} = 0.08 \div 0.20$
Prismatic	l_y	$f^y_{bc} = 5.00 \div 10.63 \rightarrow 7.08/s=1.75$	$E^y_{bo} = 2500 \div 6000$	$v^y_{bo} = 0.12 \div 0.20$

The type of failure was characterized by the formation of inclined cracks and crushing at the core of the cubic specimens, while the prismatic ones failed by cracking along the loading direction (fig. 1). The typical stress-strain curves of a prismatic specimen are shown in the fig. 2. The significant difference between the deformation on opposite surfaces of the specimen reveals flexure due to eccentricity of the loading.
- Tensile strength : Two triads of prismatic specimens 4x4x20cm were cut from uncracked bricks, along the l_x and l_y direction respectively. The ends of the specimens were glued with epoxy resin into special metallic holders and direct tension was applied. The mean values of tensile strength were found: $f^x_{bt} = 1.80$, $f^y_{bt} = 1.29$MPa respectively.

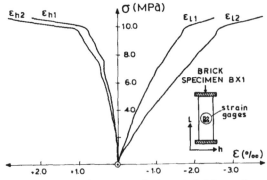

Fig.1 Prismatic brick specimen under compression

Fig.2 Stress-strain curves of prismatic brick specimen under compression

3.2 Mortar

The original mortars of Roman and Byzantine Monuments are composed of lime, pozzolana, crushed bricks, sand and water. The mortars used for restoration always contain some cement so that their strength is increased and the hardening process is accelerated. In the R/C Lab. of Aristotle Univ. of Thessaloniki a great variety of mortar compositions have been developed and tested during the last fifteen years, to support the restoration of many important monuments. The composition and the expected mechanical properties at the age of 28 days of the mortar chosen for the present research program are the following [3] :

- Composition by the weight : Lime/Pozzolana/Cement/Crashed bricks/sand/water = 1.0/0.2/0.8/3.0/3.0/2.1
- Tensile strength (flexure test according to DIN standards) :$f^f_{mt} = 0.84$MPa
- Compressive strength (according to DIN standards) : $f_{mcc} = 3.43$MPa
- Dynamic Young Modulus (ultrasonic tests) : $E^d_m = 4600$MPa

In order to test the mechanical properties of the mortar, six prismatic samples 4x4x16xcm were taken for every one of the six masonry piers constructed. The results from the 36 samples are as follows :

- $f^f_{mt} = 0.75 \div 1.75$MPa, mean value : 1.23MPa, standard deviation : s = 0.31MPa
- $f_{mcc} = 3.56 \div 4.45$MPa, mean value : 3.96MPa, standard deviation : s = 0.80MPa

3.3 Masonry Piers

In the first part of the research project presented here, six masonry piers 90x45x18cm were constructed, divided into three couples of twin piers. The only parameter was the bed joint thickness taking the values of 20, 35 and 55mm, while the brick height and the thickness of the vertical joints were 50 and 30mm correspondingly for all the piers. The piers were constructed horizontally in wooden mould. The bricks were fixed at the proper places into the mould using small nails (fig.3). Then mortar was poured to fill the joints. This method can ensure the complete filling of the joints and has the advantage of correct geometry. On the other hand it has the disadvantage of bad cohesion between mortar and bricks due to shrinkage of the mortar and the absence of the gravity pressure.

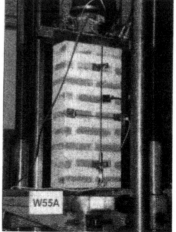

Fig.3 Wooden mould for masonry piers with the
bricks fixed at their positions

Fig.4 Instrumentation of masonry specimen

This could affect seriously the shear strength but has negligible influence on compression strength. The piers were tested under compression in vertical position. A force controlled testing machine having a capacity of 1000KN was used.

Instrumentation - monitoring : Five displacement sensors (LVDT) were mounted on each pier. Two of them were fixed longitudinally and horizontally on each side of the pier, while the fifth sensor measured the thickness deformation (ε_t) in the center of the pier (fig. 4). A load cell was used for the monitoring of the compressive force. To collect and store the data, a portable computer SANYO LT16, connected to an A/D card having 16 input channels was used. To reduce the "noise" a regulating condenser was connected to every input channel. In spite of this, "gravity functions" were used in order to smoothing the resulting stress-strain curves.

Experimental results : The compressive strength (f_{wc}), the initial value of Young Modulus (E_{wo}) and the longitudinal deformation at failure (ε'_{wu}) for the six piers are shown in the following table :

Specimen	W20			W35			W55		
	A	B	Mean	A	B	Mean	A	B	Mean
f_{wc}(MPa)	6.752	6.281	6.517	4.870	5.310	5.090	4.510	3.827	4.169
E_{wo}(MPa)	6580	6230	6405	3490	4630	4060	2460	2860	2660
ε'_{wu} (‰)	2.60	2.10	2.35	3.95	3.55	3.75	8.05	6.35	7.20

The stress-strain curves (σ - ε_l, ε_h, ε_t) for the piers W20B, W35B and W55A are shown in fig.5, 6 and 7.The mean values of the LVDT's on each of the opposite sides of the pier were used for the (σ-ε_l) and (σ-ε_h) curves. About ε_t it must be pointed out that it represents the local transversal deformation at the point of measurement. The stress-volumetric strain curve (σ-ε_v, $\varepsilon_v = \varepsilon_l + \varepsilon_h + \varepsilon_t$) is also plotted for each pier. The following remarks can be made :

• The compressive strength of masonry, as it was expected, decreases for increasing joint thickness.

• The Young Modulus follows the same rule but for the thinner joint it is greater than E_{bo}, while for the thicker joint it is smaller than E_{mo}. These inconsistencies could be attributed to many reasons such as higher E_{bo} along the thickness of the bricks (E_{bo} was measured along the two other directions of the bricks), higher E_{mo} due to the confinement of the mortar between brick layers, hair cracks due to mortar shrinkage, imperfect measurements etc.

• The ultimate deformations and the nonlinearity of (σ-ε_l) curve increase rapidly with the joint thickness.

• The shape of (σ-ε_v) curves is the typical one for brittle materials showing high dilatancy near failure and apparent instant values of Poisson ratio much greater than 0.5.

• Type of failure : Short vertical hair cracks are observed at the main sides of the piers under a compressive stress $\sigma_{cr} \approx (0.50 \div 0.65)f_{wc}$. Suddenly a major vertical crack is formatted $\sigma^m_{cr} \approx (0.75 \div 0.85)f_{wc}$, the crack is widened rapidly and the pier fails. The major crack appears either on the main sides of the specimen (pier W35B:fig. 9) or on the one or both the lateral sides of the specimen causing the splitting of the pier out of its plane (piers W20B and W55A:fig. 8,10).

4. ANALYTICAL RESEARCH

In the Reinforced Concrete Laboratory of Aristotle University of Thessaloniki a specific F.E. micromodel has been developed, for the in-plane nonlinear analysis of unreinforced masonry under monotonic loading until failure (Computer code "MAFEA" : MAsonry Finite Element Analysis) [4]. It must be pointed out that the computer program, despite its 2D character, calculates and takes rationally into account the out-of-plane principal stresses that developed in bricks and mortar joints under in-plane loading of the masonry. The model is capable of predicting cracking, crushing or out-of-plane splitting of bricks and mortar, as well as sliding or unsticking at the joints and finally the propagation of damages until failure. Using this powerful F.E. program an enlargement of the experimental findings has been attempted. At first three F.E. models (M20, M35, M55), identical with the experimentally tested masonry piers, were created and analyzed until failure in order to verificate the analytical model. The necessary inputs for the MAFEA computer code, apart from the geometry of the masonry models, are the mechanical properties of bricks, mortars and joints given in the following table. Most of them are determined by the experimental research already mentioned.

Materials	Dimensions	f_c(MPa)	f_t(MPa)	E_o(MPa)	v_o	ε_u ‰
Bricks	28x18x5cm	f_{bc}=9.00	f_{bt}=1.55	6000	0.15	3.25
Mortar	t_{bed}=1.0÷7.5	f_{mc}=3.00[A]	f_{mt}=0.70[B]	3000[C]	0.20[D]	3.25[D]
Joints	t_{vert}=3.0cm	Shear, Tensile strength: f_{jso}=0.30[D], f_{jt}=0.15[D] Mpa Friction coefficient: μ=0.75[D]				

[A]: Prismatic strength converted from the cubic strength
[B]: Direct tension strength converted from the flexural strength
[C]: E^s_{mo} under static loading converted from the dynamic Young Modulus E^d_{mo}
[D]: From the relative literature

The strength of the models was in very good agreement with the respective experimental ones (fig. 11) taking into account the wide scattering of the material properties. To enlarge the experimental results nine masonry models, with bed joint thickness varying from 1.0 to 7.5cm and the material properties of the table above, have been created and analyzed until failure. Figure 11 shows the compressive strength of masonry models (f_{wc}) versus joint thickness to brick height ratio (t_m / t_b). It is clear that f_{wc} decreases with increasing joint thickness tending asymptotically to the lower limit equal to f_{mc} and to the upper limit equal to f_{bc}, for very thick and very thin joint thickness respectively.

The damages of the masonry pier models begin with unstinkings at the vertical joints under a compressive stress $\sigma_{cr} \approx (0.60÷0.70)f_{wc}$. For higher loading the unstikings spread and propagate vertically cutting the bricks above and below the vertical joints. Failure occurs in a rather brittle manner with crushing of bed joints and cracking or out of plane splitting of bricks. The out of plane splitting appears mainly for thin joints (fig. 12 A,B=M10 model) while major vertical cracks through bricks and joints appear mainly for thick joints (fig. 12 C,D=M60 model).

5. CONCLUSIONS

It is well known that joint thickness is one of the governing parameters for the compressive strength of masonry. This is also proved to be valid for Byzantine or

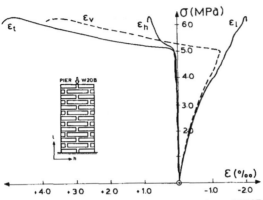

Fig.5 Stress-strain curves of masonry specimen W20B

Fig.8 Masonry specimen
W20B at failure

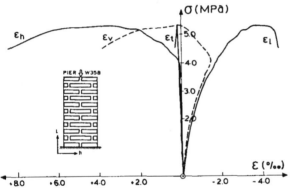

Fig.6 Stress-strain curves of masonry specimen W35B

Fig.9 Masonry specimen
W35B at failure

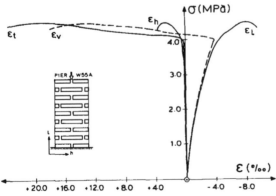

Fig.7 Stress-strain curves of masonry specimen W55A

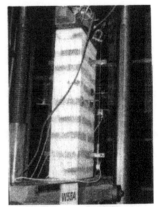

Fig.10 Masonry specimen
W55A at failure

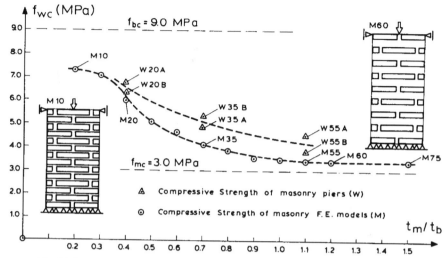

Fig.11 Compressive strength of masonry specimens and F.E. models versus
joint thickness to brick height ratio (t_m / t_b)

Roman masonry with flat bricks and relatively weak mortar. The very thick joints of this type of masonry cause a low compressive strength (despite the relatively strong bricks), nonlinear behaviour and high ultimate deformations. This deformability could be very useful under foundation settlements which are very frequent to heavy monumental buildings. Computer code MAFEA is capable of predicting the behaviour of masonry and could be very useful to enlarge and enrich the expensive experimental research.

6. ACKNOWLEDGMENTS

The authors wish to thank 9th Ephorate of Byzantine and Post-Byzantine Antiquities for the generous offering of the materials for the research program.

7. REFERENCES

1. Penelis G., Papayianni I. and Karaveziroglou M., "Pozzolanic Mortars for Repair of Masonry Structures", 1st Int. Conf. on Structural Repair and Maintenance of Historical Buildings, Florence, Italy, 1989, pp. 101-109.
2. Karaveziroglou M. and Papayianni I. "Compressive strength of masonry with thick mortar joints", Int. Conf. of RILEM on Conservation of Stone and other materials, 1993, Vol. 1, pp. 212-219.
3. Karaveziroglou M., Koulikas P., Panagiotopoulos P., Dimitreli C. and Triantafillou G., "Creep behaviour of mortars used in restoration", 4th Int. Conf. on Structural Studies of Historical Buildings, Chania, Crete, Greece, 1995, pp. 231-238.
4. Ignatakis C., Stavrakakis E. and Penelis G., "The in-plane behaviour of masonry, an analytical finite element model", Int. Conf. of ICCROM on Structural Conservation of Stone Masonry, Athens, Greece, 1989, pp. 123-130.

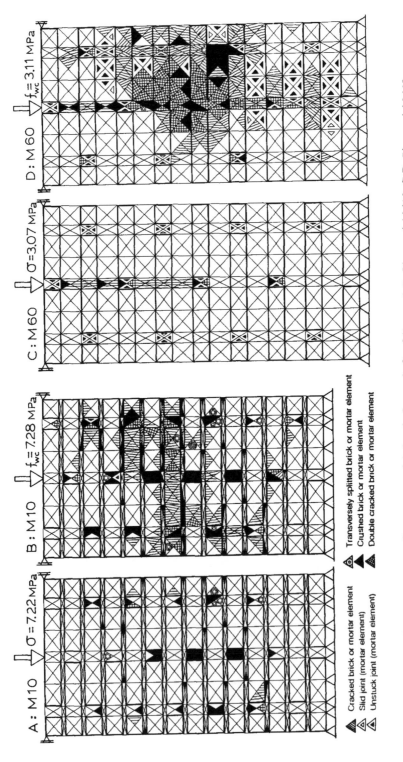

Fig.12 Damage patterns of masonry models just before and after failure : A,B: Pier model M10, C,D: Pier model M60

◤ Cracked brick or mortar element
◭ Slid joint (mortar element)
◮ Unstuck joint (mortar element)

◭ Transversely splitted brick or mortar element
◭ Crushed brick or mortar element
◭ Double cracked brick or mortar element

A : M10 ⬇ σ=7.22MPa

B : M10 ⬇ f_wc = 7.28 MPa

C : M60 ⬇ σ=3.07 MPa

D : M60 ⬇ f_wc = 3.11 MPa

STRUCTURAL CHARACTERISATION OF REINFORCED MASONRY STRUCTURES

C. Modena Professor, **D. Sonda** PhD Student and **D. Zonta** PhD Student
Department of Construction and Transport, University of Padova, Via Marzolo 9, Padova, Italy

1. ABSTRACT

The general features and the first results are presented of a research program based on dynamic experimentations on a two full scale prototype buildings, one made of reinforced masonry and the second of r.c. infilled frames, and on a reduced scale model of the reinforced masonry prototype. The scopes of the research are the calibration of theoretical models used for the seismic analyses of the structures and for the identification of structural damages induced by horizontal actions of given intensity. Identification techniques are used to these scopes. The programs allows also the comparison between different dynamic experimentation techniques and between a full scale prototype structure and the corresponding reduced scale model.

2. INTRODUCTION

For this research program based financed by the Industry and European Community (BRITE-EURAM), two identical buildings were built (Fig. 1(a)), the first made of reinforced masonry and the second of r.c. infilled frames in order to compare both costs of construction and the level of living comfort (capacity of heat insulation and sound proofing) and the behaviour under static and dynamic horizontal actions of the two buildings. In the meantime, a 1:3 scale model of reinforced masonry building was realised in order to study the seismic behaviour on a vibrating table, at the Enea centre of research of Casaccia.

This research, that involved Italian and Greek industrial partners and 4 European universities (University of Athens, Universities of Padua and Pavia, University of

Keywords: Reinforced masonry, Modal testing, Structural identification

Computer Methods in Structural Masonry – 4, edited by G.N. Pande, J. Middleton and B. Kralj.
Published in 1998 by E & FN Spon, 11 New Fetter Lane, London EC4P 4EE, UK. ISBN: 0 419 23540 X

Darmstadt), was promoted to develop a new reinforced masonry building system, based on the use of new types of blocs suited to be produced and used in different European countries and specifically studied in order to have the maximum efficiency as to the capacity of heat insulation and sound proofing, the easiness of use and the structural performance on seismic actions.

Fig. 1. (a) Overall view of the two experimental buildings, at Trento. (b) Reduced scale model of the reinforced masonry prototype, on test on the vibrating table at the Enea research centre.

But, before the realisation of the buildings and the reduced scale model, we made the complete characterisation of the used materials and the experimental individualisation of the laws of behaviour under horizontal cyclical actions of the most significative reinforced masonry structural components (that is to say panels of different size and shape) made with the type of bloc developed by the Italian industry.

In this research, the progress of the dynamic experimentation done on the two building prototypes, and on the reduced scale model is presented, illustrating the first results of the tests and numerical processing.

3. DESIGN AND CONSTRUCTION OF THE BUILDINGS

3.1. Full scale buildings

During the design of the two buildings, we tried to realise some typical constructions of our country.
The constructions (Fig. 2) were designed and realised placing them side by side in order to permit the interposition of the required elements to impose the horizontal forces to maximise the mutual contrast offered by the two structures.
Of course, we tried to create the easiest as possible the structural behaviour, to facilitate the interpretation of the tests results. This is why the staircase, common to the two buildings, is structurally independent from them, and the experimentation is done in the first phase without internal partitions.
The dimension of the structures and the correspondent structural checks were made supposing that the buildings were built in a locality characterised by a high seismicity and using the support of the regulations given by the European regulations.

REINFORCED MASONRY BUILDING   R.C.INFILLED FRAME BUILDING

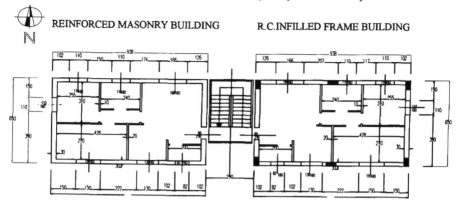

Fig. 2. Plan of full scale buildings

3.2. Reduced scale model

The realisation of the reduced scale model has been made keeping the modulus of elasticity and the density of the materials with which the prototype was built and using the scale ratio of the lengths (equal 1/3) maximum in relation to the capacity of dimensions and the load of the vibrating table, in order to make less important the effect of the factor of the scale in the behaviour of the structure.

Having defined in this way the factors of the scale in accordance with the 3 main sizes then the others were calculated (Tab. 1) in accordance with the hypothesis of elastic linear behaviour of the materials.

The main constructive characteristics of the model, built on a basic concrete plate which is 12 cm thick to allow the conveyance and the union to the vibrating table (secured by log bolts of 20mm of diameter) are indicated in the Fig.3.

The elements of brick are obtained by the same blocks used with the prototypes doing the reductions of the dimension for the scale ratio chosen through some proper cuts.

physic size	Factor of scale
density	$m_r = 1$
modulus of elasticity	$E_r = 1$
length	$l_r = 3$
time	$t_r = l_r (E/m)_r^{-1/2} = 3$
frequency	$f_r = l_r^{-1}(E/m)_r^{1/2} = 1/3$
velocity	$v_r = (E/m)_r^{1/2} = 1$
acceleration	$a_r = l_r^{-1} (E/m)_r = 1/3$
tension	$s_r = E_r = 1$
shift	$u_r = l_r = 3$
force	$F_r = E_r/l_r^2 = 9$

Table 1. Factors of scale of the modelling.

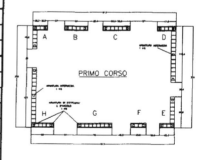

Fig. 3. Plan of the model.

As the adopted criteria in the model have a factor of scale for the accelerations different than 1, the effect of gravity can not be reproduced (neglected gravity modelling). We find only in part a solution for this inconvenient pre-pressing the masonry, as to reproduce the same state of tension of the prototype at the base of the model. For this reason, we put 8 metal turnbuckles of 12 mm of diameter, fixed on the plate of the base and blocked on the top of the masonry thank to pre-pressed spring, that can guarantee a reliable constancy of the load even used for little shifts of the structure.

Then to compensate the differences of the own weight, we added to the floor some concrete masses created with concrete plates fixed with threaded screws placed during the phase of the casting.

Obviously, the adopted constructive procedure has some inevitable differences between the full scale reinforced masonry building

4.1. DYNAMIC EXPERIMENTATION

The whole program of dynamic experimentation provides, for each building, a first phase of characterisation in elastic field; and a second one of test in non linear field (seismic test on the vibrating table for the model).

The first phase of experimentation has the following targets:

(a) to define the correspondence in the mechanic behaviour of the structures

(b) to calibrate the numerical models

(c) to define the seismic loads of the design estimated by the regulation

As second aims:

(a) to compare the working order and the correspondence of the results of the different techniques of characterisation

(b) to examine the little non linearity in the dynamic answer with regard to the structural anomalies.

4.1. Placement of transducer and techniques of experimentation

In both cases, the first of the prototype and the second of the model, the motion of the most significative points of the raising structure was measured through 8 piezoelectric accelerometers, placed as shown in the Fig. 4. Then 4 horizontal accelerometers (shown in the picture) and three verticals (Z3, Z6, Z9) were placed on the base to control its movement. Different techniques of experimentation were used for the dynamic characterisation, from AVT to shock tests and stepped sine tests.

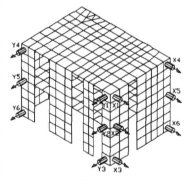

Fig. 4.

4.2. Full scale building

A dynamic exciter of big dimensions was created to allow the application of harmonic forces, up and more than the elastic region of the structure, in the Laboratory of the Department of Constructions and Transports at the University of Padua.

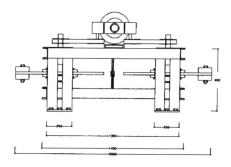

Fig. 5

The dynamic exciter (Fig. 5) is composed of two rotating steel discs of 90 cm on which the eccentric masses varying from 5 kilograms to a maximum of 40 kilograms; the force generated depends on the eccentric mass and on the frequency of the rotation, up to a maximum estimated during the design equal to 140 KN in a frequency range from 1 to 20 Hz.

The typical frequency responce functions, expressed in acceleration, got with the stationary harmonic excitation are reported in Fig. 6, and compared with the results of AVT, for the accelerometers X1 and X2, placed in correspondence with the second and the first floor respectively, in the same direction of the force. The use of the technique of the slow linear sweeping gave the same results obtained with the stationary excitation.

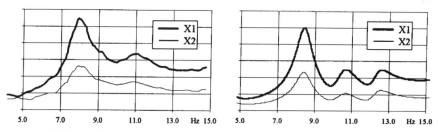

Fig. 6. FRF got (b) through the stationary harmonic excitation with the dynamic exciter placed in the centre of the second floor in the direction and (a) through AVT (regenerated)

But the use of the hammer with specified instruments and other percussion instruments gave medium results because only the high frequencies of the structure had been excited. The main results of the characterisation are reported in the Tables 2 e 3.

	AVT		stepped sine	
	f	ξ	f	ξ
I mode	6.17 Hz	4.1%	-	-
II mode	8.45 Hz	3.8%	8.01%	7.9%

Table 2.

	X1	X2	Y1	Y2	Z3	Z9
II mode	1	0.48	0.01	-	-0.27	0.42

Table 3. Modal components based on the tests with dynamic exciter.

It is important to remark that we cannot neglect the importance of the motion on the base, because the deformability of the ground produces the inversion of the first two frequencies when compared to the frequencies expected from the numerical model, where we suppose the firm ground condition on the base.

The problem of the motion on the base can be theoretically solved considering the raising structure as a substructure of the system building+ground, which can be completely characterised by the dynamic experimentation and described through the FRFs matrix.

Then, once defined the substructure as the subject-matter of our study, its state will be defined by a vector of variables. Among them, we can differentiate the variables that described only the state of the substructure from the others which describe the state of points of contact from the substructure and the rest of the system. The substructure response to an harmonic forcing of frequency equal to ω, is entirely explained in the following complex relation:

$$\begin{bmatrix} \mathbf{x} \\ - \\ \mathbf{y} \end{bmatrix} = \begin{bmatrix} \mathbf{H}_{xx} & | & \mathbf{H}_{xy} \\ - & + & - \\ \mathbf{H}_{yx} & | & \mathbf{H}_{yy} \end{bmatrix} \begin{bmatrix} \mathbf{f} \\ - \\ \mathbf{r} \end{bmatrix}$$

where the substructures $\mathbf{H}$ represent the FRFs (expressed inreceptance, mobility or inertance); $\mathbf{x}$ and $\mathbf{y}$ the vectors of the complex amplitude response (expressed in displacement, velocity or acceleration) in the own points of the structure and in the points of contact respectively; $\mathbf{f}$ e $\mathbf{r}$ the vectors of the complex components of the force applicated in correspondence with $\mathbf{x}$ e $\mathbf{y}$ respectively.

Each component $h_{ij}(\omega)$ of each submatrix can be achieved through the dynamic testing

Our purpose is to deduce the h_{ij}' FRFs, components of matrix $\mathbf{H}$', that relate the amplitude of the motion of each degree of freedom to the amplitude of the forces in the hypothesis in which there was no motion of the points of contact (grounded subsystem). Defining as it is shown $\mathbf{y} = \mathbf{0}$:

$$\mathbf{x} = \left[\mathbf{H}_{xx} - \mathbf{H}_{xy}\mathbf{H}_{yy}^{-1}\mathbf{H}_{yx} \right] \mathbf{f}$$

and then the matrix of the FRFs of the grounded substructure become:

$$\mathbf{H}'_{xx} = \mathbf{H}_{xx} - \mathbf{H}_{xy}\mathbf{H}_{yy}^{-1}\mathbf{H}_{yx}$$

This relation is particularly useful, as it allow as to exclude the contribution of the motion of the ground simply working with experimental FRFs, so that we don't need to make any hypothesis on geometry or mass distribution in the structure.

The complete characterisation of the substructure will be obtain only at the end of the first experimental phase, as it requires (for the construction of the submatrix $\mathbf{H}_{yy}$) dynamic tests with the forcing applied on many points of the base.

4.3. Reduced scale building

At the ENEA centre of Casaccia, the first characterisation has been done binding the model on the ground of the laboratory room, which can be considered rigid in the range of frequencies of interest. The characterisation, made with shock tests, confirmed the rigidity of the plan, and showd an anomalous behaviour of the raising structure (Fig. 7(a)), as modal frequencies resulted lower that it was expected ($f_1=17.8$ Hz), and an unexpected high damping. The stepped sine tests with different degrees of intensity showed a strong non linear behaviour of the structure. A careful visual check of the model pointed out some crakings between the basement and the base of some masonry panels, probably formed during the anchorage of the model, because of the imperfect union between the floor of the laboratory and the basement. Some other characterisations made with the model fixedon the vibrating table confirmed these results.

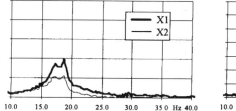 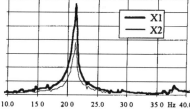

Fig. 7. FRF got through the shock test before (a) and after the intervention of the reinstatement (b)

So, a repair intervention was decided, to close such crakings, made by using filling resins. The importance of the intervention come out with the results of the further characterisations. The Fig 6(b) shows that FRF got with the shock test, forcing the structure in correspondence of the second floor and in direction X.

The Fig. 8 compares the results characterisations achieved by using the capability of the shaking table, and particularly by using (a) a slow logarithmic sweeping (0.1 oct/min) and (b) a random excitation (pink noise 5-50 Hz).

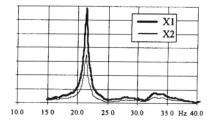 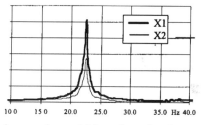

Fig. 8. FRFs got through (a) slow logarithmic sweeping and (b) random vibrations.

In the tables 4 and 5, the results are reassumed and compared in terms of frequency of the different characterisations. The shock test and the sweeping have the same results but the random excitation has frequencies slightly higher. The explanation can be find looking at the different intensity of the forces:

	hammer (before repair)		hammer (after repair)		sweeping		random	
	f (Hz)	ξ	f (Hz)	ξ	f (Hz)	ξ	f (Hz)	ξ
I mode	18.14	[1]	21.16	1.2%	21.36	1.2%	22.20	1.2%
II mode	21.16	[1]	27.22	1.2%	27.28	1.9%	29.24	1.9%

[1] Non proportional damping

Table 4.

	X1	X2	Y1	Y2	Z3	Z9
I mode	1.00	053	0.03	0.01	-	-
II mode	0.04	0.02	1	0.64	-	-

Table 5. Modal components based on the shock tests.

5. MODELLING

Numerical models with 3D elements (brick) were created from the structures under discussion before the execution of the experimentation to estimate the variability field of the own structural frequencies and then to plan the tests.

After the first results of the experimentation, a model of the structure made of plates was created in which the property of the materials were included with their measured estimations (Fig. 9).

Some slabs with anisotropic behaviour with

Fig. 9.

uniform thickness are modelled for the floors. The modulus of elasticity in the two directions is estimated as the product of the modulus of elasticity for the moment of inertia of the section is equal to the real one. For the direction of the frame of the rafters, we consider the inertia of the cope of the floor and the stringcourse of the connections between the walls

The specific weight attributed to the slab with which the floor was created considers the real own weight of the floor and the added masses to the reduced scale model to carry the ratios between mass of the walls and of the floor to the estimations of the real structure.

The identification of the model, based on the two first frequencies and relative modal components, given from the shock test and sweeping, gave, besides the correct definition of the equivalent stiffness of the floor, modelled as orthotropic slab, an estimation of the modulus of elasticity of the masonry equal to 4300 MPa, in other words lower than the medium estimation given from the results of the tests of compression.

REFERENCES

1. Bernardini, A., Modena, C., Lazzaro, G., Valluzzi, M.R., "Comportamento di pannelli sollecitati ciclicamente nel proprio piano." *D.RE.MA.B. Project*, Padova, I
2. Tomazevic, M., Modena, C., Petcovic, L., Velechovsky, T. (1989), "Shaking table study of an unreinforced- masonry building model with an internal cross-shaped wall - Test results", Institute for Testing and Research in Materials and Structures, Ljubljana, SLO
3. Modena, C., La Mendola, P., Terrusi, A., "Shaking table study of a reinforced masonry building model", *Tenth World Conference on Earthquake Engineering*, 3523-3526
4. Ewin, D. J. (1986). *Modal Testing*. Joon Wiley & Sons Inc., London, UK
5. Vestroni, F. (1995), "L'identificazione strutturale nell'ingegneria civile. Impiego di modelli agli elementi finiti", 1° Convegno nazionale di Identificazione Strutturale.

NON-LINEAR ANALYSIS OF UNREINFORCED MASONRY BUILDINGS UNDER HORIZONTAL LOADS

R. Barsotti PhD and **S. Ligarò** Assistant Professor
Department of Structural Engineering, University of Pisa, Pisa, Italy

1. ABSTRACT

An incremental-iterative procedure is presented for approximate assessment of the response of unreinforced masonry buildings to seismic loads. Working within the framework of the shear-type structural behavior, we propose a constitutive model for the resisting elements able to account for both the variation in lateral stiffness induced by cracking, and their sort of failure (i.e. ductile or brittle). Application to a practical case demonstrates how, when modeling walls with generic windowing, the manner in which they are subdivided into smaller resisting elements strongly influences the predicted structural response and the manner of failure of each wall.

2. INTRODUCTION

In order to assess the seismic response of simple masonry buildings, one of the most widely used structural models is the shear-type, in which floors behave as rigid diaphragms elastically connected by interposed walls. If the vertical displacements of the structure induced by horizontal actions are ignored, the model's degrees of freedom are extremely reduced. In such a case, the quality of the information obtained by applying the model depends upon the accuracy of the assumptions regarding the behavior of the single walls. In this regard, satisfactory descriptions can be provided of the walls' behavior during the brief initial loading stage (phase-I), where each results uncracked and behaves elastically (Fig. 1, line O-A), as well as when they reach the corresponding limit condition (line A-B). On the other hand, very little is known about the subsequent stage (phase II).

Keywords: Masonry buildings, Phase-two behavior, No-tension materials

Computer Methods in Structural Masonry – 4, edited by G.N. Pande, J. Middleton and B. Kralj.
Published in 1998 by E & FN Spon, 11 New Fetter Lane, London EC4P 4EE, UK. ISBN: 0 419 23540 X

If the resisting element (either a wall or a masonry element of walls with openings) is restrained well enough to its bases and is not excessively thin, the cracking which forms where tensile stress is at a maximum ceases almost immediately, and a new elastic phase may precede ultimate failure (line B-C). Under such conditions, the state of stress begins to grow again and reach values well above those of the preceding stage. Finally, collapse of the element may come about through brittle failure (line C-D), if, that is, the compression exceeds the material's strength, or, more frequently, through slow frictionful slipping of one part over another (line C-E), in which case, the failure turns out to be ductile.

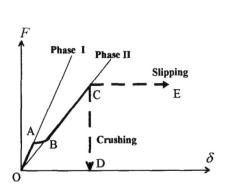

Figure 1: Two-phase behavior

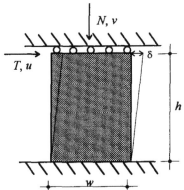

Figure 2: A masonry panel under compression and shear

In the transition from phase I to phase II, the element's stiffness undergoes an abrupt decline; thus, the stress distribution among the elements may vary considerably, with consequent worsening of the state of stress in the as yet uncracked elements. Such a situation favors a more uniform stress distribution throughout the entire structure while this is still behaving elastically and wall cracking is not evident. Greater uniformity of the stress distribution may delay failure of the elements. In any case, their manner of ultimate failure, either ductile or brittle, depends on the material's strength parameters and on the panel-aspect ratio [ ].

With the aim of accounting for the essential features of both phenomena described in the foregoing, in the following we propose a simple mechanical model able to reproduce the phase-II elastic equilibrium state attained in rectangular masonry elements. This permits the overall seismic response of a building to be deduced *via* standard incremental procedures. In doing so, each of the walls with generic windowing is treated as an assemblage of elements, some of which may be cracked. In conclusion, the main characteristics of the resulting algorithm are demonstrated in an example application.

3. THE BEHAVIOR OF A MASONRY PANEL SUBJECTED TO SHEAR

A rectangular panel of unreinforced masonry is fixed at its bases, the other sides being load-free (See Fig. 2). Let w be the width, h the height and t the thickness, assumed constant and small with respect to w and h. The panel is subjected to a relative displacement δ between its bases and parallel to them.

Phase I - In the absence of perceptible cracking, the stress distribution in the panel can be determined, within a technically acceptable margin of error, through linear elastic analysis by disregarding the effects of the material's dishomogeneities and anisotropies. In particular, by extending the beam model to the panel [1] and denoting by E and v the material's Young's modulus and Poisson's ratio, the lateral stiffness results to be

$$K^I = \frac{Et}{\left[2\chi(1+v)+\eta^2\right]\eta}, \tag{1}$$

where χ is the shear factor (6/5 for a rectangular-shaped cross-section) and $\eta = h/w$. In order to account to some extent for the actual degree of constraint, a reduced value of K^I is usually assumed for masonry elements located near the top of the building [2].

Phase II - When cracking becomes evident, the simplest mechanical model for assessing the stress state in the panel is represented by the graph in figure 3 below.

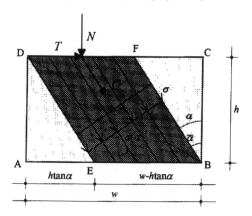

Figure 3: The load-bearing zone

In particular, it is assumed that the reactive elastic zone is made up of the parallelogram E-B-F-D, whose dimensions for given values of w and h are determined solely by angle α. Within the zone, the state of stress is constituted by a uniform compression σ, with resultant $C(\alpha)=\sigma t\, d(\alpha)=\sigma t\, (w \cos\alpha - h \sin\alpha)$. The triangular regions A-E-D and B-C-F are instead held to be stress-free. The lower the material's tensile strength, the closer such a model will conform to reality. In order to determine α, we assume a value such that the lateral stiffness of the reactive zone is at a maximum:

$$K^{II}(\alpha) = Et \sin^2\alpha \cos\alpha(\cos\alpha - \eta\sin\alpha)/\eta. \tag{2}$$

Thus, angle $\hat{\alpha}$ coincides with the maximum point of the function $f(\alpha) = \sin^2\alpha \cos\alpha(\cos\alpha - \eta\sin\alpha)/\eta$, with $0 \le \alpha \le \bar{\alpha}$, where $\bar{\alpha} = \arctan w/h$. Figure 4 shows how $\hat{\alpha}(\eta)$ is monotone decreasing between values 45° and 0°. Figure 5 instead plots the stiffness of the wall elements in phases I and II, from which it can be seen how the loss of stiffness is always significant, as long as $\eta < 2$.

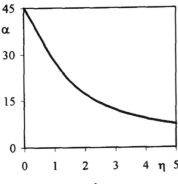

Figure 4: Angle $\hat{\alpha}$ versus the ratio h/w

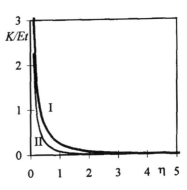

Figure 5: Panel stiffness

4. THE LIMIT STRENGTH DOMAIN

In deducing the interaction curves for the panel, we consider only cases in which failure occurs either by crushing of the weaker masonry components, or by slipping.

Denoting by $N = C \cos \alpha$ and $T = C \sin \alpha$, the vertical and horizontal components of C, respectively, and by v and u, the corresponding displacements (See Fig. 2), the interaction curves traced in the N-T plane are deduced by recourse to the static theorem of limit analysis. In doing so, we first deal with elements for which N can be considered a constant, and then elements for which v is zero and N is unknown.

4.1 Masonry elements with constant N.

Failure by crushing: the interaction curve is obtained by looking for the angle α at which the component T is a maximum under the constraint $\sigma \le \bar{\sigma}$, where σ is the maximum normal stress and $\bar{\sigma}$ is the characteristic compressive strength of the weaker component, brick or mortar, determined by standard uniaxial tests. The solution to the maximum problem

$$N \tan \alpha = \max, \qquad \text{subjected to} \qquad \frac{N}{wt} \frac{1}{\cos^2 \alpha - \eta \sin \alpha \cos \alpha} \le \bar{\sigma}, \qquad (3a,b)$$

is given in terms of the dimensionless quantities $\mathbf{n} = N/\bar{\sigma}wt$ and $\mathbf{t} = T/\bar{\sigma}wt$, by the curves depicted in Figure 6 relative to the values 0.5, 1.0 and 2.0 of the panel aspect ratio $\eta = h/w$.

Failure by slipping: the interaction curve is obtained by looking for the angle α at which T is a maximum under the constraint $|\tau| \le \bar{\tau} + \sigma \tan \bar{\varphi}$, where $\bar{\tau}$ and $\bar{\varphi}$ are the characteristic values of the material's strength determined by standard compression and shear tests on masonry panels. The solution to the maximum problem

$$N \tan \alpha = \max, \qquad \text{subjected to} \qquad N \tan \alpha \le \bar{\tau} t (w - h \tan \alpha) + N \tan \bar{\varphi}, \qquad (4a,b)$$

is given in terms of $\mathbf{n}$ and $\mathbf{t}$ by the curves shown in Figure 7 for the same values of

$\eta = h/w$, and assuming $\overline{\sigma} = 5.0N / mm^2$, $\overline{\tau} = 0.2N / mm^2$ and $\tan\overline{\varphi} = 0.4$. Such curves turn out to be weakly sensitive to variations in η and, for **n** greater than 0.05, they appear to be straight lines.

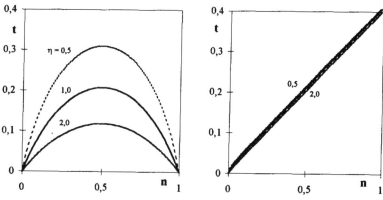

Figure 6: Interaction curves for crushing   Figure 7: Interaction curves for slipping

By superimposing the two types of curves, the limit domains showed in Figure 8 are finally obtained.

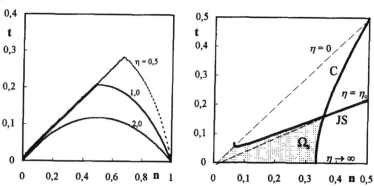

Figure 8: The limit strength domain for   Figure 9: The limit strength domain for
generic masonry panels   maximum-stiffness panels

4.2 Masonry elements operating under conditions of maximum stiffness

Failure by crushing: by substituting $C = \overline{\sigma}(w\cos\alpha - h\sin\alpha)t$ into expressions $N = C\cos\alpha$ and $T = C\sin\alpha$, and considering $\alpha = \hat{\alpha}$, we obtain

$$n = \cos\hat{\alpha}(\cos\hat{\alpha} - \eta\sin\hat{\alpha}), \qquad t = \sin\hat{\alpha}(\cos\hat{\alpha} - \eta\sin\hat{\alpha}), \qquad (5a, b)$$

which yield the limit condition for crushing (See Fig. 9, curve C).

Failure by slipping: if one admits that slipping can occur along a generic plane (here mortar is considered to be more resistant than bricks), the Mohr-Coulomb criterion yields the limit condition

$$\mathbf{n} = \frac{2\bar{\tau}\cos\bar{\varphi}}{\bar{\sigma}(1-\sin\bar{\varphi})}\cos\hat{\alpha}(\cos\hat{\alpha}-\eta\sin\hat{\alpha}), \quad \mathbf{t} = \frac{2\bar{\tau}\cos\bar{\varphi}}{\bar{\sigma}(1-\sin\bar{\varphi})}\sin\hat{\alpha}(\cos\hat{\alpha}-\eta\sin\hat{\alpha}). \quad \text{(6a, b)}$$

For each η, the values given by (6a,b) are proportional to those given by (5a,b). Therefore, panel failure will be brittle (crushing), if $\bar{\sigma} < 2\bar{\tau}\cos\bar{\varphi}/(1-\sin\bar{\varphi})$, otherwise it will be of the ductile sort. If the mortar is weaker than the brick, we may also consider the possibility of failure occurring by joint separation. Thus, by disregarding the contribution of the mortar head joints to strength, the Mohr-Coulomb criterion becomes (See Fig. 9 curve JS)

$$\mathbf{n} = \frac{\bar{\tau}}{\bar{\sigma}}\frac{(1-\eta\tan\hat{\alpha})}{\tan\hat{\alpha}-\tan\bar{\varphi}}, \qquad \mathbf{t} = \frac{\bar{\tau}}{\bar{\sigma}}\frac{(1-\eta\tan\hat{\alpha})\tan\hat{\alpha}}{\tan\hat{\alpha}-\tan\bar{\varphi}} \qquad \text{(7a,b)}$$

Coupling curves C and JS, the non-convex limit strength domain Ω_s is finally obtained; this is presented in Figure 9, where all curves are plotted using the previously given values of $\bar{\sigma}$, $\bar{\tau}$ and $\tan\bar{\varphi}$. Again, because the ratio $\mathbf{t}/\mathbf{n} = \tan\hat{a}$ depends solely on η, it is possible to deduce a critical ratio η_c for which $\tan\hat{a} = \bar{\sigma}/2\bar{\tau}(1-\sqrt{1-4\bar{\tau}/\bar{\sigma}(\bar{\tau}/\bar{\sigma}+\tan\bar{\varphi})})$; so, slender panels ($\eta>\eta_c$) will fail by crushing, if $\bar{\sigma} < 2\bar{\tau}\cos\bar{\varphi}/(1-\sin\bar{\varphi})$, or slipping along a generic plane, while wider ones ($\eta<\eta_c$) will always fail by joint separation.

7. NON-LINEAR ANALYSIS OF A MASONRY BUILDING

By applying the shear-type model, determination of the structural response of a building to seismic loads can be performed *via* sub-structuring techniques, by which the main structure is comprised of floors and the walls are secondary [3]. In such a framework, in order to account for possible openings in the walls, each wall can be thought of as an assemblage of smaller rectangular elements, each of which satisfies the constitutive equation illustrated in Figure 1 and the limit strength conditions deduced in the foregoing. Finally, with regard to the transition between the two phases, although the cessation of uncracked stage should more properly be evaluated in the context of fracture mechanics theory (in order to better account for failure of the brick-mortar interface), here for the sake of simplicity, considering the extremely low levels of stress, it was assumed that cracking starts when the maximum tensile stress component reaches the material's flexural strength.

Under the hypothesis that the loads increase proportionally to a single multiplier λ, the structural response can be obtained by proceeding incrementally [4]. In each step, an iterative procedure must be used in order to assure that the stiffness considered for each resisting element corresponds to its actual state. The resulting algorithm can be easily implemented.

As an example application of the foregoing procedure, we consider a masonry building with three above-ground storeys, used as a reference problem for comparative studies among several codes on unreinforced masonry buildings [5]. Figure 10 shows the floor-plan of the building. Details of the elements' dimensions and the materials' mechanical characteristics are available in the original publication from which the values utilized here have been drawn.

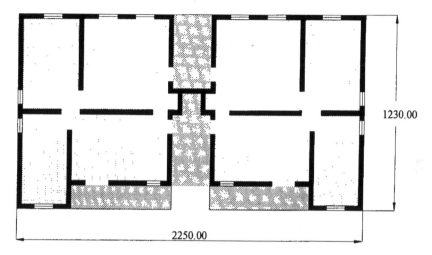

Figure 10: Floor-plan of the building considered

In particular, each of 30cm-wide wall is made of unreinforced brickwork masonry with the following mechanical properties:

$$E = 4570N / mm^2, \qquad G = 0.4E = 1828N / mm^2,$$

$$\overline{\sigma} = 5.0N / mm^2, \ \overline{\tau} = 0.2N / mm^2, \ \tan\overline{\varphi} = 0.4.$$

The values used for flexural strength and maximum shearing strain are:

$$\overline{\sigma}_t = 0.3N / mm^2, \qquad \gamma_{max} = \delta_{max}/h = 0.2,$$

The forces acting on the floors are:

$$F_1 = 441870N, \quad F_2 = 915230N, \quad F_3 = 681835N,$$

and they are applied separately along the main directions of the building plane.
The results obtained in the two cases are presented in Figures 11 and 12, which show the value of the load multiplier as a function of the displacement of the center of mass of each floor. The two elastic phases of the building's response are highlighted in figures 11b and 12b, where the relative displacements of each floor are furnished. The stiffness values, however, are strongly dependent upon the aspect ratio of the resisting elements of each walls. The case considered follows the specifications contained in the code EC8, by which the height of each resisting element is set equal to the distance between two floors. However, when a different subdivision of each wall in smaller elements was used,

the overall structural behavior and the wall's manner of failure turned out to be appreciably different. Finally, the maximum value of the load multiplier depends heavily upon that of the assumed maximum shearing strain.

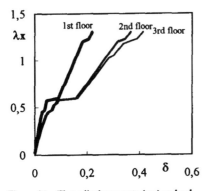

Figure 11a: Floor displacements (cm) vs load amplification factor λ_x

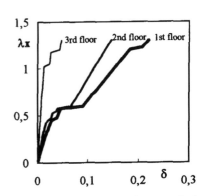

Figure 11b: Relative floor displacements (cm) vs load amplification factor λ_x

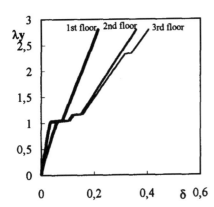

Figure 12a: Floor displacements (cm) vs load amplification factor λ_y

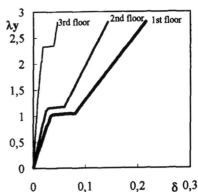

Figure 12b: Relative floor displacements (cm) vs load amplification factor λ_y

6. REFERENCES

[1] HENDRY, A. W.: "Reinforced & Prestressed Masonry", Longman, Harlow, 1991.

[2] PAULAY, T. M. and PRIESTLEY, J. N.: "Seismic Design of Reinforced Concrete and Masonry Buildings", John Wiley & Sons, New York, 1992.

[3] FAVILLI, A., LIGARO', S.:Un modello semplificato per l'analisi lineare di edifici in muratura soggetti al sisma, Atti XII Congr. AIMETA '95, Napoli, 1995, pp. 223-228

[4] MENGI, A. *et al.*: "A Model for Nonlinear Earthquake Analysis of Unreinforced Brick Masonry Buildings", Computers & Structures, 41, 1991, pp. 801-812.

[5] MODENA, C. *et al*: "Esame comparativo della normativa europea e della normativa italiana sulle strutture in muratura portante", ANDIL, Ed. Lambda, Padova, 1992.

MODELLING OF STONE MASONRY WALLS STRENGTHENED BY RC JACKETS

C. Modena Professor, **G. Zavarise** Researcher and
M.R. Valluzzi Structural Engineer
*Department of Construction and Transport, University of Padua,
Via Marzolo 9, I-35131 Padova, Italy*

1. ABSTRACT

In this paper we illustrate the research activity carried out to set up diagnostic methods for the determination of mechanical parameters of the walls and to check the effectiveness of repairing and reinforcement techniques. A series of experimental tests carried out on old buildings before and after restoration are discussed, and comparisons with numerical analyses are shown.

2. INTRODUCTION

Stone masonry walls constitute one of the most common construction typologies that have been used all over Europe and the necessity to define reliable criteria for repairing and restoring them is actually taking more and more importance. Such walls are in fact quite often not responding to the current criteria of comfort and safety.

One of the most diffused techniques for consolidation is based on the application of a thin concrete slab on both the surfaces of the wall; these layers are reinforced with a welded net and connected between them by transversal ties across the wall. A noticeable experimental activity has been devoted to the analysis of the reparation with this technology and comparisons with FE numerical methods have also considered. These methods have actually the possibility to deal with various kinds of non-linearity, hence it can be used also to integrate and partially replace experimental tests. However the current experience in using numerical method in this field is still limited, hence results should be evaluated with attention.

Keywords: Stone masonry walls, Wall reinforcement, R.C. Jackets, Numerical tests

Computer Methods in Structural Masonry – 4, edited by G.N. Pande, J. Middleton and B. Kralj.
Published in 1998 by E & FN Spon, 11 New Fetter Lane, London EC4P 4EE, UK. ISBN: 0 419 23540 X

3. EXPERIMENTAL TESTS

The walls have been tested before and after restoration using both large compression and shear tests, and flat jack tests, which allowed to determine the state of stress and the deformability. The necessity to check a representative area has forced the employment of special very large flat jacks. The scheme of the test with the check points is reported in Fig. 1 [3]. In-situ compression tests have been carried out checking large panels which dimension is 1.5×1 m. A concrete beam have been created both on the top and on the bottom of the panels to distribute the applied load and a series of checking point on the sides of the panel have been placed. The tests have been carried out both for original and restored walls, see Fig. 2. A comparison of the results obtained with flat jack tests and compression tests is reported in Fig. 3 [4], which shows a good agreement between the procedures. The efficacy of the restoration with concrete slabs is evidenced in Fig. 4, where the completely different behaviour before and after restoration is depicted.

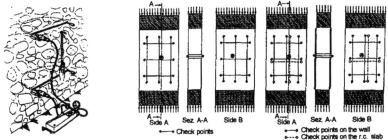

Figure 1: Flat jack test scheme   Figure 2: Compression test for original and reinforced walls.

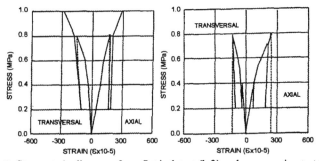

Figure 3: Stress-strain diagrams from flat jack test (left) and compression test (right).

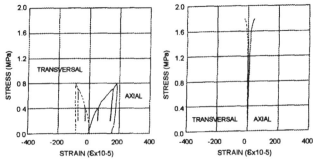

Figure 4: Stress-strain diagrams for unconsolidated (left) and consolidated wall (right).

6. NUMERICAL MODELS

The development of a methodology for the numerical modelling of the repair of existing buildings can give a considerable help to the designer in the choice of the best repairing technique. Of course, due to the high difficulty related to the characterisation of the mechanical parameters, the numerical results, at list at this stage of the research, should be used mainly for the comprehension of the global behaviour of the structure. The local state of stress depends strongly on the local characteristics of the existing walls, and it requires a different analysis carried out using a local scale.

The numerical modelling to test the effectiveness of the reinforcements of masonry walls with thin concrete slabs on the surface presents problems related to the characterisation of the shear transmission between the original walls and the new concrete. In fact, a wall reinforced with concrete slabs on the surfaces presents the same structure of a sandwich panel. The core material of the sandwich, i.e. the original wall, in reality is a non-homogeneous anysotropic material characterised by a strongly non-linear mechanical behaviour. Moreover the behaviour of the joint between the core and the outer layers presents different characteristics from the simple adhesion between two flat surfaces of homogeneous materials. In this case the wall is characterised by a superficial macroscopic roughness, hence interlocking mechanisms takes place for the stress transmission. Finally the steel reinforcements which connect the two outer slabs crossing the inner wall have a strong influence, at least on the ultimate load, because the wall benefits of the confinement state which is generated.

The complexity of the problem requires a step-by-step procedure in the development of the numerical model. One more difficult can occur when geometrical non-linearity are involved. Due to the difficulty related to the characterisation of the input parameters we start with simple models to check the basic behaviour.

6.1 Basic Model

The basic model, taken as reference for the subsequent modelling, is a 1.5 m high wall consolidated with two r.c. slabs 5 cm thick connected by $5\varnothing6/m^2$ tie-rods.

The specimen involves four different materials, i.e. the concrete at the top and at the bottom, placed to distribute the applied load, the steel reinforcement, which cross the original masonry wall, the special concrete used for reinforcement and the wall itself. We adopted a two dimensional-plane strain model and the evaluation of input data describing mechanical characteristics has been carried out on the base of specific tests. Material parameters for both concrete and steel can be determined with a low effort whereas in-situ compressive tests have been used for the characterization of the existing wall. A curve fitting of experimental results using Drucker-Prager constitutive law permitted to find the following values: elastic modulus E_m = 400 MPa, Poisson coefficient v = 0.2, friction φ = 45°, cohesion c = 1.5. Concrete for reinforcement has been modelled using a linear elastic law with E_b = 3200 MPa and Poisson coefficient v = 0.2. Steel reinforcements have been modelled using an elasto-plastic law with E_s = 210000 MPa and yield stress σ_s = 500 MPa.

A crucial task concerns the modelling of the interface between the existing wall and the new concrete. The adopted scheme of the joint (Fig. 5) permits the transfer of shear forces due to the asperities interlocking and a contemporary presence of dilatancy effects. Local effects due to the interaction of steel reinforcements with the masonry have been neglected. However the presence of steel reinforcements has been taken into account and modelled with rods connected to the external sides of the added concrete.

Moreover, the existence of horizontal and vertical symmetry has permitted to discretize only a quarter of the whole specimen, as reported in Fig. 6.

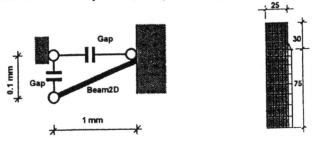

Fig. 5: Concrete-wall interface discretization.   Fig. 6: Concrete and wall discretization.

Load conditions have been applied with a step-by-step procedure, with increments of P=0.1 MPa. Transversal displacements in the middle of the specimen versus the applied vertical pressure due to the load conditions have been compared with experimental tests. Comparisons reported in Fig. 7, evidence a good agreement for the linear part of the diagrams both for original and reinforced walls. The correspondence of the non-linear behaviour is also in good agreement in the first part. For the unreinforced wall the almost perfect correspondence between the curves disappear when the experimental curve present a jump, probably due to a collapse of a brittle masonry element. The numerical model adopted is not able to take into account such effects. However it is noticeable that after the jump the two curves present the same trend till the total collapse is achieved.

Numerical results evidence also that the transmission of the shear forces which takes place at the interface between the masonry and the concrete slabs are concentrated at the top. For any applied load, more than the 90% of the shear between the wall and the reinforcement is concentrated at the top in an area which length is 40 cm.

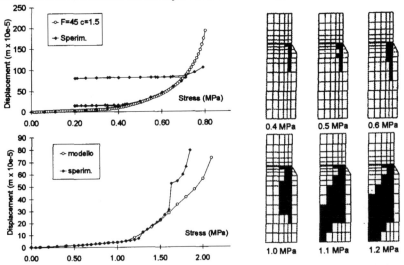

Fig. 7: Comparison of numerical and experimental stress - displacement diagrams. Original wall (top), reinforced wall (bottom).

Fig. 8: Plastic zone evolution of the consolidated wall.

It have to be remarked the fact that the reinforcement sustain more than 90% of the applied load, hence it gives a great contribution to the original wall. Even when the system is close to the collapse limit such contribution is greater than 80%. This behaviour is evidenced also in Fig. 8, where the evolution of the plastic zone is represented.

6.2 Variations on the basic model

On the basis of the numerical model previously described, the effects have been investigated of varying some characteristic geometrical parameters, such as:
- area of the connecting steel bars (A), by considering 3, 8 and 10 rods $\varnothing6/m^2$;
- thickness of r.c. slabs (S), taken equal to 3, 4, 6, 7 and 8 cm;
- wall's height (H), considered equal to 3 and 4 m.

Tab. 1 shows the combinations of the investigated parameters; the figures where the results are presented as vertical compressive stress versus transversal displacement diagrams are referred to in the lost column.

Table 1: Geometrical variations of the numerical models and numbers of the related figures.

Variations compared to the reference model	Models	Number of tie-rods	Slab's thickness (cm)	Wall's height (m)	Figures
	Reference	$5\varnothing6/m^2$	5	1.5	
Number of tie-rods	A3	$3\varnothing6/m^2$	5	1.5	
	A8	$8\varnothing6/m^2$	5	1.5	Fig. 9
	A10	$10\varnothing6/m^2$	5	1.5	
R.C. slabs' thickness	S3	$5\varnothing6/m^2$	3	1.5	
	S4	$5\varnothing6/m^2$	4	1.5	
	S6	$5\varnothing6/m^2$	6	1.5	Fig. 10
	S7	$5\varnothing6/m^2$	7	1.5	
	S8	$5\varnothing6/m^2$	8	1.5	
Wall's height	H3	$5\varnothing6/m^2$	5	3	Fig. 13
	H4	$5\varnothing6/m^2$	5	4	
Tie-rods and slabs' thickness	S3A3	$3\varnothing6/m^2$	3	1.5	Fig.12
	S8A10	$10\varnothing6/m^2$	8	1.5	
Tie-rods and wall's height	H3A3	$3\varnothing6/m^2$	5	3	
	H3A10	$10\varnothing6/m^2$	5	3	Fig. 15
	H4A3	$3\varnothing6/m^2$	5	4	
	H4A10	$10\varnothing6/m^2$	5	4	
Slabs' thickness and wall's height	H3S3	$5\varnothing6/m^2$	3	3	
	H3S8	$5\varnothing6/m^2$	8	3	Fig. 16
	H4S3	$5\varnothing6/m^2$	3	4	
	H4S8	$5\varnothing6/m^2$	8	4	

The models characterized only by the variation of the steel areas showed an increasing ultimate strength as the steel area increases (Fig. 9). The first elastic branch does not significantly depend on the quantity variations of the steel; afterwards, for stress values around 1.1 MPa, the inside core of the masonry wall comes in the plastic field and the connecting rods bear higher stresses. After their yelding (in this case has been observed that this happens always for the upper rod first) the slab is not kept any more and the collapse happens for its sudden push out due to instability phenomena.

As regards the slab thickness, a slight decrement of the ultimate strength has been obsereved as the slab's thickness increases (Fig. 10), up to a value of 2.2 MPa for the thickness of 3 cm and 1.7 MPa for that of 8 cm.

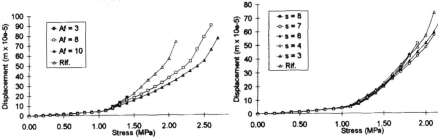

Fig. 9: Comparison among the basic model (Rif.) and the models with different steel areas.

Fig. 10: Comparison among the basic model (Rif.) and the models with different slab thicknesses.

In both the elastic and inelastic branches the global behaviour is substantially that of the masonry, thus the curves are coincident for all the models. However, the collapse happens first in the models with higher slab's thickness than in those with the lower. That is due to global instability phenomena of the slabs caused by two kinds of actions applied to the slabs themselves by the masonry. The action applied by the upper layer of the masonry causes a thrust which the two connecting rods (of which the upper is more stressed) oppose to. Consequently the slab tends to buckle as shown in Fig. 11.a. The shear load P could be considered as concentrated on the slab's upper section connected with the masonry-slab interface, so that the equilibrium of the forces causes the slab's deformation as shown in Fig. 11.b. This stresses the lower connecting rod, which is the first to yield. These two actions are combined in the loading of the slab but, while the first mechanism does not vary significantly as the thickness varies, the second tends to be prevalent over the other when the thickness increases. The results of the elaboration seem to confirm this hypothesis, since the first yelding has been found in the upper rod for the thinner walls and in the lower one in the thicker ones.

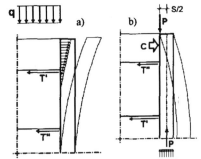

Figure 11: Masonry actions on the slab.

As regards the two numerical models obtained by combining different slabs' thickness with different steel areas (Fig. 12) a loss of stiffness has been observed in the plastic branch of the model with the lower steel area. This corresponds to a sudden instability of the slab after the simultaneous yelding of the two connecting rods. It can be noticed, however, that the failure load did increase, because of the reduction of the slab's thickness, which reduces the effects described in Fig. 11.b. Also the increase of the steel area in the model with S=8 cm produced good effects, raising the resistance from 1.7 to 2.0 MPa. These results confirm the importance of the steel quantities in the strength and stiffness increase in the plastic phase.

The height of the wall has been proved to be a very influential parameter on the global behaviour of the numerical models of the consolidated masonry. The reference model reproduces a panel resulting from an existing wall, having a reduced height (H=1.5 m) as regards those usually interested by repair interventions, since it was subjected to compressive tests. In the models with height of 3 and 4 m (Fig. 13) the ultimate strength increases from 2.1 MPa for H=1.5 m to 2.2 MPa for H=3 m, up to 2.5 MPa for the H=4 m. The increase of strength happens because of a different post-elastic phase evolution: in the reference model, in fact, the plastic zones start from the upper outer elements and join diagonally with those of the lower elements, whereas in the models of the taller panels a central core not yet in plastic phase has been found also near the failure load (Fig. 14). Hence the connecting rods set in the central zone create a restraint for the upper portion, which is more deformable and subjected to more effort. The failure happens quite in this zone and depends on a local instability phenomenon.

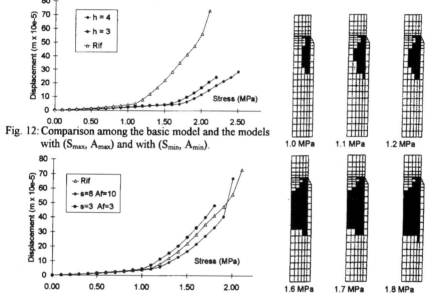

Fig. 12: Comparison among the basic model and the models with (S_{max}, A_{max}) and with (S_{min}, A_{min}).

Fig. 13: Comparison among the basic model and the models with diffrent height.

Fig. 14: Plastic zone evolution of the model H3.

Finally, in the tall panels, it has been observed that the increase of steel produces an increase of strength and stiffness (Fig. 15) and that the increase of slab's thickness implies a remarkable increase on strength (Fig. 16) due to a collapse mechanism that happens for local flexure of the slab.

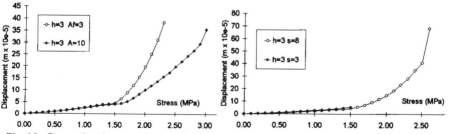

Fig. 15: Comparison between the models H3A3 and H3A10.

Fig. 16: Comparison between the models H3S8 and H3S3.

7. CONCLUSIONS

The research activity presented in this paper concerns experimental and numerical investigations on old stone masonry walls strengthened by reinforced concrete jackets. The experimental results have been used to verify the effectiveness of numerical models and their capacity to estimate the resistance of original and reinforced walls. Such numerical models, even though quite simple, confirmed the real possibility to integrate and complete the experimental tests.

From the results, some qualitative useful indications concerning the design of the strengthening intervention (number of tie-rods and slab thickness) can be drawn. The improvement of the mechanical characteristics of the masonry as the area of the connecting steel rods increases have been roughly quantified and the behaviour in the post-elastic phase up to the failure have been clarified. Tall masonry walls, for example, have a good behaviour when the tie-rods are closer one another in the upper and lower zones rather than in the central one. Also the slab's thickness has an important role in these walls because their flexural stiffness is increased. Decreasing the wall's height, it would be better not to increase the slab thickness, to avoid a collapse for instability rather than for flexure.

8. REFERENCES

1. Oden J.T. and Pires E.B., 'Nonlocal and non-linear friction laws and variational principles for contact problems in elasticity', J. Appl. Mech., vol. 50, 67-76, 1983.
2. Zavarise G., 'Problemi termomeccanici di contatto - Aspetti fisici e computazionali', Ph.D. Thesis, Istituto di Scienza e Tecnica delle Costruzioni, University of Padua, Padua, Italy, 1991.
3. Gelmi A., Modena C., Rossi P.P. and Zaninetti A., 'Mechanical characterization of stone masonry structures in old urban nuclei', 6th North American Masonry Conference, Philadelphia, Pennsylvania, U.S.A., 1993.
4. Binda L., Modena C., Baronio G. and Abbaneo S., 'Repair and investigation techniques for stone masonry walls', 6th Int. Conf. Extending the Lifes of Bridges, Civil + Building Structures, London, England, 1995.
5. Modena C. and Bettio C., 'Experimental characterization and modelling of injected and jacketed masonry walls', Proc. Italian-French Symposium Strengthening and Repair of Structures in Seismic Area, Nizza, France, 1994.

CRACK PATTERNS IN MASONRY INFILLED RC FRAMES UNDER HORIZONTAL LOADING

R. Barsotti PhD[1], **S. Ligarò** Assistant Professor[1] and
G. Royer-Carfagni Assistant Professor[2]
[1]*Department of Structural Engineering, University of Pisa, Pisa,
Italy and* [2]*Department of Civil Engineering, University of Parma,
Parma, Italy*

1. ABSTRACT

An attempt is made to determine the equilibrium state attained in masonry-infilled RC frames subjected to in-plane shear forces once the material's tensile strength has been overcome. The non-linear equilibrium problem is solved by making use of a simple mechanical model and resorting to the principle of stationary total potential energy. Numerical solutions are obtained by an in-house F.E. code that combines standard linear elastic elements with no-tension ones. Thus, the influence of geometrical and loading parameters on the system equilibrium can be assessed.

2. INTRODUCTION

Masonry is quite strong and behaves elastically, as long as it is subjected to moderate pressure forces, but turns out to be extremely weak if one tries to pull it apart. In fact, rather small tensions are sufficient to produce fractures. Such behavior remains unchanged also when masonry interacts with other, different materials.

When testing infilled RC frames by means of in-plane shear forces (See Fig. 1), three different phases can be recognized in the system's behavior. In the very earliest, the response of both components (RC frame and masonry) is practically elastic and linear, and thus so is that of the whole system until failure at their interface occurs. At the end of this phase, only compressive stress can be transmitted between the RC frame and the masonry panel. The second phase is instead characterized by a drastic change in the type of stress in the masonry which, under significant load, tends to be without tensile components. This change occurs through the growth of ever more marked cracking in the infill, even though the whole structure remains far from its ultimate state.

Keywords: Infilled RC frames, Phase-two behavior, No-tension materials

Computer Methods in Structural Masonry – 4, edited by G.N. Pande, J. Middleton and B. Kralj.
Published in 1998 by E & FN Spon, 11 New Fetter Lane, London EC4P 4EE, UK. ISBN: 0 419 23540 X

The crack pattern develops in such a way that masonry struts are still able to strengthen the RC frame. Lastly, phase three can be thought of as coinciding with failure of the as-yet resisting arrangement. Failure occurs first in the masonry (usually by head-joint separation, bed-joint slipping or crushing or out-of-plane buckling of struts) and finally in the frame (RC collapse). Therefore, a better understanding of phase-two behavior appears fundamental to assessing the actual load-bearing capacity of the composite system.

In this paper, based on the conjecture that the aforementioned resisting scheme can be deduced approximately by analyzing masonry cracking patterns, we propose a simple theoretical two-dimensional model able to numerically reproduce the crack patterns usually observed in experimental tests.

3. PROBLEM FORMULATION

Let us consider the masonry-infilled RC frame shown in Figure 1, subjected to two independent uniform loads: q, horizontally and p, vertically.

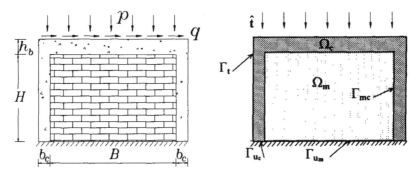

Figure 1: The panel layout   Figure 2: The problem formulation

To determine the phase-two equilibrium state, some assumptions need to be made. In particular, masonry is considered an elastic, homogeneous and isotropic no-tension material [1, 2], while the RC frame is taken to be indefinitely elastic, homogeneous and isotropic. With the notation presented in Figure 2, the problems can be stated as follows. The field equations are:

$$\text{in } \Omega_c : \begin{cases} \mathbf{E} = \frac{1}{2}\left(\nabla\mathbf{u} + \nabla\mathbf{u}^T\right), \\ \mathbf{T} = \mathbf{C}^c\mathbf{E}, \\ \nabla\cdot\mathbf{T} = 0, \end{cases} \quad \text{and} \quad \text{in } \Omega_m : \begin{cases} \mathbf{E} = \frac{1}{2}\left(\nabla\mathbf{u} + \nabla\mathbf{u}^T\right), & \mathbf{E}^w\mathbf{n}\cdot\mathbf{n} \geq 0 \,\forall\mathbf{n}, \\ \mathbf{T} = \mathbf{C}^m(\mathbf{E} - \mathbf{E}^w), & \mathbf{T}\mathbf{n}\cdot\mathbf{n} \leq 0 \,\forall\mathbf{n}, \\ \nabla\cdot\mathbf{T} = 0, & \mathbf{T}\cdot\mathbf{E}^w = 0, \end{cases}$$

where Ω_c and Ω_m represent the undistorted reference configurations of the concrete-frame and masonry-panel, respectively, $\mathbf{u}$ is the displacement field, $\mathbf{C}^c$ and $\mathbf{C}^m$ the elasticity tensors for concrete and masonry, $\mathbf{T}$ the Cauchy stress, $\mathbf{E}$ the infinitesimal strain tensor, and $\mathbf{E}^w$ denotes the inelastic strain tensor, also known as "cracking strain" [3]. Supposing both frame and panel to be clamped to a rigid support, the resulting boundary conditions are

$$\mathbf{u} = \mathbf{\hat{u}} = 0 \text{ on } \Gamma_{uc} \cup \Gamma_{um}, \quad \mathbf{Tm} = \mathbf{\hat{t}} \text{ on } \Gamma_t,$$

with $\mathbf{\hat{u}}$ imposed displacements, $\mathbf{\hat{t}}$ applied loads and $\mathbf{m}$ the outward unit normal to the external contour. In addition, the jump conditions

$$\mathbf{u}^+ = \mathbf{u}^-, \quad (\mathbf{Tn})^+ = (\mathbf{Tn})^- \text{ on } \Gamma_{mc},$$

with $\mathbf{n}$ the unit normal to the interface between concrete and masonry, indicate perfect adherence between the two materials. The solution to this problem has been determined through the principle of stationary total potential energy [4]. In particular, the search for stationary points is easily performed *via* the Lagrange multipliers method. The resulting non-linear system of equilibrium equations has been solved numerically by means of a specifically developed finite element code [5] employing simple constant stress-constant strain four nodes isoparametric elements.

4. NUMERICAL ANALYSIS

To assess the influence of: *i*) the panel aspect ratio *B/H*, *ii*) the beam and column dimensions, *iii*) the ratio *q/p* between horizontal and vertical load magnitudes, we have considered 56 different cases, summarized in Table 1, plus other intermediate situations. In all cases, the elastic properties of both the masonry and RC frames were kept fixed: a Young's modulus $E_c = 28500$ MPa and a Poisson's ratio $v_c = 0.15$ were considered for concrete while, a Young's modulus $E_m = 6000$ MPa and a Poisson's ratio $v_m = 0.18$ were assumed for masonry. It is worth noting that it is the load ratio *q/p*, rather than the magnitude of each single load, that proved to be the decisive term in defining the cracking pattern.

Table 1

B/H	1/3, 1/2, 1, 3/2, 2, 7/2, 5	
q/p	1/10, 1, 3, ∞	
Frame	Strong: beam 50 cm x 30 cm, column 40 cm x 30 cm	Weak: beam 24 cm x 60 cm, column 30 cm x 30 cm

A sample case is depicted in Figure 3.

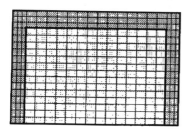

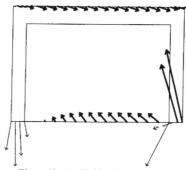

Figure 3a: The mesh utilized

Figure 3b: Applied loads and reactions

Figure 3a shows the relative mesh. The applied loads and resulting reactions are reported in Figure 3b. Figure 3c gives the deformed configuration. The directions and values of principal stress are instead represented in Figure 3d; here, thicker lines correspond to compression, while thinner ones to tension; moreover, the length of the segments defining the stress directions have been drawn proportional to the stress values at that point.

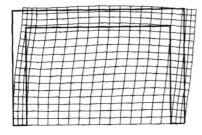

Figure 3c: The deformed configuration (magnification factor 100:1)

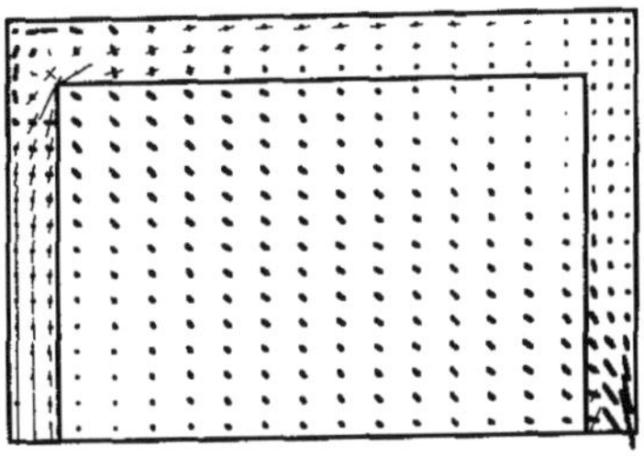

Figure 3d: Principal stresses

It can be easily verified that no tensile stress components are present in the masonry panel. Two zones appear almost inactive in the neighborhood of two opposite corners of the frame, while the masonry struts can be identified by the directions of the thicker segments. The RC column on the left-hand side acts as a tie, while the one on the right-hand side is almost inactive except for the portion closest to the clamped end, where pronounced flexure is discernible.

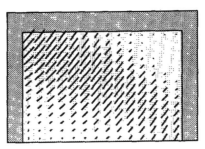

Figure 3e: Cracking strain.

Finally, Figure 3e shows the so called "crack pattern". Obviously, the model does not predict the location, length and width of each single crack, but rather gives pointwise information about the inelastic strain necessary to make the no-tension stress state compatible; otherwise, it would only be equilibrated. Thus, at each point, the illustrated segments are traced along the direction of maximum in-plane inelastic strain, and their length rendered proportional to the value of the inelastic strain expected in that region.

5. THE EFFECT OF CRACKING ON OVERALL STIFFNESS

An important parameter to consider is the ratio K_{II}/K_0, where K_{II} is the phase-two stiffness offered by the system "panel + frame" (with cracked masonry), while K_0 refers to the bare frame (i.e. without infill). Both K_{II} and K_0 have been calculated by dividing the resultant of the horizontal loads ($q \cdot L$) by the mean horizontal displacement of the RC beam. The graphs in Figure 4 show the ratio K_{II}/K_0 as a function of q/p for the strong frame case. Each line refers to different aspect ratios of the panel. In particular, q/p tending to zero corresponds to the case in which the panel is very highly compressed by vertical loads. Since compression prevents crack formation, the greatest stiffness corresponds to the largest vertical loads.

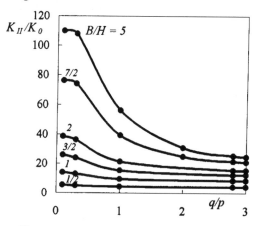

Figure 4: Stiffness ratio K_{II}/K_0 versus load ratio q/p.

In the limit case of $q/p \to 0$, the various curves tend, with horizontal tangent, to the situation of a panel with no cracks, when both panel and frame behave as indefinitely elastic also in tension. On the right-hand side, each curve tends instead to the asymptote corresponding to the case of horizontal load alone ($q/p = \infty$). In any case, whatever the B/H and q/p values are, the loss of stiffness consequent to the transition from phase-one to phase-two results always relevant.

In Figure 5, K_{II}/K_0 is now plotted as a function of the aspect ratio B/H.

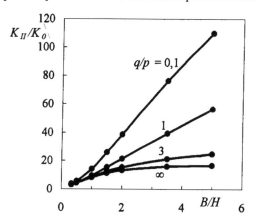

Figure 5: Stiffness ratio K_{II}/K_0 versus aspect ratio B/H.

A few things are apparent. The increase in stiffness due to increasing vertical load is much more marked for large, than for small *B/H*. When vertical loads are important (i.e. *q/p*=0.1), increasing the length of the panel increases system stiffness. On the other hand, when vertical loads are small (*q/p* → ∞), enlarging the width of the panel does not produce a corresponding increase in stiffness, as might be expected. This means that only a limited part of the composite system resists the applied loads; material that is added when the aspect ratio of the panel is enlarged is likely to remain inactive. Finally, in the case of *q/p*=1, the increase in stiffness is almost linear, suggesting that all the added material contributes in the same way to the overall stiffness of the composite system.

6. CRACK PATTERNS

The foregoing considerations can be better explained by referring to the following Figures.

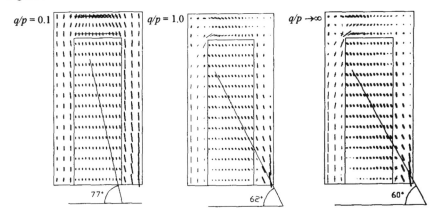

Figure 6: Principal stress for *B/H*=1/3 and various load conditions.

Figure 6 corresponds to the case of *B/H*=1/3 and gives the directions and values of principal stresses for three values of the ratio *q/p*. For high vertical loads (*q/p*=0.1), both panel and frame are highly compressed, and bending is evident in the horizontal beam. When the vertical loads are diminished, struts in the masonry appear closer to the horizontal (*q/p*=1). In addition, the RC column on the left-hand side shows a tie-like behavior, while the one on the right-hand side results almost inactive except for its portion closest to the clamped end which sustains high flexure. Even in the limit case *q/p*→∞, the inactive regions remain very limited because the frame is so narrow that it can efficiently contain the masonry panel.

The static role of each structural member is made even more evident in Figure 7, where the case of long panels (*B/H*=5) is considered.

Again in the case of *q/p*=0.1, the masonry is highly compressed but no appreciable bending is experienced by the horizontal beam, signifying that the vertical loads are supported mainly by the masonry. For *q/p*=1, the principal stress directions tend to be inclined 45° from the horizontal. In this particular case, both horizontal and vertical loads are supported mainly by the masonry struts, with very little contribution from the concrete frame.

The horizontal trend of the graph in Figure 5, corresponding to the case of $q/p=1$, is thus explained. As the horizontal loads are increased, the masonry struts draw nearer to the horizontal, and when q/p tends to infinity, most of the load-bearing capacity is given by a system composed of the left-hand column (which acts as a tie), the masonry struts (inclined about 35° from the horizontal) and the upper beam (which also acts as a tie transmitting the loads to the upper left-hand corner). In other words, the resistant structural scheme is, in a certain sense, localized. In fact, that part of the structure which lies on the right-hand side remains almost inactive, except for the horizontal beam that transmits the load to the active zone. This explains the horizontal branches in the graph presented in Figure 5: by increasing B/H, material is added, but this does not contribute to withstanding the applied loads.

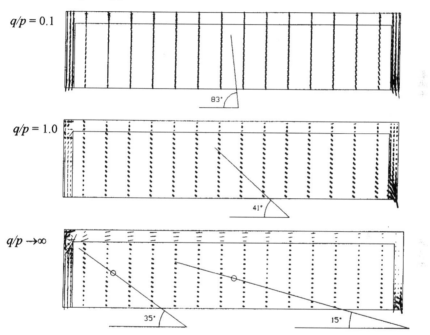

Figure 7. Principal stress for $B/H=5$ and various load conditions.

In the intermediate case of $B/H=1.5$ (See Figure 8), the behavior of the system is, as expected, midway between those of the two extreme cases discussed.

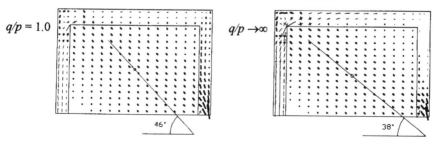

Figure 8. Principal stress for $B/H=1.5$ and various load conditions.

In order to point out the contribution of each part of the panel to supporting the applied loads, Figure 9 presents the elastic strain energy density for the whole system in the case of a strong frame, with B/H = 1,5 and q/p = 3. The level curves clearly reveal the formation of an active, though cracked, zone leading from one corner to the opposite one, and two nearly inactive zones.

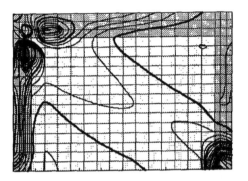

Figure 9. Strain energy density in the masonry panel.

All the features described in the foregoing remain qualitatively invariable for weak frames as well.

9. CONCLUSIONS

In conclusion, we have provided an account of the loss of stiffness in infilled RC frames due to cracking in the masonry alone and analyzed the phenomena underlying it. It is evident from the analysis that such loss is ruled by the values of ratios B/H and q/p. Consequently, the common practice of considering the presence of uncracked panels in standard static or dynamic FEM analysis leads to overestimating the system's stiffness.

10. REFERENCES

1. Di Pasquale, S., "Questioni di meccanica dei solidi non reagenti a trazione", VI Congresso nazionale AIMETA, Genova (1982), 251-263
2. Del Piero, "Constitutive equations and compatibility of the external loads for linear elastic masonry-like materials", Meccanica, 24 (1989), 150-162
3. Panzeca, T. and Polizzotto, C., "Constitutive equations for no-tension materials", Meccanica, 23 (1988), 88-93
4. Grimaldi, A., Luciano, R. and Sacco, E., "Computational problems for masonry-like materials", in "New developments in structural mechanics", Catania (1990), 51-67
5. Barsotti, R., La teoria delle membrane corrugate e le sue applicazioni nell'ingegneria strutturale e nella bioingegneria, Ph. D. Thesis, Florence (1997)

DYNAMIC RESPONSE OF MULTI-STOREY RC FRAMED STRUCTURES WITH MASONRY INFILLS: LABORATORY AND IN-SITU INVESTIGATIONS

G.C. Manos Professor[1], **B. Yasin** Research Assistant, **J. Thawambteh**
Research Assistant and **M. Triamataki** Research Assistant
Laboratory Strength of Materials, Department of Civil Engineering,
Aristotle University, Thessaloniki, Greece
[1]*E-mail: gcmanos@civil.auth.gr*

1. ABSTRACT

This paper presents a brief summary of the results of current investigation that studies the influence of masonry infills on the dynamic response of multi-story reinforced concrete (R.C.) frame structures. This investigation is two-fold. In the first part, a scaled 7-story planar R.C. frame model with or without masonry infills is studied at the Earthquake Simulator Facility of Aristotle University, being subjected to a variety of dynamic and simulated earthquake excitations. The second part of this investigation focuses on the dynamic response of a 5-story prototype R.C. building with masonry infills, which was built for this purpose at the Volvi European Test site, located near Thessaloniki.

2. THE 7-STORY PLANAR FRAME MODEL

This structure can be considered to be a 1 : 12.5 small scale model of a 7-story one-bay framed prototype structure. The amplitude and distribution of the extra weight along the height together with the basic overall dimensions are depicted in figures 1a and 1b. The concrete strength for the frame was 138kg/cm2; the yield stress for the longitudinal reinforcement for the frame was 3170kg/cm2. Model brick masonry infill panels, were incorporated within the frame structure at specific stages of the test sequence, as will be explained in what follows. The ultimate shear stress of these panels, as found from diagonal tension tests was equal to 1.8kg/cm2. The earthquake simulated tests were based on the Kern County 1957 (Taft) horizontal ground acceleration record. These tests were of progressively increasing intensity with the development of structural damage at certain stages of the testing sequence. Following each one of the simulated earthquake series, dynamic excitation tests with various intensity levels were utilized to determine the eigen-modes and eigen-frequencies at each stage of this experimental sequence.

2.1. SEQUENCE OF TESTS.

The testing sequence for the frame included dynamic excitation tests as well as simulated earthquake tests, which are depicted in figure 2. The dynamic excitation tests employed low-intensity impulse tests and frequency banded white noise tests; no use was made of frequency banded sinusoidal sweep tests. The earthquake simulated tests were based on the Taft horizontal ground acceleration record

Computer Methods in Structural Masonry – 4, edited by G.N. Pande, J. Middleton and B. Kralj.
Published in 1998 by E & FN Spon, 11 New Fetter Lane, London EC4P 4EE, UK. ISBN: 0 419 23540 X

The sequential number for each test is plotted in the abscissae of figure 2 whereas the ordinates represent the corresponding value of the fundamental eigen-frequency.

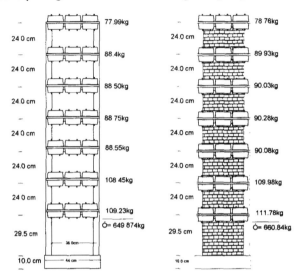

Figure 1a,b. 7-story "frame" model and extra mass distribution

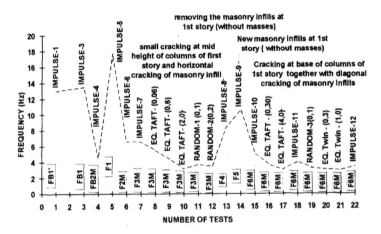

Figure 2 Sequence of Tests for the 7-story Frame Model at Aristotle University.

Thus, the 1st series of simulated earthquake tests included three tests; e.g. Taft-0.06 (Test No. 8, low-intensity), Taft-0.6 (Test No. 9, low-to-moderate intensity) and Taft- 2.0 (Test No. 10, moderate to high intensity). During this last test the masonry infill at the 1st story developed clear signs of distress in the form of horizontal cracking. This masonry infill was demolished and a new masonry panel was rebuilt before continuing with the testing sequence. The 2nd series of simulated earthquakes included two tests; Taft-0.3 (Test No 16) was again of low intensity whereas Taft-4.0 (Test No. 17) moderate to high intensity. Following each one of these simulated earthquake series, dynamic excitation tests with various intensity levels were again utilized to determine the eigen-modes and eigen-frequencies at each stage of the test sequence. The plot of figure 2 must be viewed together with Table 1 whereby brief details of the various structural configurations of the 7-story frame are listed.

Table 1   Description of the various **"7-story frame"** structural   configurations

Frame **FB1** Bare frame without extra mass	Frame **F1** Infills in all but 1st Story without extra mass	Frame **F4** Infills in all but 1st Story without extra mass
Frame **FB2M** Bare frame with extra mass	Frame **F2M** Infills in all but 1st Story with extra mass	Frame **F5** Infills in all Stories without extra mass
	Frame **F3M** Infills in all Stories with extra mass	Frame **F6M** Infills in all Stories with extra mass

2.2. SELECTED RESULTS FROM THE DYNAMIC TESTS

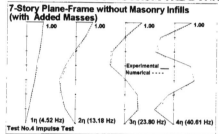

Figure   3   Test No 4 (Structure **FB2M**)

The objective here was to examine the variation of the fundamental dynamic structural parameters during the testing sequence. This variation with respect to the fundamental translational eigen-frequency is depicted in figure 2. The value of the fundamental eigen-frequency was 13Hz at the beginning of the test sequence (Test No. 1, impulse), when the model corresponded to a virgin structure completely un-cracked and without masonry infills or extra mass (Frame **FB1**). When the mass is added this frequency became 4.52Hz (Frame **FB2M**, figure 3). Next, when the extra mass was removed and the masonry infills were built in all but the 1st story they introduced a considerable increase in the stiffness (fundamental frequency 17.7Hz, Frame **F1** and 6.59Hz when the extra mass was added for Frame **F2M**). At this stage the construction of the 1st story infill did not show any increase in

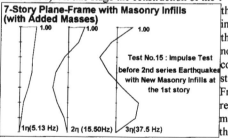

Figure 4   Test No. 15 **F6M**

the stiffness. Due to the damage that was inflicted to the masonry panels, particularly that of the 1st story , as well as to mostly non-visible   micro-cracking   there   was considerable drop in the stiffness of the structure (fundamental frequency 3.5Hz, Frame **F3M**). A part of the stiffness is regained by demolishing the damaged masonry infill and building a new one in the sequence that was described above (Test 15, fundamental frequency 5.13Hz, Frame **F6M**, Figure 4). However, during the 2nd series of simulated earthquake tests a significant loss of stiffness is again observed (Test 19, fundamental frequency 3.26Hz, Frame **F6M**).

2.3. SELECTED RESULTS FROM THE SIMULATED EARTHQUAKE TESTS.

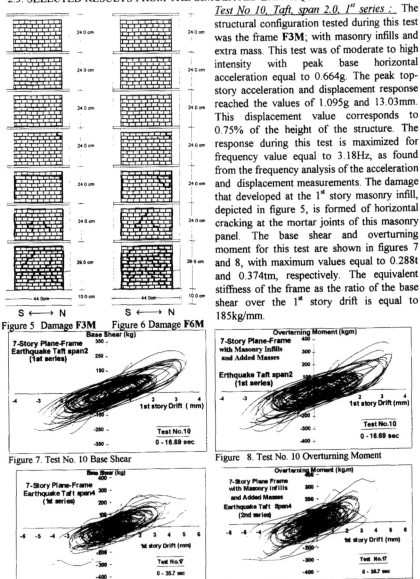

Test No 10, Taft, span 2.0, 1st series : The structural configuration tested during this test was the frame **F3M**; with masonry infills and extra mass. This test was of moderate to high intensity with peak base horizontal acceleration equal to 0.664g. The peak top-story acceleration and displacement response reached the values of 1.095g and 13.03mm. This displacement value corresponds to 0.75% of the height of the structure. The response during this test is maximized for frequency value equal to 3.18Hz, as found from the frequency analysis of the acceleration and displacement measurements. The damage that developed at the 1st story masonry infill, depicted in figure 5, is formed of horizontal cracking at the mortar joints of this masonry panel. The base shear and overturning moment for this test are shown in figures 7 and 8, with maximum values equal to 0.288t and 0.374tm, respectively. The equivalent stiffness of the frame as the ratio of the base shear over the 1st story drift is equal to 185kg/mm.

Figure 5 Damage **F3M**   Figure 6 Damage **F6M**

Figure 7. Test No. 10 Base Shear

Figure 8. Test No. 10 Overturning Moment

Figure 9. Test No. 17 Base Shear

Figure 10. Test No. 17 Overturning Moment

Test No 17, Taft, span 4.0, 1st series : Frame **F6M** was tested here, having masonry infills and extra mass, under moderate to high intensity excitation with peak base horizontal acceleration equal to 0.772g. The peak top-story acceleration and displacement response reached the values of 0.904g and 14.4mm. This displacement value corresponds to 0.82% of the height of the structure. The maximum diagonal displacement response for test No. 17 is 3.725mm, nearly twice as much as that during test No. 10 (1.893mm). The response during this test is maximized for frequency value equal to 3.05Hz. The damage that developed at the 1st story masonry infill is depicted in figure 6. The base shear and overturning moment for this test are shown in figures 9 and 10, with maximum values equal to 0.305t and 0.361tm, respectively. The equivalent stiffness of the frame during this test as the ratio of the base shear over the 1st story drift is initially equal to 235kg/mm. We can observe that this stiffness value is larger than the corresponding value during test No. 10 (185kg/mm). However, the rate of stiffness degradation for test No 17 is also larger than that of test No 10, due to the damage, in the form of diagonal cracks at the mortar joints of the masonry infills at the bottom part of the structure.

2.4. CONCLUSIVE REMARKS.
1. From the low amplitude dynamic excitation tests the damping ratio for the frame structure was found to be in the range of 5.6% to 6.05% of critical.
2.. The initial stiffness of the frame model was 235kg/mm; however, a large part of this stiffness was due to the presence of the masonry infills and degraded at an impressive rate by the diagonal cracking of the infill initially to values of the order of 100kg/mm reaching at the end of the response values of the order of 44-30kg/mm. This fact portrays the important difference in the behavior of the masonry infilled frames. That for moderate earthquake loads may retain their stiffness and thus participate up to a degree in the load bearing capacity; however, due to their brittle behavior this participation ceases to exist after the appearance of damage, usually in the form of diagonal cracks.
3. The maximum base shear and overturning moment that were sustained by the 7-story infilled frame have values equal to 0.305t and 0.374tm, respectively. These values are more than 50% the corresponding maximum values of a "similar" 7-story shear-wall model structure "(see reference Manos 1998), that was also subjected to "similar" tests.

3. 5-STORY BUILDING AND IN-SITU TESTING AT THE EUROPEAN VOLVI SITE
This 5-story structure must be considered in the four year period of its existence in the following four basic structural configurations
a. Reinforced concrete structure without added weight and without any masonry infills ("Virgin" structure, September - November, 1994
b. Reinforced concrete structure with 5 tons added weight but without any masonry infills ("Bare" structure, November 1994 - June, 1995).
c. Reinforced concrete structure with 5 tons added weight and with masonry infills in all but the ground floor (Masonry scheme 1, July 1995 - December 1996).
d. Reinforced concrete structure with 5 tons added weight and with masonry infills in all floors (Masonry scheme 2, January 1997 - today).

3.1. NUMERICAL SIMULATIONS.
The measured at various stages response was utilized together with a 3-D numerical simulation to form realistic models of this structure. The measured values of the Young's modulus for the concrete were adjusted so that their maximum becomes equal to 2984ksi (21420Mpa); moreover, the cross-sections were assumed un-cracked but taking into account the reinforcement These numerical approximation yields fundamental frequency values that agree well with the measured values that are listed together in Table 2.

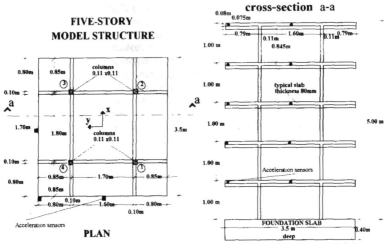

Figure 11 Dimensions of the five story Reinforced Concrete Framed Structure at Volvi.

Table 2. Summary of measured and predicted eigen-frequencies for the 5-story structure.
()* Assumed experimental value. No y-y translational pull-out was performed. In italics   *{}*** are the corresponding values obtained from the numerical simulations. *** Strong Translational (x and y) or Torsional (x and φ and / or y and φ) is observed for these modes are coupled.

Description of the 5-story Structural Configurations*	1st Transl. x-x (Hz)	1st Transl. y-y (Hz)	1st Tors. δ (Hz)	Measuring Procedure
"Virgin" structure, without added mass, without diagonals, September 1994	2.875 *{2.875}***	2.875 *{2.85}***	2.75 *{2.71}***	Real-time Analyzer
"Bare" structure, with added mass without diagonals, November 1994	2.375 *{2.40}***	2.375 *{2.38}***	2.50 *{2.30}***	Real-time Analyzer
"Bare" structure, with added mass, with diagonals ,November 1994	2.625 *{2.605}***	2.625 *{2.625}***	2.50 *{2.532}***	Real-time Analyzer
"Bare" structure, with added mass, without diagonals May 1995	2.375 *{2.40}***	(2.375)* *{2.38}***	2.375 *{2.30}***	Real-time Analyzer
"Bare" structure, with added mass, without diagonals May 1995	2.44 *{2.40}***	(2.44)* *{2.38}***	2.44 *{2.30}***	Permanent Instruments
"Bare" structure, with added mass, with diagonals May 1995	2.563 *{2.605}***	(2.563)* *{2.625}***	2.563 *{2.532}***	Permanent Instruments
Masonry scheme 1, October 1995 with diagonals, with added mass masonry infills in all but ground floor	4.15 *{4.00}***	(4.15)* *{3.94}***	*{4.20}***	Permanent Instruments
Masonry scheme 1, January 1997 with diagonals, with added mass masonry infills in all but ground floor	4.15 - 4.27 *{4.00}***	4.15 - 4.27 *{3.94}***	4.52 *{4.20}***	Permanent Instruments
Masonry scheme 1a, January 1997 with 1ˢᵗ story diagonals with added mass masonry infills in all but ground floor	4.15	(4.15)*	4.52	Permanent Instruments & R.T. A.
Masonry scheme 1b, February 1997 with no diagonals, with added mass masonry infills in all but ground floor	4.03	4.03	4.27	Permanent Instruments
Masonry scheme 2, February 1997 with added mass and masonry infills in all stories (with no diagonals)	4.395*** (5.127*** tran. coupling)	4.395*** (5.127*** tran. coupling)	4.639*** (torsional coupling)	Permanent Instruments

3.1.1. Predicted Dynamic Response Characteristics of the "Bare" Structure with added masses and Diagonal Cables : The 5-story building with added masses and without infills, but with the diagonal cables considered as being effective by small pre-stressing, was numerically simulated, in the way previously described. The material properties for the R.C. cross-sections were the ones found from correlation studies with the structure in the "bare" configuration without diagonals, as was mentioned before. The cross-sectional properties of the diagonal cables were adjusted so as to have the first three eigen-frequencies, e.g. 1st translational x-x, 1st translational y-y, and first torsional, with values equal to the ones measured in-situ (see Table 2). The axial stiffness of the diagonal cables was kept the same for all stories (EA=90t). The above adjustments were made based on the measured response during the symmetric cases with all diagonals in place. The same numerical study was also repeated for the asymmetric case.

3.1.2. Structure with added masses and Masonry as in scheme 1: The numerical model derived for the framed structure was complemented with linear panel elements, functioning mainly in shear, in order to simulate the influence of the masonry infills (Manos, 1991 and 1995). These panel elements were positioned at the location of the masonry infills that currently exists in the masonry scheme 1. A Young's modulus equal to 1712.5 Mpa (239ksi) was adopted during this numerical simulation of the masonry infills that is in accordance with the results obtained from the diagonal tension test of the masonry samples.

4.0 MEASURED AND PREDICTED RESPONSE DURING THE 4TH MAY 1995 EARTHQUAKE: Two earthquakes occurred on the 4th April, and 4th May 1995, with an 40km epicentral distance from the test site. During this earthquake sequence the 5-story structure had added weights in place, no masonry infill had been built as yet, and all diagonals were pre-stressed in all floors. Despite the fact that these earthquakes generated relatively low-intensity ground motion at the test-site, the 5-story structure was sufficiently excited with peak horizontal acceleration at the top slab just exceeding the level of 0.1g. The response of all floors was recorded and stored by the permanent instrumentation scheme.

4.1. Numerical Simulation of the Observed Earthquake Response : The numerical simulation described in section 3 before was employed in order to predict the earthquake response of this structure. The earthquake motion recorded at a distance of 12m from the foundation block of the 5-story structure was used as input in this numerical simulation. From the frequency domain study of the measured during this earthquake dynamic response and a parametric numerical study an adjustment was introduced in the stiffness properties as these were determined by the analysis described in paragraph 3.1.1. This resulted in 1st translational x-x frequency 2.496Hz (instead of 2.603Hz of Table 2); 1st translational y-y frequency 2.512Hz (instead of 2.62Hz) and 1st torsional frequency 2.50Hz (instead of 2.53Hz). Initially, a value was assigned to damping ratio equal to 2.2%, based on the study of the results obtained during the free vibration tests. However, a better agreement could be obtained between the observed maximum earthquake response at the top slab of the 5-story structure and the corresponding results obtained from this numerical simulation when this damping ratio value was taken equal to 3.5%. This increase in the damping ratio value is well justified from the fact that the structure during the free vibration tests was forced at lower levels that during the earthquake sequence of 5th May, 1995. The predicted in this way 5th story acceleration response, when compared with the one observed, exhibits good agreement.

5. MEASURED RESPONSE WITH INFILLS IN ALL STORIES
Recently (April 1997), an extensive low intensity vibration survey was performed with the masonry infills built also in the ground floor. From the analysis of in-situ measurements three frequency bands are clearly recognized from as corresponding to the first three "generalized"

modal shapes; that is 4.395 - 5.493 Hz , 14.41 - 16.73 Hz, and 19.9 - 21.25. Following the numerical simulation techniques described before the agreement that could be achieved this time between predicted and measured values is less satisfactory. This dictates further investigation that is planned for October 1997.

6. DISCUSSION OF THE RESULTS
- The numerical simulation of the dynamic characteristics of the 5-story Volvi building without masonry infills is very successful. This fact must be attributed to the very effective control of micro-cracking as well as to the almost exact estimation of the dimensions of the various structural elements and he very accurate estimation of the mass of the system. However, it must be born in mind that the measurements used and the assumptions employed in the simulations are based on linear-elastic response.
- The successful subsequent numerical simulation of the 4th of May, 1995 recorded earthquake response must also be seen under the light of the validity of the linear-elastic response assumptions, mentioned above. Despite this, a small adjustment was necessary in the stiffness and damping derived during the free-vibration in-situ test sequences.
- In studying numerically the influence of the masonry infills, the employed numerical technique with panel elements was quite successful for the case with the ground floor left without infills; for the case with masonry infills also in the ground floor certain discrepancies could be observed between measured and predicted response.

7. CONCLUSIONS
- The laboratory tests on the shaking table as well as the in-situ measurements at Volvi demonstrated the effectiveness of the masonry infills in increasing significantly the stiffness of the structure.
- The shaking table tests also demonstrated that due to cracking of the masonry infills the stiffness of the structure degraded at an impressive rate when the simulated earthquake motions became of higher intensity.
- The numerical simulation techniques employed to predict the dynamic response of the 5-story structure with masonry infills was successful for the case with the ground floor left without infills; for the case with masonry infills also in the ground floor certain discrepancies could be observed between measured and predicted response; this fact dictates the need for further investigation.

REFERENCES.
-G.C. Manos, M. Triamataki, B. Yasin, "Experimental and Numerical Simulation of the influence of masonry infills on the seismic response of reinforced concrete framed structures", *Proc. 10ECEE*, Vol. 3, pp. 1513-1518, ISBN 905410 528 3, 1995α, Balkema.
- G.C. Manos, M. Triamataki, B. Yasin, "Experimental-Numerical Study of the influence of masonry infills on the response of a two-storey RC model Subjected to Simulated Earthquakes",*Proc., 3rd Intern. Symposium on computer methods in Structural Masonry*, April 1995β, Lisbon.
- G.C. Manos, et.al. "Experimental and Numerical Simulation of the Influence of Masonry Infills on the Seismic Response of Reinforced Concrete Framed Structures", 4th Intern. Masonry Conf., London, England, 1995c.
-G.C. Manos, D. Mpoufudis, M. Demosthenous, B. Yasin, M. Triamataki, P. Skalkos, J. Thawapta et.al. "Ôhe Seismic Response of Reinforced Concrete Multistory buildings with Masonry Infills", to be published, *Proc. 2nd Egyptian Earthquake Engineering Conference*, Nov. 1997, Egypt.
- G.C. Manos et.al. "The Simulated Earthquake Response of a two 7-story R.C. Planar Model Structure - A Shear Wall and a Frame with Masonry Infills", to be presented at the 5th U.S. National Conference on Earthquake Engineering, Seattle, 1998.

A COMPUTER PROGRAM FOR THE SEISMIC ANALYSIS OF COMPLEX MASONRY BUILDINGS

F. Braga Professor, **D. Liberatore** Research Assistant and
G. Spera Doctoral Student
DiSGG, Università degli Studi della Basilicata, Via della Tecnica 3, 85100 Potenza, Italy

1. ABSTRACT

The computer program MAS3D, for the calculation of the response of masonry buildings under seismic actions, is presented. The program is based on the FEM and on the panel element developed in previous studies. It can be suitably employed to study the behaviour of complex masonry aggregates accounting for their construction history which normally gives origin to discontinuities in the masonry and to offsets between adjacent floors. The program is applied to study the seismic response of a set of masonry buildings in a row.

2. INTRODUCTION

Developing numerical models, which are both accurate and usable in practical design, for the seismic analysis of existing masonry structures is one of the major issues for the evaluation and reduction of seismic risk in Italy.

A large number of numerical models, based on the classical FEM formulation, is presently available for the seismic analysis of masonry structures [1]. Since the constitutive relations are only satisfied at Gauss points, fine meshes are normally required in order to achieve a good agreement with experimental results, and this restricts the applicability of these models to plane walls or to small wall assemblages in 3D.

Keywords: Seismic analysis, Panel element

Computer Methods in Structural Masonry – 4, edited by G.N. Pande, J. Middleton and B. Kralj.
Published in 1998 by E & FN Spon, 11 New Fetter Lane, London EC4P 4EE, UK. ISBN: 0 419 23540 X

On the opposite side, there are methods which drastically simplify the behaviour of the material and/or of the global structure. The most used among them is certainly the POR method, which models the shear response of each panel through an elastic-plastic law with limited ductility. The POR method is very efficient from the computational point of view, but presents severe inaccuracies. In particular, it neglects the flexibility of floors, the rocking effect of walls, the flexibility and failures of floor strips, flexural and sliding failures of panels. In addition, it cannot account for offsets of adjacent floors, which are very common in existing buildings. Owing to its limitations, the POR method usually leads to considerable overevaluations of the lateral strength, as pointed out through comparisons with experimental results [1].

The limitations of the POR method clearly show that the FEM is a compulsory approach for the seismic analysis of masonry buildings, especially existing ones, which present strong irregularities, both in plan and elevation. On the other hand, it is necessary to drastically reduce the number of DOF's of FE models in order to effectively analyse complex assemblages in 3D. This can be done modelling each panel in the structure through a single element. Since classical FE are very inaccurate when used to model a whole panel, alternative formulations have been proposed. These are based on some kind of simplification on material behaviour and/or on the stress field inside the panel, and lead to panel elements which represent a compromise between classical FE and simplified methods. They present a satisfactory degree of accuracy at the global level, and, at the same time, are efficient enough, from the computational point of view, to analyse complex assemblages in 3D.

In the following, a panel element developed by the first two authors in previous studies will be described, the computer program MAS3D based on this element will be illustrated and, finally, the program is applied to study the response of a complex set of masonry buildings in a row.

3. THE COMPUTER PROGRAM MAS3D

The formulation of the panel element implemented in MAS3D assumes that the stress field of the panel follows a multi-fan pattern (fig. 1) [2]. The material behaviour is assumed linear elastic in compression [3] and non-reacting in tension. The occurrence of crushing can be detected on the basis of the maximum compressive stress in the panel. In addition, it is assumed that: (a) the upper and lower faces of the panel are rigid, and (b) there is no interaction in the circumferential direction between the infinitesimal fans. Under these hypotheses, the circumferential

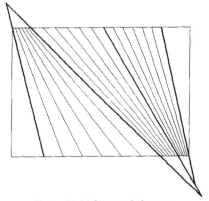

Fig. 1 - Multi-fan panel element.

and shear stresses are identically equal to zero, and the stress state in each point is defined by the orientation and intensity of the radial stress. The equilibrium and constitutive equations are easily satisfied, as well as the kinematic equation in the radial direction. However, it is not possible, in general, to satisfy the kinematic equation in the circumferential direction. The vertices co-ordinates of the elementary fans completely define the stress field of the panel element, for prescribed displacements of the upper and lower faces. These co-ordinates are not known *a priori* and should be determined imposing the minimum condition on the Total Complementary Energy. A parametric analysis in the strain space was carried out, which led to an approximate formulation of the Elastic Potential Energy as function of the strain components (axial, shear and bending). The first and second derivatives of this function permit to calculate the force vector and the stiffness matrix, respectively, thus avoiding the minimization procedure of the Elastic Complementary Energy [4]. Although the formulation of the element is stress-based, the direct calculation of its force vector and stiffness matrix permitted its implementation into the displacement-based non-linear FE program MAS3D.

MAS3D has been purposely developed by the authors in Fortran PowerStation® and runs under the Windows® 95 environment on a PC Pentium™ platform. It is complemented by a pre-processor which generates the FE model from the essential geometrical and mechanical data, and by a post-processor which provides a number of plot

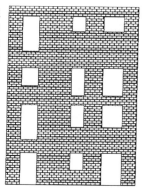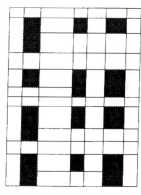

Fig. 2 - Discretization through panel elements.

files in the DXF format. The masonry structure is modelled as an assemblage of plane walls connected each other and with the floors. The FE discretization through panel elements is carried out wall by wall (fig. 2). Panel elements which are located over and below the openings have their multi-fan pattern rotated by 90 degrees. The walls are assembled in 3D in order to compose the global structure, which can present different heights in its parts. It is possible to account for the actual constraint conditions between walls and floors, as well as between different walls, in relation to the construction history which normally consists of several enlargements and raisings which took place at different times and gave origin to discontinuities in the masonry and in the floors. The floors are assumed rigid in their own plane and can be connected to only some of the walls they intersect. In fact, it is very common that floors are not connected to the pre-existent walls which are parallel to their lintels (fig. 3). The connection between contiguous walls built at the same time is taken into account imposing the equality of their interface displacements at the floor level. On the contrary, walls built at different times are not connected each other, and no such constraint is to be specified.

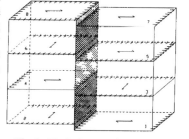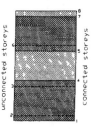

Fig. 3 - Typical constraint conditions between floors and walls.

MAS3D is also able to overcome a problem particularly important for buildings located in historical centres: in sets of buildings in a row, which result from a long evolution and are usually founded on slopes, offsets between adjacent floors are very common. They cause strong stress localization and severe damage of the walls. The FE formulation of MAS3D can easily account for offsets between floors (fig. 3). The self-weight of masonry is automatically calculated by the pre-processor and concentrated into the nodes of the FE discretization. The vertical loading applied by the floors to the walls is specified through sets of trapezoidal loads and again the pre-

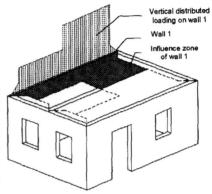

Fig. 4 - Vertical loading applied by the floor to a wall.

processor converts them in concentrated loads (fig. 4). The seismic lateral loads are concentrated at the floor levels.

The panel element showed to be able to model the local structural features with acceptable accuracy. Since it cannot reproduce the hysteresis loops, the curves of lateral load vs. lateral displacement — both at the element level and at the global level — should be interpreted as the skeleton curves of the real hysteresis loops. The ability of the panel element to accurately reproduce the skeleton curves of masonry panels under cyclic lateral loads, up to the maximum lateral load, was verified through comparisons with the experimental response of panels tested at ELSA and at the University of Pavia [5,6]. The small discrepancies found between the calculated and the experimental skeleton curves emanate from the assumptions on material behaviour (no-tension and linear elastic in compression); in particular, the panel element did not permit to evaluate the starting and the development of material degradation. Another comparison with experimental results is represented by the prediction of the response of the full-scale brick masonry prototype tested at the University of Pavia. Modelling this prototype involved two additional sources of inaccuracy: the assumption that the upper and lower faces of the panels are rigid, and the uncertainty about the interaction between orthogonal walls. Notwithstanding these limitations, at the lateral displacement corresponding to the maximum base shear, this was predicted with errors ranging from -5% to 17%, depending on the elastic modulus adopted in the analysis and the assumptions about the interaction between orthogonal walls. This results is broadly satisfactory, considering the complexity of material and structural behaviour [7].

4. EXAMPLE OF APPLICATION

MAS3D has been used to analyse a set of masonry buildings in a row 65 m long and 12 m wide (figs. 5,6) located in Tolve, near Potenza, falling in the 2nd Italian Seismic Category, for which the lateral seismic load is equal to 0.28 times the vertical load. The row, built in the sixties, was accurately surveyed in order to determine the structural properties, as well as the enlargements and raisings carried out subsequently to the original construction (fig. 7)

The perimetrical walls are 60 cm thick at the base and taper 10 cm each floor. The outer facing is made by bricks and the inner by tuff blocks; the filling consists of cobblestones and concrete. The inner transverse walls consist of single wythe tuff masonry with constant thickness of 30 cm on the whole height, except for 5 walls in brick masonry 13 cm thick in the first cell from the right (fig. 6). The elastic modulus of masonry in compression was assumed equal to 1500 MPa for all the walls. The floors are made of

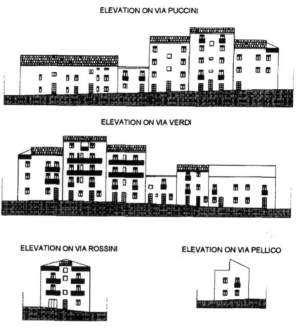

ELEVATION ON VIA PUCCINI

ELEVATION ON VIA VERDI

ELEVATION ON VIA ROSSINI

ELEVATION ON VIA PELLICO

Fig. 5 - Row of masonry buildings studied.

RC lintels and tiles. The lintels lie in the transverse direction and have span length ranging from 4 to 5 m. The floors of different cells in the row are often offset, owing to the soil slope. The roofs are partly in wooden beams, partly in steel lintels and tiles. The roof beams, or lintels, lie in the transverse direction. As an effect of enlargements and raisings carried out at different times, many contiguous walls are not connected each other, as well as many floors are not connected to the pre-existent walls parallel to their lintels. All these features have been taken into account by the numerical model.

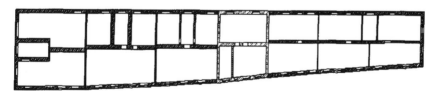

Fig. 6 - Plan of ground floor.

The numerical model of the row of buildings results in 2364 equations, with mean half bandwidth equal to 220. The lateral loads were applied separately in the longitudinal and transverse directions, with positive and negative signs, for a total of four analyses. The analyses were performed through a step-by-step procedure. At the first step the vertical load was applied, and at each of the subsequent steps the lateral loads were increased with increments of 1/25 of the maximum lateral load. At each step, equilibrium iterations were performed through the classical Newton-Raphson method. Lack of convergence indicated exceeding of the lateral strength. The number of steps, together with the calculation times on a Pentium™ 133 computer, including I/O operations, are listed in tab. 1.

The main results are the following.

1. *Overall lateral strengths.* The overall lateral strengths, divided by the total vertical load, are listed in tab. 1 for the different cases analysed. They are considerably lower than the value prescribed by the Italian seismic regulations (0.28), especially in the transverse direction, thus indicating the need to strengthen the structure.

2. *Plots of displaced floors.* They provide important information on the response of the assemblage. For instance, the plot of displaced floors for lateral loads in the transverse positive direction at step 10 (fig. 8) shows a strong displacement localization, together with a marked torsional rotation, in the 3rd cell from the left, because of the lack of connection of its floors with the pre-existent transverse walls. The small lateral strength in the transverse direction is mainly due to this effect.

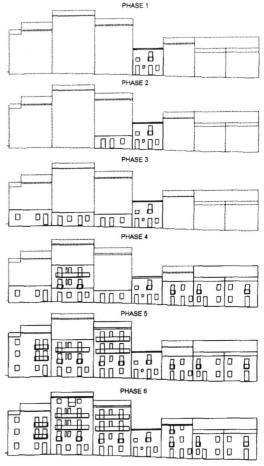

Fig. 7 - Construction phases (elevation on Via Verdi).

3. *Deflection plots of walls.* They provide essential information on the weak points of the walls. For instance, the deflection plot of the 2nd wall from the left on Via Verdi for lateral loads in the longitudinal positive direction at step 20 (fig. 9) clearly shows the low degree of coupling exerted by the narrow floor strips over the openings.

4. *Panel forces.* The plot of the forces at the center point of each panel permits a rapid understanding of the main stress flow in each wall. This plot for the wall mentioned above is shown in fig. 10.

Tab. 1 - No. of steps, calculation times and lateral strengths.

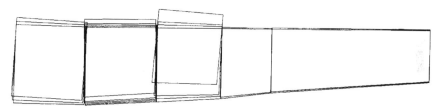

Direction	Longitudinal		Transverse	
Sign of lateral loads	Positive (left to right, elevation on Via Verdi)	Negative (right to left, elevation on Via Verdi)	Positive (left to right, elevation on Via Pellico)	Negative (right to left, elevation on Via Pellico)
No. of steps	21	19	13	12
Calculation time (s)	2558	3033	1808	2231
Lateral strength /vertical load	0.2128	0.1904	0.1232	0.1120

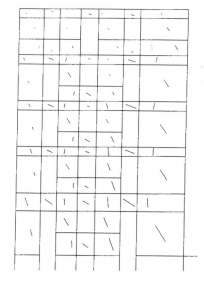

Fig. 8 - Plot of displaced floors (lateral loads in the transverse positive direction, step 10).

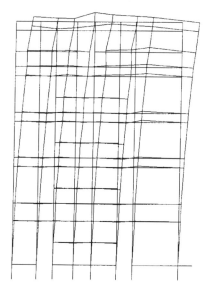

Fig. 9 - Deflection plot of the 2nd wall from the left on Via Verdi (lateral loads in the longitudinal positive direction, step 20).

Fig. 10 - Panel forces of the 2nd wall from the left on Via Verdi (lateral loads in the longitudinal positive direction, step 20).

5. CONCLUSIONS AND FUTURE DEVELOPMENTS

The computer program illustrated in this study discretizes masonry buildings through panel elements. It exploits the modelling versatility of the FEM and, at the same time, results in computational efficiency and simplicity of use. The program can account for the actual constraint conditions between walls and floors, as well as between different walls, in relation to the construction history of the assemblage. The total system of equations usually consists of a few thousands of equations, with mean half bandwidth of a few hundreds, as established through the analysis of a complex set of buildings in a row. This permits to perform a static incremental analysis on a PC computer as a routine task in practical design.

Future developments will improve the modelling capabilities of the program. In particular, models of the most common used strengthening techniques will be implemented and the in-plane flexibility of floors will be taken into account, in order to suitably model wooden floors and floors made of steel lintels and tiles.

6. ACKNOWLEDGEMENTS

The research reported in this study was partially supported by the National Group for Protection against Earthquakes (GNDT). The example of application was performed as part of the graduation thesis of Mr. C. Lucia, whose contribution is gratefully acknowledged.

7. REFERENCES

1. 'Numerical Prediction of the Experiment', Experimental and Numerical Investigation on a Brick Masonry Building Prototype, Report 3.0, CNR-GNDT, 1995.
2. Braga F., Liberatore D., 'A Finite Element for the Analysis of the Response of Masonry Buildings under Seismic Actions', 5th North American Masonry Conference, Urbana, U.S.A., 1990, pp 201-212.
3. Dhanasekar M., Kleeman P.W., Page A.W., 'Biaxial Stress-Strain Relations for Brick Masonry', ASCE, J. Struct. Div., 111, ST5, 1985, pp 1085-1100.
4. Braga F., Liberatore D., 'Elastic Potential Energy, Force Vector and Stiffness Matrix of a No-Tension Panel', Atti dell'Istituto di Scienza e Tecnica delle Costruzioni, n. 18, Potenza, Italy, 1990 (in Italian).
5. Anthoine A., Magonette G., Magenes G., 'Shear-Compression Testing and Analysis of Brick Masonry Walls', 10th European Conference on Earthquake Engineering, Vienna, Austria, 1994, pp 1657-1662.
6. Magenes G., 'Seismic Behaviour of Brick Masonry: Strength and Failure Mechanisms of Panels', Ph.D. Thesis, 1992 (in Italian).
7. Braga F., Liberatore D., Mancinelli E., 'Numerical Simulation of the Experimental Bahaviour of Masonry Panels and Walls under Horizontal Loads', 11th World Conference on Earthquake Engineering, Acapulco, Mexico, 1996.

THE MEASURED RESPONSE OF PARTIALLY REINFORCED MASONRY PIERS SUBJECTED TO COMBINED HORIZONTAL CYCLIC AND COMPRESSIVE LOADS

G.C. Manos Professor
Earthquake Simulator Facility, Laboratory Strength of Materials,
Department of Civil Engineering, Aristotle University, Thessaloniki,
Greece. E-mail: gcmanos@civil.auth.gr

1. ABSTRACT

The earthquake performance of partially reinforced masonry piers when subjected to various types of loading has been a topic of research for some time. Extensive experimental programs have been performed both in the U.S.A. as well as in various European Institutions in order to examine the performance of structural elements of this type of construction, when subjected to various types of loading. Most of these testing sequences employed masonry piers with various reinforcing arrangements, which were subjected to combined constant compression together with horizontal load reversal of varying amplitude (cyclic loading) in order to investigate the influence of certain parameters on the response. The most significant parameters that were examined are the type and strength of the materials (mortar and masonry units), the geometry and the reinforcement (in quantity and structural details), and the level of axial compression This paper includes some of the results from a current experimental investigation with partially reinforced piers employing a "Greek" type brick. This research is currently under way at the Earthquake Simulator Facility, Laboratory of Strength of Materials, Aristotle University and is under the financial support of the Greek Ministry of Energy and Industry, General Secretariat of Research and Technology.

2. INTRODUCTION

A number of researchers in the past studied experimentally the behavior of masonry piers for seismic loads by subjecting these piers simultaneously to combined horizontal and vertical loads (racking tests, Hidalgo et.al.). The influence of various reinforcing arrangements on the pier's behavior was also studied under these loading conditions (Hidalgo et.al., Tomazevic et.al.). In order to achieve a reasonable simulation of the actual earthquake loading conditions, the horizontal loads are applied in a cyclic manner, in order to represent the alternating nature of the seismic loads; the gravity action, simulated by the vertical load is assumed to remain almost constant, an assumption that can be considered reasonably valid for low-rise buildings that represent the majority for this type of construction in seismically active regions. The basic structural components of this type of masonry building are connected together with diaphragms (with a varying degree of flexibility) so that each one of the basic masonry components is subjected to simultaneously in-plane and out-of-plane horizontal seismic actions on top of the gravity loads. Despite this, a large number of experimental research has focused on the in-plane behavior separately from the out-of-plane behavior. As was shown by Manos and co-workers (1990), this can be a reasonable assumption under certain circumstances.

Computer Methods in Structural Masonry – 4, edited by G.N. Pande, J. Middleton and B. Kralj.
Published in 1998 by E & FN Spon, 11 New Fetter Lane, London EC4P 4EE, UK. ISBN: 0 419 23540 X

The dominant role in the earthquake performance of masonry buildings is played by the in-plane satisfactory performance of masonry piers that are distributed in such a way as to form the shell of a masonry building and to provide its earthquake resistance by in-plane actions in both horizontal directions. As a consequence, the importance of investigating the in-plane behavior of masonry piers and the influence that a reinforcing arrangement may have on this behavior becomes quite obvious. The current research effort belongs to this category of investigation by studying the in-plane behavior of masonry piers when they are subjected to in-plane loads simulating the combined earthquake and gravitational actions.

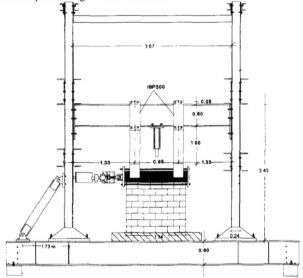

Figure 1. Masonry pier being subjected to the racking test in the strong steel reaction frame.

3. TESTING ARRANGEMENT

Figure 1 depicts the testing layout whereby the masonry pier specimen is placed within a steel reaction frame with its foundation being anchored to that frame. This reaction frame also provides the support for the horizontal servo-hydraulic actuator, which has a capacity of 250KNt and a stroke of ± 50mm and a capability of displacement control with a good fidelity in its response in the frequency range from 0 to 50Hz. Moreover, the same reaction frame also provides support for the vertical hydraulic jack that has a capacity of 200KNt and a stroke of ± 200mm, with the force being applied statically. Because this vertical jack is not displacement-controlled a system of accumulators was added in order to avoid variations of the vertical force, when the specimen develops at the post-cracking stage excessive deformations that also include significant vertical displacements. Thus, whereas the vertical load is kept almost constant at a predetermined level, the horizontal force is varied in a cyclic manner, this results from controlling the imposed horizontal displacement at this point of the masonry pier in a predetermined way. The imposed cyclic displacement time history is depicted in figure 2, in terms of displacement amplitude versus number of cycles. The frequency of this cyclic loading is also one of the parameters to be studied and it can be specified at the beginning of the test for each pier. The prescribed horizontal displacement depicted in figure 2 has been already applied with two distinctly different frequencies, the first being a rather slow variation of the horizontal forces at 0 01Hz whereas the second is a rather fast variation of the horizontal forces at 1 00Hz. The former simulates the prototype earthquake forces only in the reversible (cyclic) nature of the loading whereas the latter both in the cyclic nature as well as in the frequency content, which is quite representative of the dominant frequency content that is expected to develop in the dynamic response of such masonry components being part of a low-rise masonry building The validity of this assumption will be further investigated numerically in the framework of this project

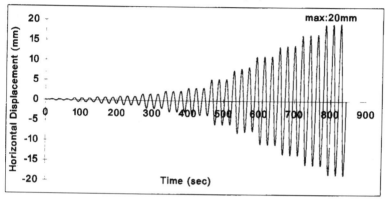

Figure 2. Applied horizontal cyclic displacement.

4. MASONRY SPECIMENS

All masonry specimens will be constructed with special "Greek" brick with vertical holes depicted in figure 3 This masonry unit was initially developed as a pilot brick unit in the framework of a Brite-Uram project by the industrialists Filippou Structural Clay Products in cooperation with the Technical University of Athens under the leadership of Professor T Tasios (Psylla et.al.) Moreover, during this preliminary research effort, the mechanical properties of the masonry unit in itself as well as that of masonry piers built with it, similar to the ones of the current project and under similar loading conditions were also examined. The whole Brite-Euram research effort was related to the use of reinforced masonry for building in seismic zones (Modena et.al.). As a result of this preliminary project the geometry of this pilot brick unit underwent certain modifications with regards to its height as well as the dimensions of its vertical holes. Moreover, the reinforcing arrangements that are studied during the current project are also different as what is investigated now is the applicability of this type of construction for low-rise housing (1 or 2-story buildings) in moderate hazard seismic zones of Greece (zones 1 and 2 of the 1992 Greek Seismic Code).

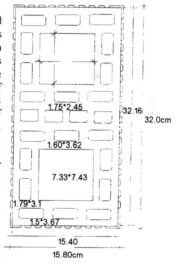

Figure 3. "Greek" brick unit
Cross-section

The tests that have been performed so far and will also be performed in the future include partially reinforced piers of the following geometry

- Length 1330mm , height 1330mm and thickness 154mm
- Length 660mm , height 1330mm and thickness 154mm

Apart from the above geometry, which must be considered as ½ scaled specimens and represent the bulk of the testing sequence, a limited number of specimens near to prototype scale will also be tested These specimens will be of the following geometry

- Length 2700mm , height 2475mm and thickness 320mm.
- Length 2700mm, height 2475mm and thickness 154mm.

These later specimens will be tested under the conditions similar to the ones described in section 3, however, only the slow variation of the horizontal forces at 0.01Hz will be applied in this case

All test specimens will be constructed with hollow brick masonry units with vertical holes; these were initially developed in the framework of a Brite - Euram project by Filipou Structural Clay Products industry. These units were modified in terms of their basic dimensions and composition of their ceramic material in the framework of the current project; moreover they are currently produced by Filipou industry under full production conditions that included certain modifications in the furnace and dryer conditions. .

The compressive strength of the mortar employed in the construction of these piers is aimed to have the following strengths:

- Category O, target compressive strength 2.5Mpa
- Category N, target compressive strength 5 0Mpa.

Finally, various longitudinal and transverse reinforcing arrangements are examined together with a variation of the level of uniform compression

5 STANDARD TESTS

A number of standard tests are carried out prior to the complex racking tests. These include the following

- Simple compression tests of the masonry units.
- Simple compression tests of cubes taken from mortar during the construction of the racking specimens
- Simple compression tests of cubes taken from the grout during the construction of the racking specimens.
- Simple tension tests of the reinforcement used in the construction of the racking specimens
- Simple pull-out tests of the reinforcement used in the construction of the racking specimens from cubes of grout.

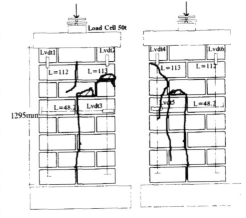

Figure 4 Standard compression test of unreinforced masonry pier.

- Simple compression tests of masonry piers, without reinforcement. These piers have 660mm length, 1330mm height and 154mm thickness; they are constructed at the same time and with the same materials used in the construction of a group of racking specimens.
- Simple diagonal tension tests of masonry specimens without reinforcement These specimens have length 99.5mm, height 106mm and thickness 106mm; they are constructed at the same time and with the same materials used in the construction of a group of racking specimens.

Instrumentation is provided so that during some of these standard masonry tests the variation of the applied load together with certain important deformation levels is recorded.
This is depicted in figures 4 and 5, for the simple compression and figure 6 for the diagonal tension tests, respectively.

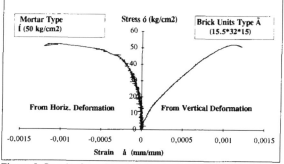

Figure 5. Results from Compression

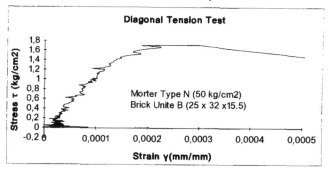

Figure 6. Standard diagonal tension test of masonry specimen

6. INSTRUMENTATION FOR THE CYCLIC TESTS

For the racking tests, the specimens are instrumented with a number of displacement transducers in order to record the peak displacement response at the top of the pier as well as a number of deformations at critical points together with the maximum compression and horizontal cyclic loads. Figure 7 depicts the used instrumentation scheme with 14 displacement transducers and two load cells in order to monitor the variation of the horizontal and vertical load. As already mentioned, a system of accumulators was attached to the vertical loading system in order to minimize the fluctuations of the vertical load during the racking test from a

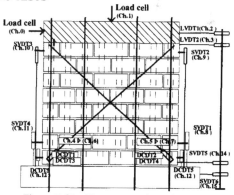

Figure 7. Used Instrumentation scheme

set value selected at the beginning of the test. The load cell that monitors the vertical load is intended to check whether this system operates satisfactorily with the fluctuations of the vertical load kept to a minimum.

7 TESTING PARAMETERS

A number of specimens were already tested and the measured response has been treated in such a way as to deduce the most significant state of stress. In general the following dominant response modes are expected to develop in these masonry specimens

- A predominantly rocking mode at the foundation level that is due to the development of a limit flexural state of stress at the foundation level under the combined action of the horizontal and vertical loads.
- A predominantly flexural mode for the rest of the masonry pier, apart from the foundation level due to the flexural state of stress for the rest of the masonry pier under the combined action of the horizontal and vertical loads.
- A predominantly sliding mode at the foundation level that is due to the development of a limit sliding state of stress at the foundation level under the combined action of the horizontal and vertical loads.
- A predominantly shearing mode for the rest of the masonry pier, apart from the foundation level, that is due to the development of a shearing state of stress for the rest of the masonry pier under the combined action of the horizontal and vertical loads.

The variation of the basic parameters studied in the present test sequence are expected to exert an influence in the development of the above dominant modes of response. The instrumentation scheme that is provided, and has been presented in section 6, is aimed to be able to identify the contribution of each one of the above response modes to the total response of the masonry piers during the racking tests. The basic parameters whose influence is studied in the framework of the current project are listed below:

- The geometry in terms of height over length ratio; one represents rather slender piers with height over length ratio approximately equal to 2 whereas the second addresses less slender piers with height over length ratio approximately equal to 1. This must be viewed   together with the two different types of thickness, as described in section 3.
- The type of mortar; two distinct types, O and N are examined
- The level of axial compression that is applied on these piers together with the horizontal loads This is set as a percentage of the compressive strength of the masonry specimens subjected to the standard simple compression test presented in section 5. As already mentioned, the current project is intended to investigate the performance of this type of masonry when applied in low-rise housing; for structural elements composing such low-rise housing, it is expected that their level of axial compression is rather low. Consequently, two levels of axial compression are adopted; the first is 4% of the masonry strength and the second is 8% of the masonry strength. Moreover, a limited number of specimens will be tested with no axial compression.
- The amount of horizontal (transverse) reinforcement in terms of ratio of the area of this type of reinforcement of the corresponding gross cross sectional area; thus this ratio is varied from a relatively low value, approximately equal to 0 05%, to a somewhat larger value, approximately equal to 0.150%. The amount of the vertical (longitudinal) reinforcement, in terms of ratio of the area of this type of reinforcement of the corresponding gross cross sectional area, remains constant for all the specimens, approximately equal to 0 125%.
- In addition to the above parameters the influence of the frequency of the cyclic horizontal loading, as mentioned in section 3, is also one of the parameters to be studied in the reduced scale specimens.

8. DISCUSSION OF TEST RESULTS

Some of the measured response during this test sequence, together with the observed damage patterns, are presented and discussed in what follows

Figure 7 depicts the details of a specimen with dimensions 1330mm length by 1330mm height and a thickness 15mm. The mortar used was of type N and the axial compression level was 4% of the strength of this type of masonry in compression.

The vertical and horizontal reinforcing arrangement is also shown in figure 7 whereas figure 8 depicts the specimen on the testing facility. finally, figure 9 depicts the horizontal load and horizontal displacement variation at the top of this pier whereas figure 10 portrays the damage sustained by this specimen at the end of the test.

The following observations can be made on the basis of the observed behavior.

- The tested pier exhibits an increasing horizontal load capacity up to the 9th group of cyclic loading, which corresponds to a maximum displacement of 7.5mm.

- For this level of deformation, the load degradation observed for the subsequent two cycles, which are of the same displacement level as the first cycle of this group, is rather limited.

- An increasing horizontal load capacity can also be seen even for the 10th group of cyclic loading, which corresponds to a maximum displacement of 10mm. However, for this level of deformation the load degradation observed for the subsequent two cycles becomes noticeable.

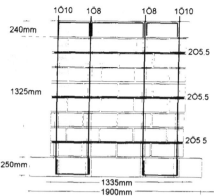

Figure 7. Reinforcing Details

Figure 8. Masonry Pier during Test

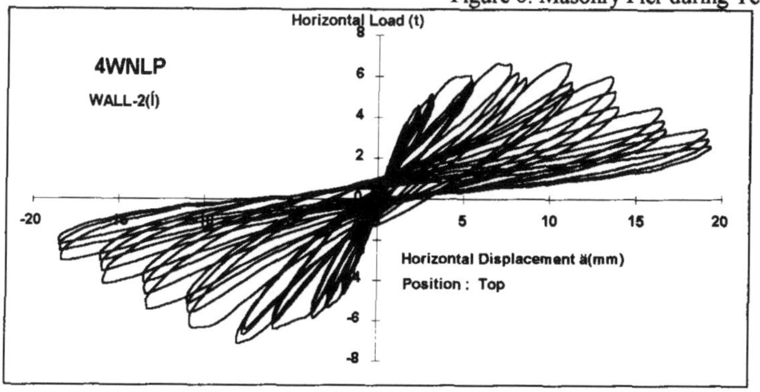

Figure 9. Load-Deformation Cyclic Behavior

- For the remaining three groups of cyclic loading from 12.5mm to 20mm the pier cannot sustain the maximum load that was measured before (just above 6t), and its capacity deteriorates in each subsequent cycle

The above observations are in agreement with the observed damage, which is of a rather shear type, accompanied with the disintegration of the central part of the pier during the last groups of cycles.

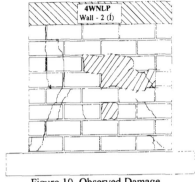

Figure 10. Observed Damage

9. CONCLUSIVE REMARKS

The outline of an ongoing experimental sequence has been presented, which is aiming to examine the seismic performance of partially reinforced brick masonry piers constructed by a newly developed hollow "Greek" brick unit with vertical holes.

This investigation is focusing on low-rise housing to be constructed in low to moderate seismic zones of Greece.

The basic parameters whose influence on the piers' performance will be studied were also outlined together with the instrumentation scheme that is employed in identifying the dominant modes of response.

Finally, typical results related to one tested masonry pier are presented and discussed. It can be seen that this pier has reached its ultimate state by the used loading system. Moreover, the employed instrumentation recorded this performance in a way that adequately describes the observed behavior.

Further testing is currently under way; from the results up to now certain important issues were identified that, when properly solved, will lead to the satisfactory exclusion of undesirable response modes from this type of structural elements.

REFERENCES.
- P. A. Hidalgo, R.L. Mayes, H. D. McNiven and R.W. Clough. "Cyclic Loading Tests of Masonry Single Piers" Volumes1,2,3, EERCE Reports No. 78/27, 78/28, 79/12, University of California at Berkeley.
- Gulkan P., Clough R.W., Manos G.C. and Mayes R.L. "Seismic Testing of Single-story Masonry Houses : Parts 1 and 2", Journal of Str. Eng. ASCE, Vol. 116, No 1, January 1990, pp. 235-274.
- T. P. Tasios "The Mechanics of Masonry", Athens, 1987 (in Greek)
- M. Tomazevic, M. Lutman, L. Petkovic "In-Plane Behavior of Reinforced Masonry Walls Subjected to Cyclic Lateral Loads", Parts 1 and 2, Reports ZRMK/PI-92/06 and 07, Institute for Testing and Research in Materials and Structures, Ljubljana, Slovenia, 1993.
- N. Psilla, E. Vintzileou, T. P. Tasios "Seismic Behavior of Reinforced Masonry", Proceedings of Greek Conference on Reinforced Concrete Structures, Vol. II, pp. 284-294, Cyprus, 1996 (in Greek)
- C. Modena, A. Bernardini et.al. "Reinforced Masonry for Buildings in Seismic Zone", Report of University of Padova in the framework of Brite Euram Project No 4001, 1996.

STATISTICAL ANALYSIS OF PARTIAL SAFETY FACTORS FOR STRUCTURAL MASONRY

M. Holický Senior Research Worker, Associate Professor[1], **J. Middleton** Reader[2] and **M. Vorlíček** Senior Research Worker, Associate Professor[1]
[1]*Klokner Institute, Czech Technical University in Prague, Solínova 7, 166 08 Praha 6, Czech Republic and* [2]*Department of Civil Engineering, University of Wales Swansea, Singleton Park, Swansea SA2 8PP, Wales, UK*

1. ABSTRACT

Presented statistical analysis of partial safety factors for structural masonry is focused on mutual dependence of quality control and partial factors of material properties. Taking the coefficient of variation as an indicator of the level of quality control, simple statistical techniques are proposed for an assessment of adequate partial safety factors using available experimental data and previous experience. It is shown that the assessed partial safety factors are significantly dependent on the coefficient of variation, number of available observations and asymmetry of probability distribution.

2. INTRODUCTION

Quality control of structural masonry is generally recognised as an important factor determining strength and safety of structural elements. Considerable variation in quality level of masonry may cause significant differences in structural safety [1]. That is why the present Eurocode 6 [2] recommends the total of six partial safety factors of material properties γ_M depending on two categories of manufacturing control of masonry units and three categories of execution. It was hoped that in this way differences in safety level of structural masonry will be even out. However, the present Eurocode 6 [2] does not provide appropriate guidance on how to classify masonry in particular conditions into proposed categories of quality level. The aim of this contribution is to propose simple statistical techniques enabling rational choice of an adequate quality level using previous experience and available experimental data.

Keywords: Partial safety factor, Masonry strength, Statistical methods, Assessment

Computer Methods in Structural Masonry – 4, edited by G.N. Pande, J. Middleton and B. Kralj.
Published in 1998 by E & FN Spon, 11 New Fetter Lane, London EC4P 4EE, UK. ISBN: 0 419 23540 X

3. PARTIAL SAFETY FACTORS IN EUROCODE 6

Partial safety factors in Eurocode 6 are dependent on category of
  - manufacturing control of masonry units,
  - execution.
The present Eurocode 6 [2] recognises two categories of manufacturing control of masonry units and three categories of execution and recommends the total of six partial safety factors γ_M indicated in Table 1 which is a part of Table 2.3 of Eurocode 6 [2].

Table 1 - Partial safety factors γ_M

Masonry		Category of execution		
		A	B	C
Category of manufacturing	I	1,7	2,2	2,7
control of masonry units	II	2,0	2,5	3,0

Thus, the partial safety factors are within a broad range from 1,7 to 3,0. Moreover, it is not clear how to classify particular site conditions. This may obviously lead to a „safe" and consequently quite uneconomical design. It is, therefore, desirable to get more precise guidance how to choose adequate categories of both aspects or of production quality or, possibly, how to determine the partial safety factor itself. This is particularly true for masonry executed under given conditions, which may be dependent on local traditions, when category of manufacturing control of masonry units and category of execution may be difficult to assess.

It was shown in a previous study [1] that the most relevant statistical technique to be used for determination of masonry strength is the classical coverage method with the confidence level 0,75. The characteristic and design strength are then given as

$$f_k = \mu_f - \kappa_k \, \sigma_f = \mu_f (1 - \kappa_k \, v_f) \qquad (1)$$
$$f_d = \mu_f - \kappa_d \, \sigma_f = \mu_f (1 - \kappa_d \, v_f) \qquad (2)$$

where μ_f denotes the mean, σ_f standard deviation, κ_k and κ_d are the fractiles of the probability distribution of a standardised random variable for $p = 0,05$ (characteristic strength) and $p = 0,001$ (design strength) respectively, $v_f = \sigma_f / \mu_f$ denotes the coefficient of variation. The partial safety factor is then

$$\gamma_M = \frac{f_k}{f_d} = \frac{\mu_f - \kappa_k \sigma_f}{\mu_f - \kappa_d \sigma_f} = \frac{1 - \kappa_k v_f}{1 - \kappa_d v_f} \qquad (3)$$

Thus, there is one to one correspondence between the partial safety factor and the coefficient of variation. Consequently, if the coefficient of variation is known (from previous experience or current experimental data), the partial factor may be calculated from equation (3).

To select appropriate quality category and appropriate partial safety factor it may be useful to know what are the coefficients of variation corresponding to the partial safety factors specified in Eurocode 6. It follows from equation (3) that for a given partial safety factor γ_M the coefficient of variation v_f may be determined using the following formula

$$v_f = \frac{\gamma_M - 1}{\kappa_d \gamma_M - \kappa_k} \qquad (4)$$

Using equation (4) and assuming the normal distribution of masonry strength (which is assumed in Eurocodes) the coefficients of variation v_f corresponding to the partial safety factors γ_M given in Table 1, are indicated in Table 2.

Table 2 - Coefficients of variation v_f corresponding to the partial safety factors γ_M given in Table 1

Masonry		Category of execution		
		A	B	C
Category of manufacturing	I	0,193	0,233	0,254
control of masonry units	II	0,220	0,247	0,262

Thus, the best quality of manufacturing control and execution, represented by the categories I and A, corresponds to the coefficient of variation 0,193, which seems to be quite high value, certainly greater than commonly observed coefficients of variability. Consequently, the partial safety factors given in Table 1 for this quality level may be rather conservative and uneconomical. One would expect a lower coefficient of variation for the „best quality" considered in Eurocode.

On the other hand, the worst quality, represented in Table 2 by categories II and C, corresponds to the coefficient of variation 0,262, which seems to be quite reasonable and realistic value. It may, however, depend on local conditions. Certainly, the range of the coefficients of variation from 0,193 to 0,262 indicated in Table 2 seems to be quite narrow. It appears, that the best quality (categories I and A) could be characterised by lower coefficient of variability than almost 0,20, say by 0,15. In the following a broad range from 0,10 to 0,30 is considered.

4. PARTIAL SAFETY FACTORS DERIVED FROM EXPERIMENTAL RESULTS

In building industry available experimental data are almost always of limited size. From the statistical point of view limited size of data causes the fundamental uncertainty in assessment of observed material property. Generally, having n experimental measurements of masonry strength the sample mean m_f, standard deviation s_f and coefficient of variation $V_f = s_f / m_f$ may be calculated. Accepting the confidence level 0,75 [1], it may be shown that lower bounds of the characteristic and design strength are given as follows

$$f_k > m_f - K_k s_f = m_f (1 - K_k V_f) \qquad (5)$$
$$f_d > m_f - K_d s_f = m_f (1 - K_d V_f) \qquad (6)$$

where the coefficient K_k and K_d are so called noncetral values [1] of normal distribution with $n-1$ degrees of freedom, corresponding to the probabilities $p = 0,05$ and $p = 0,001$. Selected values of both coefficient K_k and K_d are given in Table 3.

Table 3 - Coefficients K_k and K_d

Coefficient	Number of observations n			
	10	20	30	∞
K_k	2,10	1,93	1,87	1,64
K_d	3,86	3,56	3,45	3,09

It can be further shown that an upper bound of the partial safety factor is then

$$\gamma_{M,max} = \frac{m_f - K_k s_f}{m_f - K_d s_f} = \frac{1 - K_k V_f}{1 - K_d V_f} \tag{7}$$

Thus, using equation (7) the upper bounds $\gamma_{M,max}$ may be estimated on the bases of the coefficients of variation V_f obtained from n experimental measurements. Selected values of $\gamma_{M,max}$ are given in Table 4.

Table 4 - Upper limit of the partial safety factors $\gamma_{M,max}$

Coefficient of variation V_f	Number of observations n			
	10	20	30	∞
0,10	1,29	1,25	1,24	1,21
0,15	1,63	1,52	1,49	1,41
0,20	2,54	2,13	2,02	1,76
0,25	x	4,70	3,87	2,59
0,30	x	x	x	6,96

Note: Symbol x denotes those cases when the assumption of normal distribution fails.

It follows that the partial safety factor γ_M is dependent not only on the coefficient of variation V_f, but also on the sample size n. Note, that for higher values of coefficient of variation (0,25 and 0,30) and decreasing sample size, the assumption of normal distribution fails and more realistic probabilistic models, including asymmetric distributions (lognormal), should be considered.

There are several possible ways how to use Table 4. When available (newly obtained or previous) measurements are well defined and their size n is known, then the application of Table 4 is quite straightforward. However, in some cases, particularly when previous measurements are used and the sample characteristics m_f, s_f and $V_f = s_f / m_f$ may be available only, then number of observations n to be used in assessing the partial safety factor using Table 3, may be substituted by an estimate n' [1] obtained as

$$\sqrt{n'} = \frac{s_f}{m_f V(m_f)} = \frac{V_f}{V(m_f)} \tag{8}$$

Here $V(m_f)$ is the coefficient of variation of the mean m_f which may be estimated from the standard deviation of the mean of prior observations. The standard deviation may be even adjusted to reliability and quality of available data. Obviously this step may require collaboration with an experienced specialist. Furthermore, current observations may be effectively combined with prior information using Bayesian approach as indicated in [1].

5. EFFECT OF DISTRIBUTION ASYMMETRY

The whole procedure may become more complicated if the distribution of masonry strength is asymmetrical. Then, the coefficients κ_k and κ_d (the fractiles of the probability distribution of a standardised random variable for $p = 0,05$ and $p = 0,001$) are dependent on the coefficient of skewness α_f (a measure of asymmetry). Table 5 shows the coefficients κ_k and κ_d for selected values of α_f for lognormal distribution which seems to be a suitable model [3, 4].

Table 5 - Coefficients κ_k and κ_d

Coefficient	Coefficient of skewness α_f				
	-1,0	-0,5	0,0	0,5	1,0
κ_k	1,85	1,77	1,64	1,49	1,34
κ_d	4,70	3,86	3,09	2,46	1,98

In equations (1) to (4) the coefficients κ_k and κ_d should be generally used. It follows from Table 5 that effect of asymmetry may be quite significant. However, unless there is a convincing evidence of asymmetry and a reliable value of the coefficient α_f is available it is recommended to use the above assumption of normal distribution. Generally a positive coefficient of skewness α_f less than 1 is observed for material properties as masonry strength.

The effect of asymmetry is even more pronounced when experimental data of a limited size are used to assess characteristic and design strengths (5% and 0,1% fractiles) [1] and the partial safety factor in accordance to equation (5), (6) and (7). Then the coefficients K_k and K_d should be taken as noncentral values of lognormal distribution [3, 4], for which selected values are indicated in the following Table 6.

Table 6 - Coefficients K_k and K_d

Coefficient	Coefficient of skewness α_f	Number of observations n			
		10	20	30	∞
K_k	-1,0	2,63	2,33	2,23	1,85
	-0,5	2,36	2,14	2,05	1,77
	0,0	2,10	1,93	1,87	1,64
	0,5	1,87	1,74	1,68	1,49
	1,0	1,69	1,56	1,51	1,34
K_d	-1,0	6,40	5,74	5,49	4,70
	-0,5	5,03	4,54	4,39	3,86
	0,0	3,86	3,56	3,45	3,09
	0,5	3,06	2,81	2,75	2,46
	1,0	2,52	2,31	2,24	1,98

It should be emphasized that the design strength is derived from the characteristic value determined in accordance with the Eurocode 6 (without using experimental data) which is divided by the partial safety factor; the only statistical characteristic used in the above procedure directly is the coefficient of variation. When, however, representative experimental data are available, it is possible to assess the characteristic and design strength directly using equations (5) and (6) as described in [1] using both fundamental statistical characteristics, the mean m_f and standard deviation s_f. The final decision should be made taking into account size and quality of available data.

6. EXAMPLE

Using available information a masonry in a given conditions is classified by the category of manufacturing II and the category of execution B. In accordance with Table 1 the partial safety factor $\gamma_M = 2,5$. It follows from Table 2 the corresponding coefficient of variation $v_f = 0,247$ (which seems to be quite high).

Assume that a long term experience with a given masonry provide a convincing evidence that the coefficient of variation is only $v_f = 0,20$. Then according to equation (3) or Table 4 for $n = \infty$, $\gamma_M = 1,76$. When experimental data are used and the coefficient of variation 0,15 is determined from 20 available observations, then according to Table 4 $\gamma_{M,max} = 1,52$. Furthermore, if there is a convincing evidence that the strength has asymmetric distribution with the coefficient of skewness $\alpha_f = 0,5$, then it follows from equation (7) in which the coefficients K_k and K_d are taken from Table 6 that $\gamma_{M,max} = 1,28$. If, however, $\alpha_f = -0,5$, then $\gamma_{M,max} = 2,13$.

7. CONCLUSIONS

(1) The partial safety factors for masonry strength recommended by Eurocode 6 correspond to the coefficients of variation from 0,19 to 0,26.
(2) In some cases of a better quality of masonry production, the recommended partial safety factors may be rather conservative and uneconomical.
(3) Proposed statistical technique enable rational assessment of the partial safety factors using experimental data and previous experience.
(4) The assessment of partial safety factor is significantly dependent on number of observations, coefficient of variation and asymmetry of distribution.

8. REFERENCES

1. Holický M., Pume D. and Vorlíček M., "Masonry strength determination from tests," Proceedings Computer Methods in Structural Masonry - 3, Lisbon 1995, pp. 107 - 116.
2. ENV 1996-1-1 Eurocode 6, "Design of masonry structures- Part 1-1 General rules for buildings - Rules for reinforced and unreinforced masonry".
3. Holický M. and Vorlíček M., "General three parameter lognormal distribution in statistical quality control," Proceedings, ICASP7, Paris 1995, pp. 719 - 724.
4. Holický M. and Vorlíček M., "Fractile estimation and sampling inspection in building," Acta Polytechnica, Czech Technical University, Prague, 1992/1, pp. 87-96.

Acknowledgement. This paper presents a part of the findings of research project Copernicus - CT 94-174 "Advanced Testing of Masonry," supported by European Commission and co-ordinated by the University of Wales, Swansea, UK.

RELIABILITY ANALYSIS OF A MASONRY WALL

M. Holický Associate Professor[1], **J. Marková** Civil Engineer[1] and
G.N. Pande Professor[2]
[1]*Czech Technical University, 166 08 Prague, Czech Republic and*
[2]*Department of Civil Engineering, University of Wales Swansea,*
Singleton Park, Swansea SA2 8PP, Wales, UK

1. ABSTRACT

Reliability analysis of a masonry wall is focused on verification of safety level of a basic structural element designed in accordance with Eurocodes. It is shown that reliability of the masonry wall depends significantly on eccentricity of the load. Reliability index decreases with increasing eccentricity and in some cases may be less than recommended minimum 3.8. Further calibration of Eurocodes is needed.

2. INTRODUCTION

It is well recognised that probabilistic methods provide rational framework for reliability verification of various types of structures including structural masonry [1,2,3,4]. The design concept of limit states in conjunction with methods of partial safety factors enables relatively straightforward transformation of available research data into partial safety factors and other reliability elements. This possibility of gradual updating of reliability elements in design specification using appropriate probabilistic methods was perhaps the decisive reason for accepting methods of partial factors as a basic design format for development of present generation of Eurocodes [2,3,4].

The reliability analysis of masonry walls according to American standards was made by Ellingwood [5]; the reliability of unreinforced walls seems to be in some cases insufficient particularly for increasing eccentricity of the load effect. It is shown that uncertainties of the strength of material and workmanship could be taken into account

Keywords: Masonry column, Eccentricity, Statistical properties, Reliability index.

Computer Methods in Structural Masonry – 4, edited by G.N. Pande, J. Middleton and B. Kralj.
Published in 1998 by E & FN Spon, 11 New Fetter Lane, London EC4P 4EE, UK. ISBN: 0 419 23540 X

by adjustments of various reliability elements. However, further research is required

Submitted contribution shows reliability analysis of a typical component of structural masonry - unreinforced wall of a multi-storey structure. The aim of the probabilistic study is to verify reliability of a component designed in accordance with Eurocodes 1 and 6. Following similar study of reinforced concrete columns [6], ten study cases of the masonry wall corresponding to different structural arrangements are analysed here. Reliability analysis has been done using the programme Comrel-TI 6.0 of the Structural Reliability Analysis Programme System STRUREL [7] and the First-Order Reliability Method (FORM) of probability integration.

3. STRUCTURAL CHARACTERISTIC

A multi-storey structure shown in Fig. 1 is considered in this study. An edge unreinforced wall of an internal transversal frame having the height H and rectangular cross-section $b \times t$ is assumed in this reliability analysis. Masonry walls are located within distance a_1 in the transversal direction of the structure and within distance a_2 in the longitudinal direction. The structure has n storeys above the analysed wall where n is from 1 to 4. The heights of the storeys are $H = 4$ m and $h_s = 3$ m. The masonry walls carry the effects of permanent and imposed load. Resistance to horizontal actions is considered to be provided by a system of shear walls.

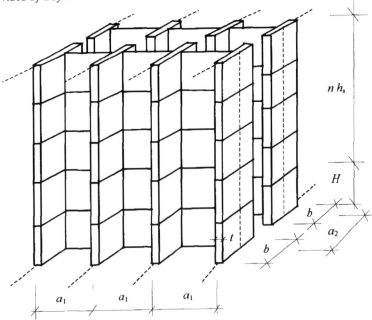

Fig. 1. A multi-storey structure with masonry walls.

One permanent load and one imposed load are considered only. Effect of actions consist of the axial force N with appropriate subscript. In the design calculation, the axial force is represented by the design value N_{Sd} as

$$N_{Sd} = \gamma_G N_w + \gamma_Q N_{imp} \qquad (1)$$

where $\gamma_G = 1.35$ is the partial safety factor for permanent actions and $\gamma_Q = 1.50$ is the partial safety factor for variable actions. N_w is the characteristic value of the axial force due to self-weight, $N_w = 0.5 (n+1) a_1 a_2 t_s \rho_c$, where ρ_c is the volume density of concrete considered as 0.024 MN/m^3. Characteristic value of permanent load is determined as a weight of a concrete floor of a uniform equivalent thickness $t_s = 0.30$ m (including weight of the concrete slab, floor, cladding and self-weight of the walls). N_{imp} is the characteristic value of axial force due to imposed load, $N_{imp} = 0.5 \, \alpha_n \, n \, a_1 \, a_2 \, p_{imp}$, where α_n is the load reduction factor according to the clause 6.3.1.2 of Eurocode 1 [3] Choosing the category B (Public Building) the characteristic value of imposed load p_{imp} equals 0.003 MN/m^2.

The following material characteristics of solid clay masonry units of Group 1 having dimensions $0.290/0.140/0.065$ m are considered in the deterministic design of a masonry wall. Normalized compressive strength of masonry units is $f_b = 15.4$ MPa. It is assumed general purpose mortar of class M 5 with specified compressive strength $f_m = 5$ MPa. The characteristic compressive strength of unreinforced masonry made with general purpose mortar with longitudinal joints (coefficient $K = 0.5$) may be calculated from the formula given in Eurocode 6 [4] as

$$f_k = K \, f_b^{0.65} \, f_m^{0.25} = 4.42 \text{ MPa}. \qquad (2)$$

4. DETERMINISTIC DESIGN

In accordance to the clause 4.4 of Eurocode 6 [4], the design vertical load on a masonry wall, N_{Sd}, should be less than the design vertical load resistance of the wall, N_{Rd}, such that $N_{Sd} \leq N_{Rd}$. The design vertical load resistance of a masonry wall of the length b is given as

$$N_{Rd} = \Phi \, t \, b \, f_k / \gamma_M \qquad (3)$$

where Φ is the reduction factor allowing for the effects of slenderness and eccentricity of loading, t is the thickness of the wall (considered 0.44 m in all the study cases), f_k is the characteristic compressive strength of the masonry (2) and γ_M is the partial safety factor from the clause 2.3.3.2 of Eurocode 6 [4]. For considered category of execution B and category of manufacturing control of masonry units II the partial safety factor is 2.5. The value of the reduction factor Φ is given as

$$\Phi = 1 - 2 \, e / t \qquad (4)$$

where total eccentricity $e = e_0 + e_a$, eccentricity of the loads $e_0 = M_d/N_{Sd}$, M_d is the design bending moment at the top or the bottom of the wall resulting from the eccentricity of the floor load at the support and is given by equation C.1 of Eurocode 6 [4], e_a is an accidental eccentricity and $e \geq 0.05\ t$. The results of the deterministic design of ten study cases are shown in Table 1. Note that in accordance with [4] the reduction factor within the middle height of the wall is greater than the reduction factor at the top of the wall. Therefore, the strength at the top of the wall is considered only.

Table 1. Deterministic design.

Study case	n	b (m)	a_1 (m)	N_w (MN)	N_{imp} (MN)	N_{Sd} (MN)	$k\,M_d$ (MNm)	e_a (m)	e_0 (m)	e (m)	Φ
1	4	3.44	8.0	0.720	0.204	1.278	0.1368	0.007	0.107	0.114	0.48
2	4	4.04	8.5	0.765	0.217	1.358	0.1571	0.009	0.116	0.125	0.44
3	4	4.79	9.0	0.810	0.230	1.438	0.1786	0.009	0.124	0.133	0.39
4	2	1.79	6.0	0.324	0.090	0.572	0.0680	0.009	0.119	0.128	0.42
5	2	2.39	6.5	0.351	0.098	0.620	0.0844	0.009	0.136	0.145	0.34
6	2	3.29	7.0	0.378	0.105	0.668	0.1020	0.009	0.153	0.162	0.27
7	2	4.79	7.5	0.405	0.113	0.716	0.1207	0.009	0.169	0.178	0.19
8	1	1.04	5.0	0.180	0.038	0.299	0.0386	0.009	0.129	0.138	0.37
9	1	1.79	5.3	0.191	0.040	0.317	0.0509	0.009	0.160	0.169	0.23
10	1	2.24	5.4	0.194	0.041	0.323	0.0549	0.009	0.170	0.179	0.19

5. RELIABILITY ANALYSIS

The limit state function g may be expressed as the difference between the design vertical load resistance of the wall N_{Rd} and the design vertical load on a masonry wall N_{Sd}, such that

$$g = \xi_R \left(t - 2\left(e_a + e_0\right)\right) b f - \xi_E\,0.5\left(\left(n+1\right) a_1 a_2\,t_s\,\rho + n\,a_1 a_2\left(p_{st} + p_{lt}\right)\right) \tag{5}$$

Comparing with equations (1) and (3) new variables are here. Coefficients of model uncertainties ξ_R and ξ_E are introduced as random variables covering imprecision and incompleteness of the relevant theoretical models for resistance and load effect and imposed load p_{imp} is described by long-term load p_{lt} and short-term load p_{st}, similarly as in the study by Vrouwenvelder T. et al. [8].

Basic variables applied in the reliability analysis are listed in Table 2. Some of the variables are assumed to be deterministic values - denoted "DET" (a_1, a_2, n, t_s), while the others are considered as random variables having the normal distribution - "N", lognormal distribution - "LN" and Gamma distribution - "GAM". Statistical properties of the random variables are further described by the moment characteristics, the mean and standard deviation, partly taken from previous study by Vrouwenvelder T. et al. [8] and from CIB Reports [9, 10].

6. RELIABILITY INDEX

Reliability index β (FORM) is determined using the software COMREL-TI 6.0 [7]. Time variant reliability analysis is not considered here. The reliability of a masonry wall is significantly affected by the eccentricity e_0 of loads. The eccentricity depends on the ratio between the bending moment and the axial force. Thus, it was important to determinate the mean and coefficient of variation $v(e_0)$. The mean of the eccentricity was considered by its nominal value. Statistical estimation of the coefficient of variation $v(e_0)$ was made using relevant relationships given in Annex C of Eurocode 6 [4] and statistical properties of appropriate basic variables. The lower bound was assessed by the value $v(e_0) = 0.03$ and the upper bound by the value $v(e_0) = 0.11$.

Table 2. Statistical properties of basic variables.

Category of basic variables	Name of basic variables	Sym-bol	Distrib. type	Dimen-sion	Mean	Standard deviation
Material properties	Compressive strength of masonry	f	LN	MPa	6.58	mean $v^{(1)}$
Geometric data	Wall distance in plane	a_1	DET	m	nom.	0
	Perpend. distance of wall	a_2	DET	m	nom.	0
	Length of the wall	b	N	m	nom.	0.02
	Thickness of the wall	t	N	m	nom.	0.01
	Number of floors	n	DET	-	nom.	0
	Uniform thickness of floor	t_s	DET	m	nom.	0
	Accidental eccentricity	e_a	N	m	0	0.01
	Eccentricity of the loads	e_0	N	m	nom.	0.11 mean
			N	m	nom.	0.03 mean
Model uncertainty	Uncertainty of resistance	ξ_R	N	-	1.1	0.11
	Uncertainty of load effect	ξ_E	N	-	1.0	0.10
Actions	Volume density of concrete	ρ	N	MN/m³	0.024	0.00192
	Imposed long-term load	p_{lt}	GAM	MN/m²	0.0006	mean $v^{(2)}$
	Imposed short-term load	p_{st}	GAM	MN/m²	0.0002	mean $v^{(3)}$

Notes:
(1) The mean and standard deviation of masonry are assessed by considering normal distribution of the compressive strength of masonry and coefficient of variation $v = 0.2$. The characteristic value of compressive strength of masonry follows from equation (2).

(2) The mean and standard deviation correspond to the distribution of 7 years maximum; the coefficient of variation v is given as $v = ((0.16 + 8/(a_1 a_2)) (1/n + r(1-1/n)))^{0.5}$ [10], where the coefficient of correlation of the long-term loads on two floors is considered as $r = 0.5$.

(3) The mean and standard deviation correspond to the distribution of the 12 hour maximum (one day); the coefficient of variation is $v = (50/a_1 a_2)^{0.5}$.

Thus, two different series of reliability indexes β were obtained for the following two probabilistic models assumed for eccentricity e_0 as indicated in Table 3:

(1) normal type of distribution of e_0, the mean equals to nominal value and the coefficient of variation $v(e_0) = 0.11$.

(2) normal type of distribution of e_0, the mean equals to nominal value and the coefficient of variation $v(e_0) = 0.03$.

Table 3. Reliability indexes β for ten considered study cases and two probabilistic models (1) and (2) assumed for eccentricity e_0.

Study case	n	b (m)	a_1 (m)	e/t	t (m)	(1) β_1	(2) β_2
1	4	3.44	8.0	0.24	0.44	5.69	6.79
2	4	4.04	8.5	0.26	0.44	5.12	6.70
3	4	4.64	9.0	0.28	0.44	4.62	6.48
4	2	1.79	6.0	0.29	0.44	4.95	6.66
5	2	2.24	6.5	0.31	0.44	3.86	5.87
6	2	3.14	7.0	0.35	0 44	2.87	4.73
7	2	4.49	7.5	0.38	0.44	2.06	3.61
8	1	1.04	5.0	0.29	0.44	4.28	6.14
9	1	2.54	5.5	0.36	0.44	2.49	4.22
10	1	3.44	5.6	0.39	0.44	1.99	3 52

Note that the reliability of the masonry column designed according to Eurocodes 1 and 6 may be in some study cases insufficient (β is less than 3.8). For example in the case of one-storey structure the reliability index β significantly decreases with increasing eccentricity e_0.

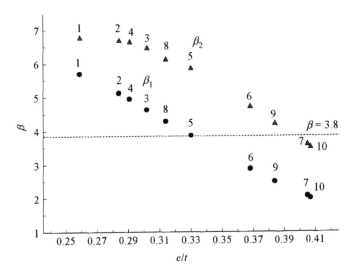

Fig. 2. Reliability indexes β_1 (for $v\,(e_0) = 0.11$) and β_2 (for $v\,(e_0) = 0.03$) versus eccentricity ratio e/t for study cases 1 to 10 given in Table 3.

An additional reliability study concerning a masonry wall of one-storey structure assuming gradually increasing eccentricity of load has been made. The distance in the transversal and longitudinal direction of the structure is assumed to be 5 m and the thickness of the wall 0.44 m. The length b of the wall is considered by theoretical values without taking into account dimensions of masonry units and disregarding the lower bound of the value of eccentricity ($e \geq 0.05\,t$). The resulting reliability indexes β as functions of the eccentricity ratio e/t are shown in Fig. 3.

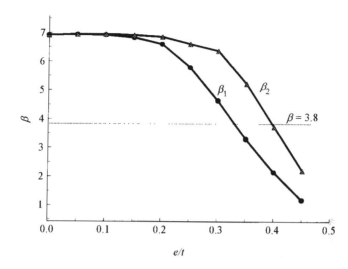

Fig. 3   Reliability indexes β_1 (for v $(e_0) = 0.11$) and β_2 (for v $(e_0) = 0.03$) versus eccentricity ratio e/t.

It appears, that for the ratio e/t greater than 0.33 and probabilistic model (1) or for the ratio e/t greater than 0.4 and probabilistic model (2), the reliability index β may be less than the recommended value 3.8. Note, that in the Czech standard [1] there is limitation of the ratio e/t which should be less than the value 0.45. Reliability analysis of reinforced masonry walls using advanced first-order, second-moment method FOSM was made by Ellingwood [5]. Fig. 3 of his study indicates that the reliability of unreinforced walls designed according to Building Code Requirements for Engineered Brick Masonry of America also rapidly decreases with increasing e/t ratio. If the ratio e/t is greater than 1/3, then the codes require designing the masonry walls using a modified section analysis. The Canadien National Building Code [11] also gives formulas for different types of eccentricity coefficients and for assessing the allowable vertical load. If the eccentricity exceeds the ratio $t/6$ or $t/3$, another type of formula is necessary to use.

Effect of the material characteristic on resulting reliability of walls were studied analysing reliability of two types of masonry walls made of two different characteristic compressive strength f_k. The masonry made of the above assumed masonry units and mortar M 10 with the characteristic compressive strength $f_k = 5.23$ MPa and masonry made of perforated clay masonry units of Group 2 b with $f_k = 3.64$ MPa were considered. It appears that the effect of characteristic compressive strength of unreinforced masonry on reliability level of the designed masonry wall is insignificant.

Detailed analysis of sensitivity of basic variables was also made. The sensitivity factors indicate that   with   increasing   ratio e/t relative   significance   of   eccentricity e_0 increases and relative significance of the strength of masonry f decreases. The reliability indexes of the wall are always higher than $\beta = 3.8$ for smaller values of the ratio e/t ($e/t < 0.25$). They can be influenced by the possible change of statistical properties of compressive strength of masonry f. Nevertheless the reliability indexes β decrease dramatically for ratio $e/t > 0.3$, while the strength f becomes less significant. Reliability of the wall turns to be insufficient.

7. CONCLUSIONS

1. Reliability of a clay brick masonry wall seems to depend significantly on the eccentricity.

2. Reliability of a masonry wall decreases with increasing eccentricity e_0 of the loads and may be insufficient in some cases (β is less than 3.8). With increasing eccentricity e_0 the reliability index β decreases from 6.9 to 1.1 if the coefficient of variation $v(e_0) = 0.11$ and from 7 to 2.1 if $v(e_0) = 0.03$.

3. It appears that the reliability index β is less than recommended minimum 3.8 if the value of the eccentricity-thickness ratio e/t is greater than 1/3.

4. Further reliability analysis of masonry walls including those having greater slenderness and calibration of reliability level provided by Eurocode 6 is needed.

8. REFERENCES

[1] ČSN 731101 'Design of masonry structures', 1980.
[2] ENV 1991-1: Eurocode 1, 'Basis of design and actions on structures - Part 1: Basis of design', 1994.
[3] ENV 1991-2-1: Eurocode 1, 'Basis of design and actions on structures - Part 2.1: Densities, self-weight and imposed loads', 1995.
[4] ENV 1996-1-1: Eurocode 6, 'Design of masonry structures - Part 1-1: General rules for buildings - Rules for reinforced and unreinforced masonry', 1995.
[5] Ellingwood B., Tallin A. 'Limit states criteria for masonry construction', Journal of Structural Engineering, Vol. 111, No. 1, January, 1985.
[6] Holický M., Vrouwenvelder T., 'Reliability analysis of a reinforced column designed according to the Eurocodes', IABSE Colloquium Basis of Design and Actions on Structures, Delft, 1996, pp 251-264.
[7] COMREL-TI 6.0 with Symbolic Processor, RCP-Reliability Consulting Programs, Munchen, 1996.
[8] Vrouwenvelder T., Neuenhofer A., Goyet J., Denver H., Holický M., 'JCSS Working document on Eurocode random variable models', 1995.
[9] CIB Report, ' Actions on structures, self weight loads', Publication 115, 1989.
[10] CIB Report, 'Actions on structures, live loads in buildings' Publication 116, 1989.
[11] Canadian Structural Design Manual, Supplement No. 4 to the National Building Code of Canada, National Research Council of Canada, 1970.

Acknowledgement. This paper presents a part of the findings of research project Copernicus-CT 94-174 "Advanced Testing of Masonry," supported by European Commission and co-ordinated by the University of Wales, Swansea, UK.

COMPUTER-AIDED DESIGN OF STRUCTURAL MASONRY ELEMENTS

J. Uzarski Partner and **A. Robinson** Associate
SOHA Structural Engineers, San Francisco, California 94108, USA

1. ABSTRACT

This paper will describe a computer program named CMDxx. The program is an aid for the structural design of reinforced concrete or clay hollow masonry unit construction. CMDxx is an easy to use tool for accurate and economical structural masonry design.

2. INTRODUCTION

CMDxx is an acronym for Concrete (or Clay) Masonry Design conforming to the 19xx edition of the Uniform Building Code (UBC). The UBC is widely used in the western United States.

The UBC is published in three year cycles. CMD88 was the first version of CMDxx to be released. CMD91 and CMD94 were next. CMD97 is the latest edition.

3. ELEMENT TYPES

The program can design beams, columns, walls subjected to in-plane loads (IPL) and walls subjected to out-of-plane loads (OOPL).

Keywords: Reinforced masonry design, Computer aided design, Seismic design

Computer Methods in Structural Masonry – 4, edited by G.N. Pande, J. Middleton and B. Kralj.
Published in 1998 by E & FN Spon, 11 New Fetter Lane, London EC4P 4EE, UK. ISBN: 0 419 23540 X

3.1 Beam elements

Beams can be designed for major axis transverse loadings including axial load. The transverse loads can be either uniformly distributed, trapezoidal or concentrated at defined points. Boundary conditions ranging from fixed to pinned can be obtained by varying the user specified fixed-end moments and shears. The program checks for longitudinal flexural reinforcement and for transverse shear reinforcement.

3.2 Column elements

Columns can be designed for axial load with biaxial moments and shears, but transverse loadings are not considered. Boundary conditions ranging from fixed to pinned can be defined by the user. Columns are checked for longitudinal and transverse reinforcement.

3.3 Walls subjected to in-plane loads

IPL walls can be designed for axial loads, shears and moments that are assumed to act in the plane of the wall. Out-of-plane loadings are not considered. IPL walls can be rectangular or they can be defined as non-symmetrical with "H", "T", "C" or "L" shaped cross sections. Boundary conditions ranging from fixed to pinned can be defined by the user. IPL walls are checked for vertical and horizontal reinforcement.

3.4 Walls subjected to out-of-plane loads

OOPL walls can be designed for uniformly distributed or trapezoidal out-of-plane loads combined with vertical axial load. In-plane moments and shears are not considered. Boundary conditions must be specified as either pinned-pinned, fixed-pinned or fixed-free. Only vertical reinforcement is checked. The shear check assumes no shear reinforcing; it assumes that the masonry alone resists the entire out-of-plane shear force.

Most practical solutions will employ the use of a single layer of vertical reinforcement placed in the center of the wall, but non-symmetrical layers of double vertical reinforcement can also be defined. However, compression reinforcement is not included in the calculations for OOPL walls.

4. DESIGN METHODS

Two different methods can be selected by the user:

- Allowable Stress Design (ASD), or
- Ultimate Strength Design (USD).

Both methods assume that:

- plane sections remain plane,
- strain varies linearly across the depth of the section, and
- masonry cannot resist tensile stress.

Elements designed using USD generally require less reinforcement and exhibit enhanced seismic performance than those designed using ASD.

4.1 Allowable stress design

The ASD option uses unfactored loads to check masonry and reinforcement stresses against allowable maximum values. This traditional procedure assumes that all stresses remain in the elastic range. It will usually produce designs with factors of safety around four or five. When the ASD option is used for OOPL walls, the maximum permitted height to thickness ratio is 30.

4.2 Ultimate strength design

The solution technique for OOPL walls is different than for beams, columns and IPL walls.

4.2.1 USD for beams, columns and IPL walls

Conventional ultimate strength design procedures, similar to reinforced concrete design, are used for beams, columns and IPL walls. Factored loads and strength reduction factors are used to compare element demands and capacities. The ultimate masonry compression strain is assumed to be 0.003 and the reinforcement is assumed to yield at its specified yield strength. A rectangular compression stress block is assumed for the masonry with a depth equal to 85% of the distance from the extreme compression fiber to the neutral axis with a uniform compression stress equal to 85% of its 28-day unconfined compression strength. For a user defined element cross section and reinforcement configuration, CMDxx generates a non-linear axial load versus moment curve by iterating on the depth of the neutral axis in 25 steps.

4.2.2 USD for OOPL walls

The solution technique for OOPL walls uses factored loads and strength reduction factors to check ultimate strength and unfactored loads to check deflections. P-delta effects are included in the strength and deflection calculations.

When this design procedure is used, practical construction considerations should keep the maximum height to thickness ratio to 40 or less.

5. BASIC INPUT INFORMATION

The user must always input the:

- cross section of the element,
- externally applied loading conditions,
- 28-day unconfined compression strength of the masonry, and
- yield strength of the reinforcement.

If USD is desired, the user must always input the configuration of the reinforcement. The program will then verify the adequacy of the member.

If ASD is desired and if the user inputs the configuration of the reinforcement, the program will calculate the resulting stresses in the masonry and reinforcement.

If ASD is desired but the user does not specify any reinforcement, the program will automatically determine the amount of required reinforcement. This option is useful for the trial design of elements where USD is desired and an initial reinforcement configuration is unknown.

6. DESIGN LOADS

With the exception of element self-weight as defined by the user, CMDxx does not compute externally applied forces such as dead, live, snow, wind, earth pressure and/or seismic loads. All externally applied axial loads, shears or moments must be independently calculated by using manual techniques or by using a finite element computer program.

7. MODULAR CONSTRUCTION

Although CMDxx is capable of designing reinforced clay masonry, only reinforced hollow concrete unit construction will be discussed here.

The most common type of hollow concrete block used for structural applications in regions of high seismicity such as California has the following nominal dimensions:

- 200 mm ($\approx$ 8 inches) thick,
- 400 mm ($\approx$ 16 inches) long, and
- 200 mm ($\approx$ 8 inches) tall.

Usually, these blocks have two cells for the placement of vertical reinforcement. Therefore, the spacing of vertical reinforcement is limited to multiples of 200 mm ($\approx$ 8 inches), or 400 mm ($\approx$ 16 inches), 600 mm ($\approx$ 24 inches), 800 mm ($\approx$ 32 inches) and so on.

Horizontal reinforcement is placed on the top of depressed vertical cross webs. For the typical 200 mm ($\approx$ 8 inches) tall hollow concrete block, the height of the cross webs are reduced to approximately 125 mm ($\approx$ 5 inches). This leaves approximately 75 mm ($\approx$ 3 inches) of clear space for placement of the horizontal reinforcement. Therefore, similarly as for vertical reinforcement, the spacing of horizontal reinforcement is also limited to multiples of 200 mm ($\approx$ 8 inches).

According to the UBC, the maximum limit for the spacing of vertical and horizontal reinforcement in regions of high seismicity is 1200 mm ($\approx$ 48 inches).

CMDxx can recognize nominal block thicknesses of 150 mm (≈ 6 inches), 200 mm (≈ 8 inches), 250 mm (≈ 10 inches), 300 mm (≈ 12 inches) and 400 mm (≈ 16 inches). However, 150 mm (≈ 6 inches) thick blocks are not considered appropriate for resisting significant seismic loadings.

8. PARTIALLY GROUTED CONSTRUCTION

The only cells that are required to receive grout are those which contain reinforcement. Partially grouted construction is usually limited to reinforced masonry walls. Partial grouting of beams or columns is generally not cost effective.

For partially grouted OOPL walls, the program automatically uses a "T-beam" analysis if the depth to the neutral axis exceeds the thickness of the face shell.

For partially grouted IPL walls, the program assumes an equivalent solid masonry section with a reduced thickness. As an example, the equivalent solid thickness of a 200 mm (≈ 8 inches) wall with vertical reinforcement at 400 mm (≈ 16 inches) centers would be 130 mm (≈ 5⅛ inches).

9. HARDWARE AND SOFTWARE REQUIREMENTS

The minimum hardware requirements are an IBM® or compatible personal computer with an Intel® 486 chip or equivalent / better and at least 8 MB of RAM. Microsoft® Windows 3.1 or 95 is also necessary.

10. QUALITY ASSURANCE

Exhaustive testing and verification have gone into the development of CMDxx to assure accurate internal calculations and conformance with the UBC. However, since use of the program requires judgment, responsibility for the interpretation of results resides with the structural engineer in-charge.

11. ACKNOWLEDGMENTS AND AVAILABILITY

Funding for the development and verification of CMDxx was provided by the Concrete Masonry Association of California and Nevada (CMACN). The program is available from CMACN for a nominal fee. CMACN is located at: 6060 Sunrise Vista Drive, Suite 1875, Citrus Heights, California, 95610, USA.

AUTHOR INDEX

SUBJECT INDEX

This index is compiled from the keywords provided by the authors of the papers, edited and extended as appropriate. The numbers refer to the first page of the relevant papers.

Printed and bound by CPI Group (UK) Ltd, Croydon, CR0 4YY

05/11/2024

01784332-0001